METHODS IN MOLECULAR BIOLOGY

Series Editor
John M. Walker
School of Life and Medical Sciences
University of Hertfordshire
Hatfield, Hertfordshire, UK

For further volumes:
http://www.springer.com/series/7651

For over 35 years, biological scientists have come to rely on the research protocols and methodologies in the critically acclaimed *Methods in Molecular Biology* series. The series was the first to introduce the step-by-step protocols approach that has become the standard in all biomedical protocol publishing. Each protocol is provided in readily-reproducible step-by-step fashion, opening with an introductory overview, a list of the materials and reagents needed to complete the experiment, and followed by a detailed procedure that is supported with a helpful notes section offering tips and tricks of the trade as well as troubleshooting advice. These hallmark features were introduced by series editor Dr. John Walker and constitute the key ingredient in each and every volume of the *Methods in Molecular Biology* series. Tested and trusted, comprehensive and reliable, all protocols from the series are indexed in PubMed.

RNA Spectroscopy

Methods and Protocols

Edited by

Véronique Arluison

Université de Paris, Paris, France
Laboratoire Léon Brillouin LLB, CEA, CNRS UMR12, Université Paris Saclay, Gif-sur-Yvette, France

Frank Wien

Synchrotron SOLEIL L'Orme des Merisiers Saint Aubin, Gif-sur-Yvette, France

Humana Press

Editors
Véronique Arluison
Université de Paris
Paris, France

Laboratoire Léon Brillouin LLB
CEA, CNRS UMR12
Université Paris Saclay
Gif-sur-Yvette, France

Frank Wien
Synchrotron SOLEIL L'Orme
des Merisiers Saint Aubin
Gif-sur-Yvette, France

ISSN 1064-3745 ISSN 1940-6029 (electronic)
Methods in Molecular Biology
ISBN 978-1-0716-0277-5 ISBN 978-1-0716-0278-2 (eBook)
https://doi.org/10.1007/978-1-0716-0278-2

This Humana imprint is published by the registered company Springer Science+Business Media, LLC, part of Springer Nature.
The registered company address is: 233 Spring Street, New York, NY 10013, U.S.A.

Preface

RNAs are ubiquitous molecules, acting at different levels of gene expression in the cell. Although significant progress has been made to understand their atomic resolution structure, analysis of their function requires in most cases the contribution of various proteins acting as co-factors. The discovery of novel RNA molecules and interaction networks provides new challenges for structural studies including, e.g., dynamic and single molecule investigations. This volume provides an introduction to different spectroscopic and biophysical methods to assess the structure and folding of RNAs and to follow their interaction with proteins. Each chapter explains how the technique works, how to set up these methods for nonexperts, and illustrates possibilities and limitations in each case. Selected examples of applications are presented.

Gif-sur-Yvette, France

Véronique Arluison
Frank Wien

Contents

Contributors

LAURA ANTOINE • *Architecture et Réactivité de l'ARN - CNRS UPR 9002, Institut de Biologie Moléculaire et Cellulaire, Université de Strasbourg, Strasbourg, France*

VÉRONIQUE ARLUISON • *Université de Paris, Paris, France; Laboratoire Léon Brillouin LLB, CEA, CNRS UMR12, Université Paris Saclay, Gif-sur-Yvette, France*

V. BALOBANOV • *Institute of Protein Research Russian Academy of Sciences, Pushchino, Moscow Region, Russia*

SERENA BERNACCHI • *Architecture et Réactivité de l'ARN - CNRS UPR 9002, Institut de Biologie Moléculaire et Cellulaire, Université de Strasbourg, Strasbourg, France*

THIERRY BIZEBARD • *Expression Génétique Microbienne, UMR8261 CNRS/Université Paris 7, IBPC, Paris, France; UMR 8104 CNRS/INSERM/Université Paris 5, Institut Cochin, Paris, France*

ALEXANDRA BLUHM • *Architecture et Réactivité de l'ARN, UPR 9002, IBMC, CNRS, Université de Strasbourg, Strasbourg, France*

ULRICH BOCKELMANN • *Nanobiophysics, ESPCI Paris, Paris, France; UMR 8104 CNRS/INSERM/Université Paris 5, Institut Cochin, Paris, France*

RICHARD BÖRNER • *Department of Chemistry, University of Zurich, Zurich, Switzerland; Laserinstitut Hochschule Mittweida, University of Applied Sciences Mittweida, Mittweida, Germany*

EMILY K. BOWMAN • *Institute for Cellular and Molecular Biology, The University of Texas at Austin, Austin, TX, USA*

JOERG E. BRAUN • *RNA Therapeutics Institute, University of Massachusetts Medical School, Worcester, MA, USA*

FLORENT BUSI • *Unité de Biologie Fonctionnelle et Adaptative, Université de Paris, CNRS UMR 8251, Paris, France*

TERESA CARLOMAGNO • *Centre for Biomolecular Drug Research, Leibniz University Hannover, Hannover, Germany; Helmholtz Centre for Infection Research, Group of Structural Chemistry, Braunschweig, Germany*

LYDIA M. CONTRERAS • *Institute for Cellular and Molecular Biology, The University of Texas at Austin, Austin, TX, USA*

RAPHAËL DE WIJN • *Architecture et Réactivité de l'ARN - CNRS UPR 9002, Institut de Biologie Moléculaire et Cellulaire, Université de Strasbourg, Strasbourg, France*

BORRIES DEMELER • *Department of Chemistry and Biochemistry, University of Lethbridge, Lethbridge, AB, Canada*

CARMELO DI PRIMO • *Laboratoire ARNA, University of Bordeaux, Bordeaux, France; INSERM U1212, CNRS UMR 5320, Institut Européen de Chimie et Biologie, Pessac, France*

ERIC ENNIFAR • *Architecture et Réactivité de l'ARN - CNRS UPR 9002, Institut de Biologie Moléculaire et Cellulaire, Université de Strasbourg, Strasbourg, France*

MAGALI FRUGIER • *Architecture et Réactivité de l'ARN - CNRS UPR 9002, Institut de Biologie Moléculaire et Cellulaire, Université de Strasbourg, Strasbourg, France*

FRANK GABEL • *Univ. Grenoble Alpes, CEA, CNRS, IBS, Grenoble, France*

HUI BON HOA GASTON • *INSERM, Fontenay Aux Roses, France*

LAURENT GEFFROY • *Nanobiophysics, ESPCI Paris, Paris, France; Department of Chemistry, University of Michigan, Ann Arbor, MI, USA*
FRÉDÉRIC GEINGUENAUD • *Plateforme CNanoMat, UFR SMBH, Université Paris 13, Sorbonne Paris Cité, Bobigny, France; INSERM, U1148, Laboratory for Vascular Translational Science, UFR SMBH, Université Paris 13, Bobigny, France*
QINGGUO GONG • *Hefei National Laboratory for Physical Sciences at Microscale and School of Life Sciences, University of Science and Technology of China, Anhui, China*
MÉLODIE C. A. S. HADZIC • *Department of Chemistry, University of Zurich, Zurich, Switzerland*
REKA A. HARASZTI • *RNA Therapeutics Institute, University of Massachusetts Medical School, Worcester, MA, USA*
LAURIANE KUHN • *Plateforme Protéomique Strasbourg Esplanade, CNRS FR1589, Strasbourg, France*
AUDRONE LAPINAITE • *Department of Molecular and Cell Biology, University of California, Berkeley, CA, USA*
CHRISTOPHE LAVELLE • *Museum National d'Histoire Naturelle, CNRS UMR 7196/ INSERM U1154, Paris, France*
ETIENNE LE BRUN • *Laboratoire Léon Brillouin LLB, CEA, CNRS UMR12, Université Paris Saclay, Gif-sur-Yvette, France*
ISABELLE LEBARS • *Architecture et Réactivité de l'ARN - CNRS UPR 9002, Institut de Biologie Moléculaire et Cellulaire, Université de Strasbourg, Strasbourg, France*
N. LEKONTSEVA • *Institute of Protein Research Russian Academy of Sciences, Pushchino, Moscow Region, Russia*
BRIDGET LI • *Institute for Cellular and Molecular Biology, The University of Texas at Austin, Austin, TX, USA*
XIAODAN LIU • *Hefei National Laboratory for Physical Sciences at Microscale and School of Life Sciences, University of Science and Technology of China, Anhui, China*
BERNARD LORBER • *Architecture et Réactivité de l'ARN - CNRS UPR 9002, Institut de Biologie Moléculaire et Cellulaire, Université de Strasbourg, Strasbourg, France*
STEFANO MARZI • *Architecture et Réactivité de l'ARN - CNRS UPR 9002, Institut de Biologie Moléculaire et Cellulaire, Université de Strasbourg, Strasbourg, France*
MIA K. MIHAILOVIC • *Institute for Cellular and Molecular Biology, The University of Texas at Austin, Austin, TX, USA*
A. MIKHAYLINA • *Institute of Protein Research Russian Academy of Sciences, Pushchino, Moscow Region, Russia*
VALERIA MILITELLO • *Department of Physics and Chemistry, University of Palermo, Viale delle Scienze, Palermo, Italy*
SOMDEB MITRA • *Department of Chemistry, New York University, New York, NY, USA*
A. NIKULIN • *Institute of Protein Research Russian Academy of Sciences, Pushchino, Moscow Region, Russia*
CAROLINE PAULUS • *Architecture et Réactivité de l'ARN - CNRS UPR 9002, Institut de Biologie Moléculaire et Cellulaire, Université de Strasbourg, Strasbourg, France*
OLIVIER PIÉTREMENT • *Laboratoire Interdisciplinaire Carnot de Bourgogne, UMR 6303 CNRS, Université de Bourgogne, Dijon Cedex, France*
JAVIER ROL-MORENO • *Architecture et Réactivité de l'ARN - CNRS UPR 9002, Institut de Biologie Moléculaire et Cellulaire, Université de Strasbourg, Strasbourg, France*
KÉVIN ROLLET • *Architecture et Réactivité de l'ARN - CNRS UPR 9002, Institut de Biologie Moléculaire et Cellulaire, Université de Strasbourg, Strasbourg, France*

JOËLLE RUDINGER-THIRION • *Architecture et Réactivité de l'ARN - CNRS UPR 9002, Institut de Biologie Moléculaire et Cellulaire, Université de Strasbourg, Strasbourg, France*
CLAUDE SAUTER • *Architecture et Réactivité de l'ARN - CNRS UPR 9002, Institut de Biologie Moléculaire et Cellulaire, Université de Strasbourg, Strasbourg, France*
YUNYU SHI • *Hefei National Laboratory for Physical Sciences at Microscale and School of Life Sciences, University of Science and Technology of China, Anhui, China*
ROLAND K. O. SIGEL • *Department of Chemistry, University of Zurich, Zurich, Switzerland*
ANGELITA SIMONETTI • *Architecture et Réactivité de l'ARN - CNRS UPR 9002, Institut de Biologie Moléculaire et Cellulaire, Université de Strasbourg, Strasbourg, France*
ANNE THÉOBALD-DIETRICH • *Architecture et Réactivité de l'ARN, UPR 9002, IBMC, CNRS, Université de Strasbourg, Strasbourg, France*
AURÉLIEN THUREAU • *Synchrotron SOLEIL, Saint-Aubin, Gif-sur-Yvette, France*
BERTRAND VILENO • *Laboratoire des Propriétés Optiques et Magnétiques des Architectures Moléculaires, Institut de Chimie (UMR7177) Université de Strasbourg/CNRS, Strasbourg, France; French EPR Federation of Research (REseau NAtional de Rpe interDisciplinaire (RENARD), Fédération IR-RPE CNRS #3443), Strasbourg, France*
FRANK WIEN • *Synchrotron SOLEIL, L'Orme des Merisiers Saint Aubin, Gif-sur-Yvette, France*
PHILIPPE WOLFF • *Architecture et Réactivité de l'ARN - CNRS UPR 9002, Institut de Biologie Moléculaire et Cellulaire, Université de Strasbourg, Strasbourg, France; Plateforme Protéomique Strasbourg Esplanade FRC1589 du CNRS, Université de Strasbourg, Strasbourg, France*
JIHUI WU • *Hefei National Laboratory for Physical Sciences at Microscale and School of Life Sciences, University of Science and Technology of China, Anhui, China*
PENGZHI WU • *Hefei National Laboratory for Physical Sciences at Microscale and School of Life Sciences, University of Science and Technology of China, Anhui, China; Department of Biology, Institute of Molecular Biology and Biophysics, ETH Zürich, Zürich, Switzerland*
SUSANN ZELGER-PAULUS • *Department of Chemistry, University of Zurich, Zurich, Switzerland*

Chapter 1

Encapsulation of Fluorescently Labeled RNAs into Surface-Tethered Vesicles for Single-Molecule FRET Studies in TIRF Microscopy

Susann Zelger-Paulus, Mélodie C. A. S. Hadzic, Roland K. O. Sigel, and Richard Börner

Abstract

Imaging fluorescently labeled biomolecules on a single-molecule level is a well-established technique to follow intra- and intermolecular processes in time, usually hidden in the ensemble average. The classical approach comprises surface immobilization of the molecule of interest, which increases the risk of restricting the natural behavior due to surface interactions. Encapsulation of such biomolecules into surface-tethered phospholipid vesicles enables to follow one molecule at a time, freely diffusing and without disturbing surface interactions. Further, the encapsulation allows to keep reaction partners (reactants and products) in close proximity and enables higher temperatures otherwise leading to desorption of the direct immobilized biomolecules.

Here, we describe a detailed protocol for the encapsulation of a catalytically active RNA starting from surface passivation over RNA encapsulation to data evaluation of single-molecule FRET experiments in TIRF microscopy. We present an optimized procedure that preserves RNA functionality and applies to investigations of, e.g., large ribozymes and RNAs, where direct immobilization is structurally not possible.

Key words Single-molecule FRET (smFRET), Lipid vesicle encapsulation, RNA labeling, RNA folding, Ribozyme, Group II intron

1 Introduction

Förster resonance energy transfer (FRET) is a non-radiative transfer of energy between two nearby fluorophores, a donor and an acceptor, overlapping emission and absorption spectra. It is a widely used spectroscopic technique to investigate intra- or intermolecular interactions such as binding events or structural rearrangements of proteins and nucleic acids [1, 2]. FRET is particularly successful on the single-molecule level as it allows to disentangle dynamic processes which obscure short-lived or rare folding intermediates usually averaged out in the molecule ensemble [3–5].

Véronique Arluison and Frank Wien (eds.), *RNA Spectroscopy: Methods and Protocols*, Methods in Molecular Biology, vol. 2113, https://doi.org/10.1007/978-1-0716-0278-2_1,

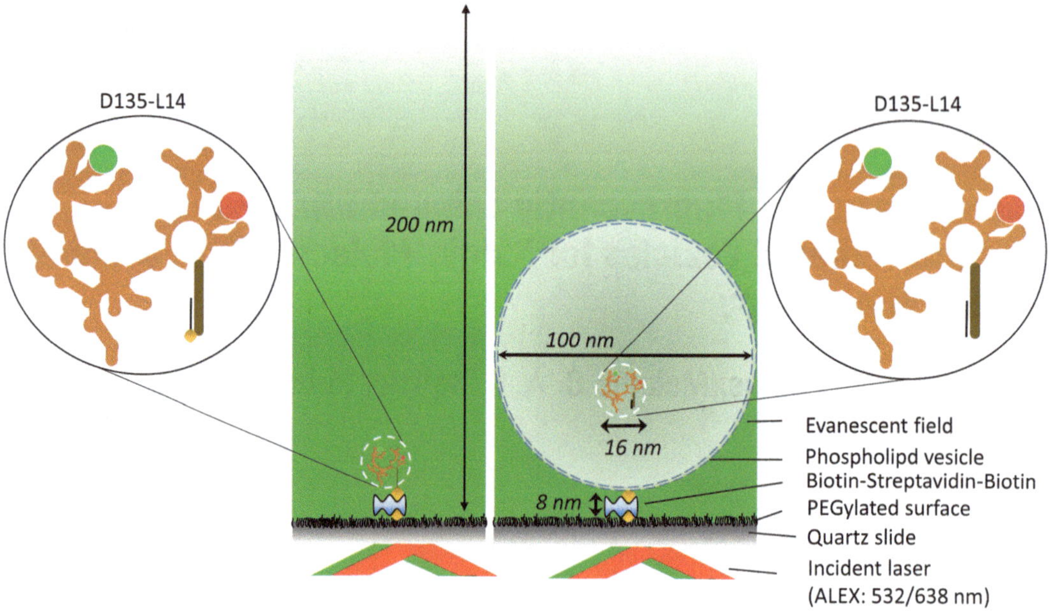

Fig. 1 To scale scheme of both immobilization techniques: classical surface immobilization of a large RNA (left) versus vesicle encapsulation (right). The group II intron ribozyme construct D135-L14 is either directly tethered to the biotin-PEG-passivated quartz surface via a biotinylated DNA oligonucleotide hybridizing to the extended 3′-end of the RNA (left) or freely diffusing in a phospholipid vesicles, which itself is tethered in the respective way (right)

smFRET measurements can be carried out by applying two different methods: confocal or wide-field microscopy. In the latter, total internal reflection (TIR) illumination is used to reduce the background signal contribution yielding a high signal-to-background ratio. In TIR fluorescence microscopy (TIRFM), single molecules are usually immobilized on a quartz slide or coverslip with a very low surface density, prior being imaged (Fig. 1) [6, 7]. This method allows to follow the folding and reaction kinetics of hundreds of single molecules in parallel over time, only limited by the frame rate (>100 Hz) of the camera and the limited observation time due to fluorophores photobleaching [8, 9]. The identification of distinct states (model selection) and their interconversion (state transition detection) allows to develop a mechanistic view of the investigated biomolecular system [10].

The immobilization to a functionalized and passivated surface via biotin-streptavidin linkage implies RNA modifications to hybridize a biotin-carrying DNA or PNA oligonucleotide [11, 12] or to covalently link a biotin moiety to the RNA [13]. Molecule immobilization can cause interactions between the molecule and the surface, which might influence the molecule's fold, and thus leads to misfolding or changes in the associated activity [14]. Further, the surface can alter the photophysical properties of the fluorophores [15]: changes in rotational freedom of the dyes promote fluorescence fluctuations that yield molecule-unrelated FRET changes [5, 16]. An elegant way to circumvent the disadvantages of a direct surface immobilization while maintaining prolonged observation periods is to use surface-attached nanoscaled liposomes which carry the molecule of interest [17, 18]. The vesicles are ~100 nm in diameter and anchored to the surface by a biotin-streptavidin linkage (Fig. 1) [17, 19–21]. Thus, the exponential decay of the evanescent excitation field within the vesicle diameter is negligible, and the excitation probability can be regarded to be nearly constant within the vesicle's diameter. Lipid vesicle membrane permeability becomes maximum near the glass transition T_g of the phospholipids they are composed of [22]. Therefore, the molecules freely diffuse inside the vesicle while the exchange of ions and small molecules is still possible through the vesicle lipid membrane [17]. Further, encapsulating catalytically active biomolecules allows to follow not only folding processes but also catalytic reactions as both, reactants and products, are trapped inside the vesicle. Recently, it was shown that vesicles have a similar effect on RNA folding as crowding reagents, which chaperone the folding and thus increase RNA fitness [14, 23].

Here, we describe a comprehensive protocol about the encapsulation of fluorescently labeled nucleic acids into phospholipid vesicles. Our model system is a 632-nt-long derivative of the wild-type group IIB intron of *Saccharomyces cerevisiae*, named D135-L14 according to the remaining domains (D) 1, 3, and 5 and two artificial loops (L) within domains 1 and 4 as labeling platforms [11, 24]. We combine established protocols for surface passivation and vesicle encapsulation and adapt them to retain RNA functionality [17, 19]. Therein, we waive the common freeze-thaw cycles, keep the time of lipid hydration relatively short, and adjust the temperature to prevent catalytic activity during the encapsulation process. Applying this protocol enables to successfully encapsulate functional RNAs with a high mono-encapsulation rate and an appropriate vesicle density on the surface. Lastly, we show a comparison of FRET results between surface-immobilized and encapsulated molecules.

2 Materials

All buffers and solutions are prepared with chemicals of at least *purissimum per analysis* quality and purified deionized water. All chemicals were purchased from Sigma-Aldrich and Thermo Fisher Scientific, unless otherwise indicated. To prevent RNA degradation, use sterile-filtered buffer (0.2 μm pore size), gel, and salt solutions.

2.1 Lipid Vesicle Preparation

1. 1 mg DMPC (1,2-dimyristoyl-sn-glycero-3-phosphocholine) : biotin cap PE (1,2-dipalmitoyl-sn-glycero-3-phosphoethanolamine-N-(cap biotinyl)) (99:1, *see* **Note 1**)
2. 1 mL Chloroform
3. Extruder and polycarbonate membrane (100 nm, e.g., Avanti Polar Lipids, USA).

2.2 Microfluidic Chamber and Surface Passivation

1. Quartz slide (7.5 × 2.5 × 0.3 cm).
2. Diamond driller (1 mm).
3. Glass coverslip (24 × 24 mm, type #1): thickness 0.13–0.16 mm, to be used with a high numerical aperture (NA) water-immersion objective (NA 1.2, magnification 60x, working distance >0.27 mm).
4. 80 mg Methoxy-poly(ethylene glycol)-succinimidyl valerate (MW 5000 g/mol, mPEG-SVA-5000).
5. 2 mg Biotin-PEG-SVA-5000 (*see* **Note 1**).
6. Imaging spacer (thickness 0.12 mm, e.g., SecureSeal™ from Grace Bio-Labs, USA).
7. Plasma cleaner (e.g., Diener Electronics, Germany).

2.3 Reagents for RNA Labeling, Encapsulation, and Imaging

1. Fluorescently labeled (and biotinylated) DNA oligonucleotides were purchased from IBA (Göttingen, Germany), and PNA oligonucleotides were synthesized as described elsewhere [12]. Any molecule carrying a dye always needs to be protected from light.
2. 3 mg/mL Trolox.
3. 20 μL/mL Streptavidin in 10 mM phosphate-buffered saline, pH 7.4 (KOH).
4. 2170 U/mL Catalase.
5. 165 U/mL Glucose oxidase.
6. T50 buffer: 50 mM NaCl, 50 mM TRIS, pH 7.5 (NaOH) [25].
7. Oxygen scavenger system (OSS) solution (100×): 22 U catalase, 1.7 U glucose oxidase, T50 buffer (*see* **Note 2**) [26, 27].
8. Standard buffer (SB) (5×) to be adapted to the optimized folding and activity conditions of the investigated RNA: 2.5 M KCl, 0.4 M MOPS, pH 6.9 (KOH), sterile filtered, store under darkness at 4 °C [28].
9. Anti-blinking buffer (AB) (10 mL): 2 mL of 5× SB, 0.1 M MgCl2, 3 mg/mL Trolox, pH 6.9 (KOH) (*see* **Note 3**) [29, 30].
10. Sugar-anti-blinking buffer (SAB): 1 mL of AB, 1 % w/v D-glucose (*see* **Note 4**).
11. Imaging buffer (IB): 495 μL SAB, 5 μL 100× OSS (*see* **Note 4**).

3 Methods

3.1 Lipid Cake Preparation

1. Dissolve 0.1 mg of phospholipids (99:1 DMPC:biotin cap PE) in 100 μL chloroform [17].
2. Prepare a thin phospholipid layer deposited on the round bottom of a 1.5 mL reaction tube, called lipid cake, by evaporating the chloroform under nitrogen flow for at least 2 h [17, 19]. Remove remaining traces of solvent under vacuum overnight (*see* **Note 1**).

3.2 Site-Specific Labeling of Long RNAs

Prior to labeling the RNA of interest, here, the D135-L14 ribozyme construct is prepared by in vitro transcription with homemade T7 RNA polymerase and purified by means of polyacrylamide gel (5 %) electrophoresis (PAGE) under denaturing conditions [31].

1. Incubate 1 μM of purified RNA in SB for 1 min at 90 °C.
2. Decrease the temperature to 42 °C for 3 min allowing secondary structures to form [12].
3. Add fluorescently labeled DNA [11] or PNA oligonucleotides [12] carrying each a sulfonated fluorophore, the donor (sCy3) and acceptor (sCy5), which are complementary to the respective labeling sites, in a 1:1 ratio (*see* **Note 5**) to the RNA for hybridization illustrated in Fig. 2 (*see* **Note 6**).
4. The folding of the RNA is induced by adding 100 mM $MgCl_2$. Please note that the specific folding conditions need to be adapted for any RNA under investigation. The ribozyme activity, here, the cleavage reaction of D135-L14, can be followed by incubating the RNA at 42 °C or stalled at 22 °C to follow the folding only.

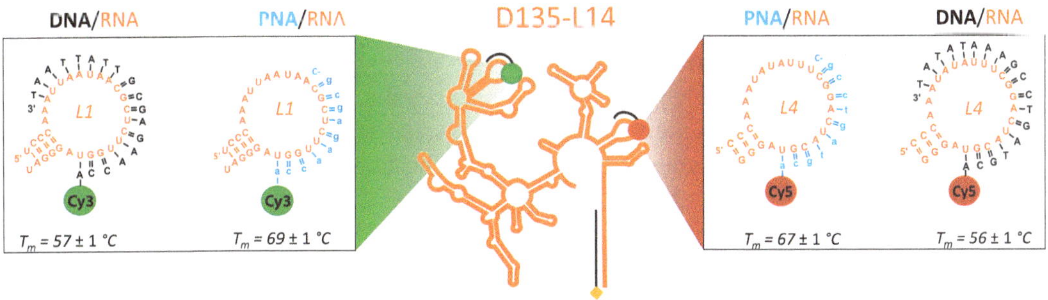

Fig. 2 Labeling scheme of D135-L14, a derivative of the wild-type group II intron *Sc*.ai5γ. The RNA is labeled via artificially introduced loops in domains D1 and D4 (L1, L4). Complementary DNA oligonucleotides (black) or PNAs (blue) carrying a fluorescent dye, here sCy3 (donor) and sCy5 (acceptor), can hybridize to these labeling platforms. The corresponding melting temperatures (T_m) of the hybrids were determined experimentally. D135-L14 can be directly immobilized on a streptavidin-coated surface via the biotinylated DNA oligonucleotide complementary to the 3′ elongation

Site-specific labeling can also be achieved by DNA-helper-strand-guided posttranscriptional base modification and subsequent bi-orthogonal RNA labeling as presented recently [32].

3.3 RNA Encapsulation into Phospholipid Vesicles

1. Prepare a 200 nM solution of fluorescently labeled RNA with a final volume of 100 μL.
2. Hydrate the prepared lipid cake with the 100 μL RNA solution; incubate both shaking for 5 min, 30 °C, i.e., above DMPC glass transition ($T_g = 23$ °C), at 1400 rpm, followed by 20 min, 30 °C at 700 rpm (*see* **Note 7**) [19].
3. Meanwhile, assemble the extruder according to the instructions of the manufactory. The two filters left and right from the membrane have to pre-equilibrate in dH_2O.
4. Equilibrate the polycarbonate membrane with 100-nm-diameter pores inside the extruder with 1 mL AB. Take up 1 mL AB with one syringe, mount the syringe accordingly, and push the buffer twice through the extruder.
5. Place the extruder on a heating block and let it warm up to 30 °C (*see* **Note 8**).
6. Centrifuge the lipid-RNA solution for 2 min at 7500 × g to remove tiny vesicles (Fig. 3).
7. Dilute the suspension with 500 μL of AB to a final RNA concentration of 33.3 nM by gently resuspending the solution (*see* **Note 9**).
8. Extrude the cloudy suspension at 30 °C 35 times through the polycarbonate membrane, which determines the final size of the vesicles (here, 100 nm). The obtained RNA-vesicle sample is ready for the immobilization step, e.g., in a microfluidic chamber.

3.4 Vesicle and Dye Characterization

Vesicle diameter (100 nm) has been chosen in such a way that the RNA of interest (here, 16 nm) can diffuse freely, the influence of the confined environment on the RNA is reduced as much as possible [14, 23], and intensity fluctuations due to the exponential decay ~ exp $(-z/d)$ of the evanescent field along the optical axis z are negligible (Fig. 3). The latter is realized by adapting the penetration depth $d = \lambda/4\pi\sqrt{n^2_{\text{quartz}}\sin^2\theta - n^2_{\text{H}_2\text{O}}} \le 70$ nm of the evanescent field which can be varied by changing the incidence angle θ (Fig. 3); the refractive indices $n_{\text{quartz}} = 1.46$ and $n_{\text{H}_2\text{O}} = 1.33$ as well as the wavelength of the incident laser beam in vacuum $\lambda = 532$ nm are determined by the experimental setup and cannot be changed.

It has been reported that the fluorophore labels of the encapsulated biomolecules interact with the phospholipid membrane, changing the fluorophore photophysics, thus yielding

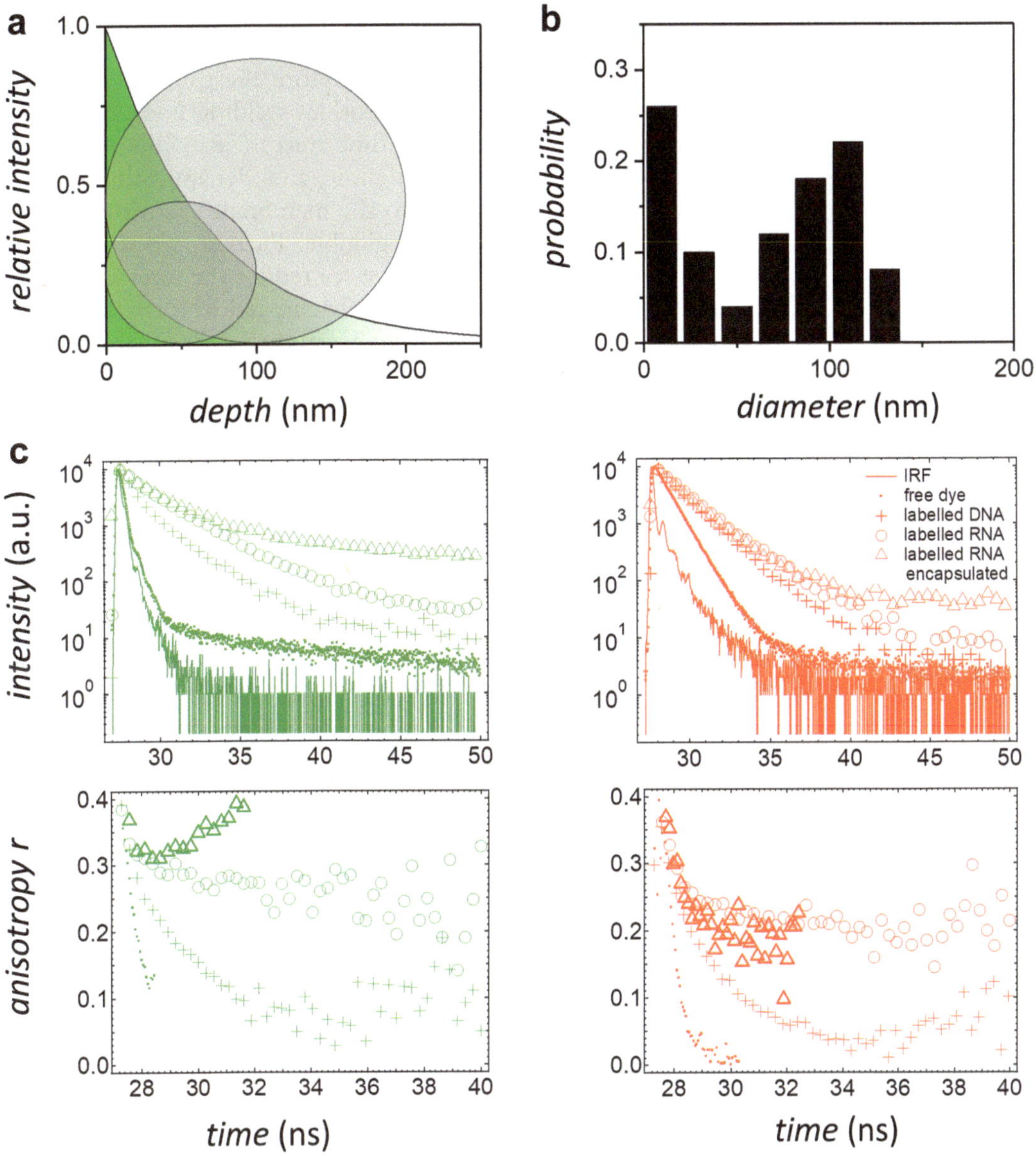

Fig. 3 Vesicle properties. (**a**) Exponentially decaying evanescent field in TIRFM in comparison to the size of the lipid vesicle. Decay of the excitation probability over the vesicle diameter is less pronounced for smaller vesicles; excitation probability fluctuations within the integration time of the camera average out due to RNA diffusion within the vesicle. (**b**) Size distribution of vesicles measured by means of dynamic light scattering (DLS) after using the extruder, but without centrifugation necessary to remove too small vesicles prior to extrusion. (**c**) Fluorophore characterization. Fluorescence lifetime (top) and dynamic anisotropy (bottom) measurements of carbocyanines, the donor sCy3 (left) and the acceptor sCy5 (right), free in solution, covalently attached to DNA oligonucleotides, DNA oligonucleotides hybridized to D135-L14, and the labeled D135-L14 encapsulated into lipid vesicles. The lifetime of both carbocyanines increases upon decreasing the cis-trans isomerization propensity in the presence of DNA oligonucleotides and RNA, respectively [33]. The decreased decay time of the dynamic anisotropy supports this finding. The lifetime and dynamic anisotropy of the labeled RNA free in solution and encapsulated are similar; thus, the encapsulation does not alter the characteristics of the fluorophores, and a possible interaction of the fluorophores and/or RNA with the lipid membrane can be neglected. Upon encapsulation of the labeled RNA, the scatter background increases drastically, limiting the observation window to a few nanoseconds

unreliable FRET traces [34]. We therefore encapsulated carbocyanine dye labeled DNA oligonucleotides yielding constant single-molecule fluorescence intensity time traces (no blinking) and no changes in the fluorescence lifetime and dynamic fluorescence anisotropy (no intercalation into the membrane) proving that the dye-membrane interaction is negligible. Further, we recommend the use of sulfonated-carbocyanines to reduce the interaction propensity of the dye with the negatively charged RNA backbone. In addition, carbocyanines show changes in their quantum yield upon interacting with the RNA due to changes of their cis-trans isomerization propensity, known as RNA-induced fluorescence enhancement (RIFE) [33]. Rigid carbocyanines, such as Cy3B, cannot undergo a cis-trans isomerization and are thus RIFE-free.

3.5 Microfluidic Chamber Preparation Including Surface Passivation for Prism-Based TIRFM

Prepare the self-made microfluidic chamber prior to the smFRET experiment [35]:

1. Drill two holes per chamber into the quartz slide according to the chamber design, e.g., provided by the imaging spacer for flushing the required solutions, using a diamond driller.
2. Rinse and sonicate the quartz slide (and the coverslip) for 20 min at r.t. successively with dH_2O, technical acetone, and dH_2O.
3. Clean and activate the surface of the quartz slide and the coverslip with a plasma cleaner for 30 min and low-pressure oxygen atmosphere according to standard protocols.
4. Both surfaces of the chamber are aminosilanized, followed by overnight PEGylation according to standard protocols [36]. Incubate the coverslip and quartz slide with 30 μL solution containing 99 mg mPEG-SVA-5000 and 1 mg biotin-PEG-SVA-5000 dissolved in 1 mL of 0.1 M sodium carbonate buffer (*see* **Note 10**).
5. After successively washing with dH_2O and drying with nitrogen gas stream, join the drilled quartz slide and the coverslip with a double-sided adhesive imaging spacer, thus building the microfluidic sample chamber (*see* **Note 11**).
6. The PEGylated chambers can be stored in falcon tubes under nitrogen atmosphere at −20 °C for 1 month without observing quality loss.

3.6 smFRET Experiment

1. Flush the chamber once with 200 μL of T50 buffer before coating the surface with 100 μL streptavidin (20 μL/mL) for 10 min at r.t.
2. Remove unbound streptavidin by a washing step with 100 μL AB.

3. Immobilize 100 μL of the lipid vesicle suspension and incubate for 10 min at r.t. In case of a direct surface immobilization (Fig. 4), the vesicle suspension is replaced by the sample solution containing the biotin-functionalized and fluorescently labeled RNA of interest with a pM concentration.
4. Meanwhile, prepare fresh IB containing SAB, the oxygen scavenger system (OSS), and Trolox to prevent early photobleaching of the fluorophores; flush the chamber finally with 200 μL IB and incubate for another 5 min prior imaging.
5. Seal the drilled holes in the quartz slide to prevent evaporation, and mount the chamber onto the water-immersion objective (*see* **Note 12**).
6. Use a drop of immersion oil before placing the quartz TIRF prism onto the quartz slide of the chamber (*see* **Note 13**).
7. The temperature of the objective and the sample holder can be controlled by a temperature controller (*see* **Note 14**).

3.7 Single-Molecule Video (SMV) Processing and Data Analysis

SMV processing for TIRF-based smFRET experiments is presented elsewhere [38]. However, a number of mandatory steps are described herein. We recommend to perform data analysis of smFRET movies by using our multifunctional analysis software for heterogeneous FRET (MASH-FRET) data (https://github.

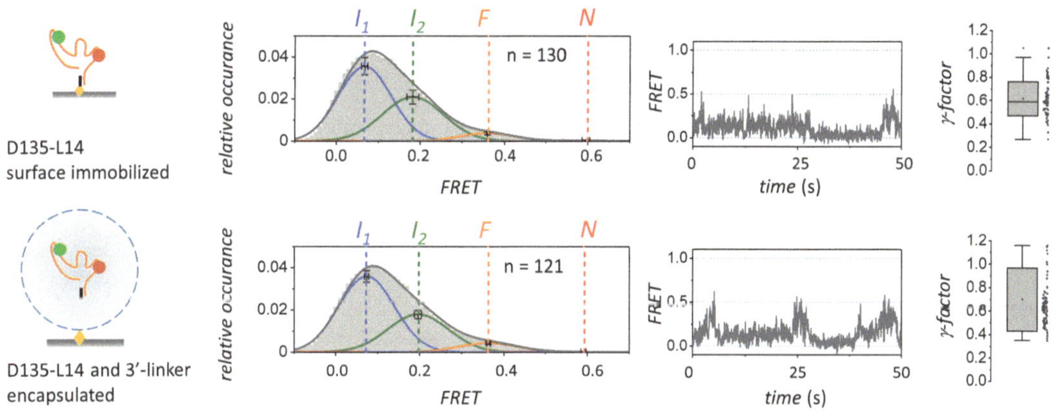

Fig. 4 Comparison of (top) the direct surface-immobilized and (bottom) encapsulated D135-L14 ribozyme with a (right) typical FRET trajectory and (middle) the ensemble FRET probability histograms of *n* molecules. Although the FRET histograms are almost identical, which demonstrates the negligible effects of the encapsulation on the RNA folding, the population of higher FRET states increases upon encapsulation ($\Delta F + \Delta N = 3\%$). Further, state transitions to higher FRET states appear more frequently upon encapsulation. These results are in line with previous observations of the ribozyme in a crowded, and thus confined, environment [23]. Error bars were obtained via bootstrapping [37]. Further, the γ-factor for individual molecules was determined according to the protocol proposed by McCann et al. [45] (very right). The characterization of the γ-factor distribution is summarized as box plot centered around the mean value with one standard deviation box height. The median value of the distribution is drawn as horizontal line (0.59 for surface immobilized and 0.61 for encapsulated molecules)

com/RNA-FRETools/MASH-FRET.git) [10, 39] or iSMS [40], both Matlab-based home-written software packages.

1. Single molecules are localized on the SMVs, and fluorophore emission intensity trajectories in imaging counts per frame (IC/frame) but usually recalculated in counts per second (a.u./s) are created as described elsewhere [39].
2. Trajectories are individually corrected from the background signal by using the "median" method [41]. Photon emission detected in the Cy5 channel is reduced by 7 % of Cy3 photon emission (bleed-through) and 2 % of Cy5 photon emission upon 638 nm excitation (direct excitation) summarized elsewhere [39, 42].
3. Single encapsulation is ensured by selecting trajectories displaying Cy3 and Cy5 emission. Single dye emission level is estimated from single bleaching step trajectories upon respective direct excitations. The fluorescence intensity trajectories selected for analysis fulfill the following criteria:
   - $I_{\mathrm{Cy5em}}^{\mathrm{Cy5ex}}(t)$ must appear stable and correspond to single Cy5 emission level along the observation time (no defocusing, blinking, photobleaching, nor inferring neighboring molecules).
   - Total photon emission upon Cy3 excitation $\left(I_{\mathrm{tot\,em}}^{\mathrm{Cy3\,ex}}(t) = I_{\mathrm{Cy3\,em}}^{\mathrm{Cy3\,ex}}(t) + I_{\mathrm{Cy5em}}^{\mathrm{Cy3ex}}(t)\right)$ must appear stable and should correspond to a single Cy3 emission level without energy transfer along the observation time.
4. The trajectories for apparent energy transfer FRET (t) and stoichiometry $S(t)$ are calculated using the respective equations:

$$\mathrm{FRET_{PR}}(t) = \frac{I_{532,\mathrm{Cy5\,em}}(t)}{I_{532,\mathrm{Cy5\,em}}(t) + I_{532,\mathrm{Cy3\,em}}(t)} \quad (1)$$

$$S(t) = \frac{I_{532,\mathrm{Cy5\,em}}(t) + I_{532,\mathrm{Cy3em}}(t)}{I_{532,\mathrm{Cy5\,em}}(t) + I_{532,\mathrm{Cy3\,em}}(t) + I_{638,\mathrm{Cy5\,em}}(t)} \quad (2)$$

   $\mathrm{FRET_{PR}}$ values can be corrected for possible variations in quantum yield and photon detection efficiency, called γ-correction [43, 44], yielding true FRET values that can be converted into absolute distances between the two labeling position L1 and L4 [42, 45]. The time traces selected for analysis exhibit both dyes within one vesicle, resulting in an average stoichiometry $S \approx 0.5$.
5. Multi- or mono-encapsulation can be distinguished by considering the fluorescent intensities, the stoichiometry, and single-step photobleaching (Fig. 5).

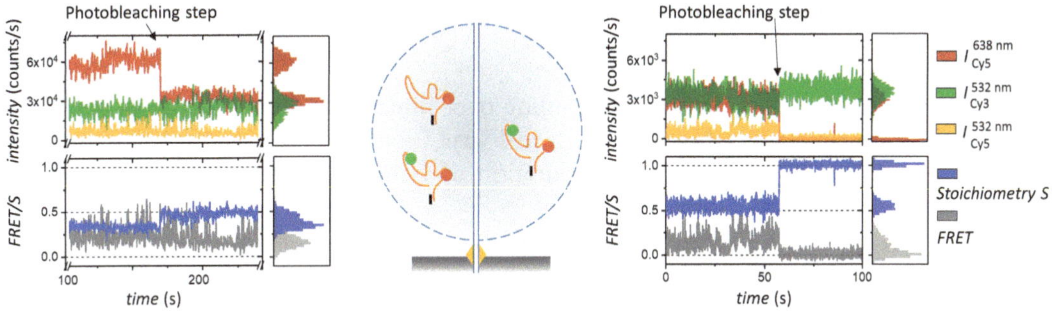

Fig. 5 Examples of smFRET trajectories (30 frames/s) using millisecond ALEX in prism-based TIRFM showing (left) double- and (right) mono-encapsulation (top). The fluorescence intensities (red, green, and yellow) as well as the calculated FRET efficiency (grey) and stoichiometry (blue) are shown together with their respective probability histograms (left). The intensity trajectory shows a bleaching step of the acceptor dye, Cy5 (red), after 175 s (arrow), indicating a double encapsulation (right). The intensity trajectory shows a bleaching step of Cy5, too (arrow). In contrast, the fluorescence of Cy5 drops to zero, while at the same time the stoichiometry goes up from 0.5 to 1, clearly indicating a mono-encapsulation

6. Selected single molecules can be classified into "dynamic" and "static" according to the presence or the absence of FRET state transitions, i.e., detectable by eye or using model selection and state transition detection methods discussed elsewhere [10].
7. Contributions of all selected trajectories are summed into a 0.01 FRET-binned histogram to obtain the ensemble FRET probability distribution (Fig. 4).
8. A stable or transient state in the histogram is modeled with a Gaussian distribution, centered on the state's FRET value. The presumed relative state populations are accurately characterized by fitting a mixture of Gaussian functions to the histograms (Fig. 4). The Gaussian mixture model (GMM) inferred via BIC selection is obtained by using an expectation-maximization (E-M) method for likelihood maximization [45].
9. In order to correct for differences in the detection efficiency of both detection channels, which enables the comparison of FRET results between different microscope setups, as well as to correct for differences of fluorophore quantum yields, which allows to use FRET as a distance measure, gamma-correction is carried out. We compared the gamma factor distribution of surface immobilized with encapsulated RNA molecules fluorescently labeled with Cy3/5 and found almost no difference besides an increasing number of outliers and thus a broadening of the distribution for encapsulated molecules (Fig. 4).

4 Notes

1. Biotin is degrading over time; the expiration date of the lipids should not exceed 1 year. Lipid cakes were stored at −20 °C for a maximum duration of 1 month, and no loss of quality was observed.
2. T50 buffer is used for the respective OSS but can be replaced by SB.
3. [Mg^{2+}] should be adapted to the optimized folding and activity conditions of the investigated RNA [11, 28].
4. Both the IB and the SAB should not be used longer than 1 week.
5. The ratio of RNA to fluorescently labeled DNA or PNA oligonucleotide is crucial for mono-encapsulation. If the label is in excess over the RNA, the probability of encapsulating DNA or PNA oligonucleotides only is very high.
6. High temperatures can destroy the fluorescent dyes; therefore, the labels should be added after the denaturation step at 90 °C. To increase the overall labeling efficiency, the design of dye-carrying DNA or PNA oligonucleotides is rationalized as follows:
   - The RNA labeling site is unique within the RNA sequence and reports on dynamics important for folding and function.
   - The size of the DNA oligonucleotides should be as short as possible to reduce necessary modifications of the RNA sequence or—if binding to the natural sequence does not influence the RNA functionality—to reduce the influence on the RNA folding. In both cases, the GC content should be chosen is such a way to yield ideally a melting temperature above 60 °C of the hybrid. In contrast to DNA, PNA is without charge and thus has a higher binding affinity toward RNA. For that reason, PNAs can be kept much shorter (down to 8–10 nt) and retain similar hybrid stability [12].
   - The labeling efficiency via hybridizing complementary oligonucleotides depends on the respective RNA and on the length and type (PNA or DNA) of the oligonucleotides [12]. Labeled and unlabeled molecules were left unseparated prior to encapsulation. Molecular sorting is further performed by means of alternating laser excitation (ALEX), which enables to identify single-labeled species [43, 46].
7. The ratio of RNA to lipids is very important to obtain mono-encapsulated RNAs.

8. We are interested to follow the functional RNA before, during, and after its catalytic step. Therefore, it is important to prevent the catalytic step (here the cleavage reaction of D135 with its substrate 17/7 in trans) during the encapsulation. This is achieved by keeping the temperature relatively low at 30 °C and adding magnesium(II) ions as late as possible; in vitro cleavage does not take place at temperatures below 30 °C and at low magnesium(II) concentrations.
9. The volume of the lipid-RNA mixture depends on the size of the syringes and the dead volume; here, we use two 1 mL syringes. The RNA concentration is crucial to ensure an appropriate density of fluorescently labeled RNAs.
10. A mixture of PEG/biotinylated PEG (99:1) is used for vesicle immobilization. The well-known surface passivation and immobilization via BSA/biotin-BSA cannot be used here as the vesicles loose surface contact and roll along the surface, in particular, upon rising the temperature above 40 °C.
11. The thickness of the imaging spacer and the coverslip must be smaller than the working distance of the used WI objective.
12. Sealing the quartz slide holes is especially important at higher temperatures to prevent evaporation of the solution inside the chamber.
13. Use immersion oil for quartz only ($n_{\text{quartz, 532 nm}} = 1.46$).
14. Wait 10 min until the system including the microscope objective, immersion water, immersion oil, and the sample chamber itself equilibrates to the set temperature. Otherwise, you will lose focus meanwhile the acquisition time. Using both an objective and a sample holder heating device, prevent for a temperature gradient within the sample chamber.

Application notes are as follows:

- The TIRFM setup design for single-molecule FRET measurements is described elsewhere [35, 47, 48].
- Encapsulation of NA [14, 17, 20] and proteins [18, 49] is used to investigate their binding interactions, folding, and chemical reactions.

Acknowledgments

Financial support from the Swiss National Science Foundation [to RKOS], the UZH Forschungskredit [FK-13-095, FK-15-09 to SZP, FK-13-091 to MCASH, FK-14-096, FK-15-095 to RB], the UZH Stiftung für wissenschaftliche Forschung [to RKOS and RB], and the University of Zurich is gratefully acknowledged.

References

1. Schuler B, Hofmann H (2013) Single-molecule spectroscopy of protein folding dynamics—expanding scope and timescales. Curr Opin Struct Biol 23(1):36–47. https://doi.org/10.1016/j.sbi.2012.10.008
2. Ha T, Enderle T, Ogletree DF et al (1996) Probing the interaction between two single molecules: fluorescence resonance energy transfer between a single donor and a single acceptor: fluorescence resonance energy transfer between a single donor and a single acceptor. Proc Natl Acad Sci U S A 93 (13):6264–6268. https://doi.org/10.1073/pnas.93.13.6264
3. Chung HS, McHale K, Louis JM et al (2012) Single-molecule fluorescence experiments determine protein folding transition path times. Science 335(6071):981–984. https://doi.org/10.1126/science.1215768
4. Schuler B, Eaton WA (2008) Protein folding studied by single-molecule FRET. Curr Opin Struct Biol 18(1):16–26. https://doi.org/10.1016/j.sbi.2007.12.003
5. Deniz AA, Dahan M, Grunwell JR et al (1999) Single-pair fluorescence resonance energy transfer on freely diffusing molecules: observation of Forster distance dependence and subpopulations. Proc Natl Acad Sci U S A 96 (7):3670–3675. https://doi.org/10.1073/pnas.96.7.3670
6. Lamichhane R, Solem A, Black W et al (2010) Single-molecule FRET of protein-nucleic acid and protein-protein complexes: surface passivation and immobilization. Methods 52 (2):192–200. https://doi.org/10.1016/j.ymeth.2010.06.010
7. Cardo L, Karunatilaka KS, Rueda D et al (2012) Single molecule FRET characterization of large ribozyme folding. Methods Mol Biol 848:227–251. https://doi.org/10.1007/978-1-61779-545-9_15
8. Gust A, Zander A, Gietl A et al (2014) A starting point for fluorescence-based single-molecule measurements in biomolecular research. Molecules 19(10):15824–15865. https://doi.org/10.3390/molecules191015824
9. Holden SJ, Uphoff S, Hohlbein J et al (2010) Defining the limits of single-molecule FRET resolution in TIRF microscopy. Biophys J 99 (9):3102–3111. https://doi.org/10.1016/j.bpj.2010.09.005
10. Hadzic MCAS, Börner R, König SLB et al (2018) Reliable state identification and state transition detection in fluorescence intensity-based single-molecule FRET data. J Phys Chem B. https://doi.org/10.1021/acs.jpcb.7b12483
11. Steiner M, Karunatilaka KS, Sigel RKO et al (2008) Single-molecule studies of group II intron ribozymes. Proc Natl Acad Sci U S A 105(37):13853–13858. https://doi.org/10.1073/pnas.0804034105
12. Schmitz AG, Zelger-Paulus S, Gasser G et al (2015) Strategy for internal labeling of large RNAs with minimal perturbation by using fluorescent PNA. Chembiochem 16 (9):1302–1306. https://doi.org/10.1002/cbic.201500180
13. Kowerko D, König SLB, Skilandat M et al (2015) Cation-induced kinetic heterogeneity of the intron-exon recognition in single group II introns. Proc Natl Acad Sci U S A 112 (11):3403–3408. https://doi.org/10.1073/pnas.1322759112
14. Saha R, Verbanic S, Chen IA (2018) Lipid vesicles chaperone an encapsulated RNA aptamer. Nat Commun 9(1):2313. https://doi.org/10.1038/s41467-018-04783-8
15. Börner R, Kowerko D, Krause S et al (2012) Efficient simultaneous fluorescence orientation, spectrum, and lifetime detection for single molecule dynamics. J Chem Phys 137 (16):164202. https://doi.org/10.1063/1.4759108
16. Schuler B, Lipman EA, Steinbach PJ et al (2005) Polyproline and the "spectroscopic ruler" revisited with single-molecule fluorescence. Proc Natl Acad Sci U S A 102 (8):2754–2759. https://doi.org/10.1073/pnas.0408164102
17. Cisse I, Okumus B, Joo C et al (2007) Fueling protein DNA interactions inside porous nanocontainers. Proc Natl Acad Sci U S A 104 (31):12646–12650. https://doi.org/10.1073/pnas.0610673104
18. Boukobza E, Sonnenfeld A, Haran G (2001) Immobilization in surface-tethered lipid vesicles as a new tool for single biomolecule spectroscopy. J Phys Chem B 105 (48):12165–12170. https://doi.org/10.1021/jp012016x
19. Liu B, Mazouchi A, Gradinaru CC (2010) Trapping single molecules in liposomes: surface interactions and freeze-thaw effects. J Phys Chem B 114(46):15191–15198. https://doi.org/10.1021/jp104614d
20. Okumus B, Wilson TJ, Lilley DMJ et al (2004) Vesicle encapsulation studies reveal that single molecule ribozyme heterogeneities are

intrinsic. Biophys J 87(4):2798–2806. https://doi.org/10.1529/biophysj.104.045971
21. Ishitsuka Y, Okumus B, Arslan S et al (2010) Temperature-independent porous nanocontainers for single-molecule fluorescence studies. Anal Chem 82(23):9694–9701. https://doi.org/10.1021/ac101714u
22. Blicher A, Wodzinska K, Fidorra M et al (2009) The temperature dependence of lipid membrane permeability, its quantized nature, and the influence of anesthetics. Biophys J 96 (11):4581–4591. https://doi.org/10.1016/j.bpj.2009.01.062
23. Paudel BP, Fiorini E, Börner R et al (2018) Optimal molecular crowding accelerates group II intron folding and maximizes catalysis. Proc Natl Acad Sci U S A 30:201806685. https://doi.org/10.1073/pnas.1806685115
24. Fiorini E, Börner R, Sigel RKO (2015) Mimicking the *in vivo* environment – the effect of crowding on RNA and biomacromolecular folding and activity. Chimia 69(4):207–212. https://doi.org/10.2533/chimia.2015.207
25. Zhao R, Rueda D (2009) RNA folding dynamics by single-molecule fluorescence resonance energy transfer. Methods 49(2):112–117. https://doi.org/10.1016/j.ymeth.2009.04.017
26. Cooper D, Uhm H, Tauzin LJ et al (2013) Photobleaching lifetimes of cyanine fluorophores used for single-molecule Förster resonance energy transfer in the presence of various photoprotection systems. Chembiochem 14 (9):1075–1080. https://doi.org/10.1002/cbic.201300030
27. Ha T (2001) Single-molecule fluorescence resonance energy transfer. Methods 25(1):78–86. https://doi.org/10.1006/meth.2001.1217
28. Pyle AM, Green JB (1994) Building a kinetic framework for group II intron ribozyme activity: quantitation of interdomain binding and reaction rate. Biochemistry 33(9):2716–2725. https://doi.org/10.1021/bi00175a047
29. Ha T (2001) Single-molecule fluorescence methods for the study of nucleic acids. Curr Opin Struct Biol 11(3):287–292. https://doi.org/10.1016/S0959-440X(00)00204-9
30. Cordes T, Vogelsang J, Tinnefeld P (2009) On the mechanism of Trolox as anti-blinking and antibleaching reagent. J Am Chem Soc 131 (14):5018–5019. https://doi.org/10.1021/ja809117z
31. Gallo S, Furler M, Sigel RKO (2005) In vitro transcription and purification of RNAs of different size. Chimia 59(11):812–816. https://doi.org/10.2533/000942905777675589
32. Zhao M, Steffen FD, Börner R et al (2017) Site-specific dual-color labeling of long RNAs for single-molecule spectroscopy. Nucleic Acids Res. https://doi.org/10.1093/nar/gkx1100
33. Steffen FD, Sigel RKO, Börner R (2016) An atomistic view on carbocyanine photophysics in the realm of RNA. Phys Chem Chem Phys 18 (42):29045–29055. https://doi.org/10.1039/c6cp04277e
34. Hughes LD, Rawle RJ, Boxer SG (2014) Choose your label wisely: water-soluble fluorophores often interact with lipid bilayers. PLoS One 9(2):e87649. https://doi.org/10.1371/journal.pone.0087649
35. Selvin PR, Ha T (eds) (2008) Single-molecule techniques: a laboratory manual. Cold Spring Harbor Laboratory Press, New York, NY
36. Chandradoss SD, Haagsma AC, Lee YK et al (2014) Surface passivation for single-molecule protein studies. J Vis Exp (86). https://doi.org/10.3791/50549
37. König SLB, Hadzic MCAS, Fiorini E et al (2013) BOBA FRET: bootstrap-based analysis of single-molecule FRET data. PLoS One 8 (12):e84157. https://doi.org/10.1371/journal.pone.0084157
38. Börner R, Kowerko D, Miserachs HG et al (2016) Metal ion induced heterogeneity in RNA folding studied by smFRET. Coord Chem Rev. https://doi.org/10.1016/j.ccr.2016.06.002
39. Börner R, Kowerko D, Hadzic MCAS et al (2018) Simulations of camera-based single-molecule fluorescence experiments. PLoS One 13(4):e0195277. https://doi.org/10.1371/journal.pone.0195277
40. Preus S, Noer SL, Hildebrandt LL et al (2015) iSMS: single-molecule FRET microscopy software. Nat Methods 12(7):593–594. https://doi.org/10.1038/nmeth.3435
41. Preus S, Hildebrandt LL, Birkedal V (2016) Optimal background estimators in single-molecule FRET microscopy. Biophys J 111 (6):1278–1286. https://doi.org/10.1016/j.bpj.2016.07.047
42. Hellenkamp B, Schmid S, Doroshenko O et al (2018) Precision and accuracy of single-molecule FRET measurements-a multi-laboratory benchmark study. Nat Methods 15 (9):669–676. http://arxiv.org/pdf/1710.03807
43. Hohlbein J, Craggs TD, Cordes T (2014) Alternating-laser excitation: single-molecule FRET and beyond. Chem Soc Rev 43 (4):1156–1171. https://doi.org/10.1039/c3cs60233h

44. Lee NK, Kapanidis AN, Wang Y et al (2005) Accurate FRET measurements within single diffusing biomolecules using alternating-laser excitation. Biophys J 88(4):2939–2953. https://doi.org/10.1529/biophysj.104.054114
45. McCann JJ, Choi UB, Zheng L et al (2010) Optimizing methods to recover absolute FRET efficiency from immobilized single molecules. Biophys J 99(3):961–970. https://doi.org/10.1016/j.bpj.2010.04.063
46. Kapanidis AN, Laurence TA, Lee NK et al (2005) Alternating-laser excitation of single molecules. Acc Chem Res 38(7):523–533. https://doi.org/10.1021/ar0401348
47. Martens KJA, van Beljouw S, van der Els S et al. (2018) An open microscopy framework suited for tracking dCas9 in live bacteria. https://hohlbeinlab.github.io/miCube. Accessed 22 Jan 2019
48. Craggs TD, Ambrose B, Cully J et al. smfBox: Open Source Single-Molecule FRET. https://benjaminambrose.github.io/smfBox/. Accessed 22 Jan 2019
49. Benítez JJ, Keller AM, Chen P (2010) Nanovesicle trapping for studying weak protein interactions by single-molecule FRET. Methods Enzymol 472:41–60. https://doi.org/10.1016/S0076-6879(10)72016-4

Chapter 2

Preparation of $SNAP_f$-Beads for Colocalization Single-Molecule Spectroscopy (CoSMoS) of RNA-Protein Complexes

Reka A. Haraszti and Joerg E. Braun

Abstract

The $SNAP_f$-tag is a chemical tag that allows rapid and highly specific covalent labeling of proteins even in the full complexity of the cellular environment. The $SNAP_f$-tag has been instrumental to study native RNA-protein complexes at single-molecule resolution in their cellular environment as efficient labeling of the RNAs and proteins of interest is essential for this colocalization single-molecule spectroscopy (CoSMoS) technique. However, removal of excessive benzylguanine dye after the labeling reaction has remained challenging. Here, we describe a strategy to remove excessive benzylguanine dye using $SNAP_f$-tag coated beads as sponges.

Key words RNA spectroscopy, CoSMoS, Colocalization single-molecule spectroscopy, RNA-protein complex, RNP, TIRF microscopy, Single-molecule spectroscopy, Single-molecule fluorescence microscopy, Protein labeling

1 Introduction

The $SNAP_f$-tag is a chemical tag that allows rapid and highly specific labeling of proteins even in the full complexity of the cellular environment [1–3]. The $SNAP_f$-tag reacts with benzylguanine dye conjugates which allow covalent attachment of a dye to the $SNAP_f$-tagged protein. The $SNAP_f$-tag has been instrumental to study native RNA-protein complexes at single-molecule resolution in their cellular environment as efficient labeling of the RNAs and proteins of interest is essential for this colocalization single-molecule spectroscopy (CoSMoS) technique [4–8]. However, removal of excessive benzylguanine dye after the labeling reaction has remained a challenge and is critical to obtain high-quality data. Here, we describe a strategy to remove excessive benzylguanine dye using $SNAP_f$-tag coated beads. $SNAP_f$-tag coated beads react with the excessive benzylguanine dye and can afterward easily be removed from the biological sample by centrifugation (*see* Fig. 1).

Véronique Arluison and Frank Wien (eds.), *RNA Spectroscopy: Methods and Protocols*, Methods in Molecular Biology, vol. 2113,
https://doi.org/10.1007/978-1-0716-0278-2_2, © Springer Science+Business Media, LLC, part of Springer Nature 2020

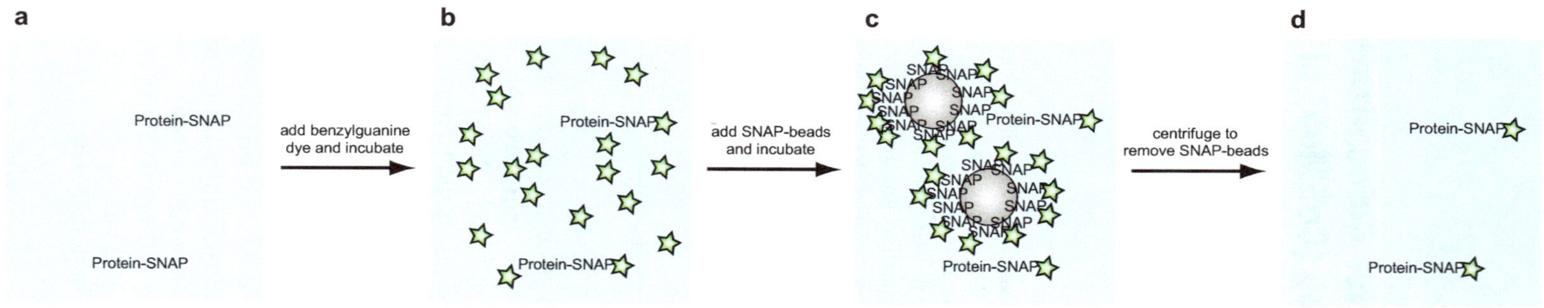

Fig. 1 (**a**) Biological sample with SNAP-tagged protein of interest. (**b**) For protein labeling, benzylguanine dye (□) is added and reacts with SNAP-tag. After labeling, the biological sample contains $SNAP_f$–tagged protein reacted with benzylguanine dye (protein–$SNAP_f$–□) and excessive free benzylguanine dye (□). (**c**) $SNAP_f$-coated beads are added and react with free benzylguanine dye. (**d**) $SNAP_f$-coated beads with excess benzylguanine dye bound on the surface are removed by centrifugation, resulting in a biological sample cleared from excess (free) benzylguanine dye

2 Materials

2.1 Expression and Purification of SNAP$_f$-Protein

1. Plasmid pET24b-6His-fSNAP (available from Addgene).
2. *E. coli* competent cells, e.g., Rosetta2(DE3) (*see* **Note 1**).
3. LB medium.
4. Kanamycin and chloramphenicol (*see* **Note 2**).
5. Isopropyl β-D-1-thiogalactopyranoside (IPTG).
6. Buffer A (50 mM HEPES buffer pH 7.5, 150 mM NaCl, 20 mM imidazole, 2 mM ß-mercaptoethanol) (*see* **Note 3**).
7. Lysozyme.
8. DNase.
9. Protease inhibitor, e.g., cOmplete EDTA-free protease inhibitor cocktail.
10. 2× Protein sample loading buffer
11. 0.45 μm Syringe filter
12. Nickel-charged chromatography column (5 ml), e.g., HisTrap.
13. Buffer B (50 mM HEPES buffer pH 7.5, 150 mM NaCl, 500 mM imidazole, 2 mM ß-mercaptoethanol).
14. SDS-PAGE gel, e.g., 12%.
15. Coomassie blue staining.
16. Centrifugal filter, 3 K.
17. Dialysis device, 3 K.
18. Glycerol.
19. Chromatography setup, e.g., ÄKTApurifier.

2.2 Immobilization of SNAP$_f$-Protein on Agarose Beads

1. RNase-free H_2O (*see* **Note 4**).
2. NHS-activated agarose dry resin.
3. PBS.
4. 1 M Tris pH 6.8
5. SDS-PAGE gel, e.g., 12%.

2.3 Depletion of SNAP-Substrates Using SNAP$_f$-Beads

1. Biological sample reacted with SNAP-substrate.
2. SDS-PAGE gel, e.g., 12%.

3 Methods

3.1 Expression and Purification of SNAP$_f$-Protein

1. Transform plasmid pET24b-6His-fSNAP into appropriate *E. coli* competent cells (*see* **Note 1**), e.g., Rosetta2(DE3) according to the manufacturer's recommendations.
2. Supplement 5 liter LB medium with respective antibiotics: kanamycin and chloramphenicol (*see* **Note 2**). Inoculate

medium with an overnight culture of transformed *E. coli* and grow bacteria at 37 °C. When OD (600 nm) reaches 0.4, shift temperature to 20 °C. Let cells cool down for 15 min, add 1/1000 volume of 0.5 M IPTG, and grow cells for another night. Harvest cells at 5000 g and freeze cell pellet at −80 °C (anticipated weight ~20 g).

3. Perform all further protein purification steps at 4 °C.
4. Suspend the pellet in buffer A (5 ml/g pellet; anticipated volume ~100 ml) supplemented with 1 g/l lysozyme, protease inhibitor, and DNase. Stir for 20 min (do not vortex) and take sample aliquot of 20 μl. Add 1 volume 2× protein sample loading buffer to sample aliquot and freeze.
5. Sonicate cell suspension from **step 4** on ice and centrifuge at 27,000 g for 45 min to clear the lysate. Filter the cleared lysate (supernatant) using a 0.45 μm syringe filter. Take sample aliquot, add 1 volume 2× protein sample loading buffer, and freeze. Discard pellet.
6. Load filtered lysate (supernatant from **step 5**) at approximately 1 ml/min on a nickel-charged chromatography column. Take sample aliquot from flow-through, add 1 volume 2× protein sample loading buffer, and freeze. Wash column with 10 column volume buffer A, and start linear gradient to buffer B from 0% to 100% in 20 column volumes. Collect 2 ml elute fractions and take sample aliquots from each fraction, add 1 volume 2× protein sample loading buffer, and freeze.
7. Analyze sample aliquots from **steps 4** to **6** by SDS-PAGE and Coomassie blue staining (follow the manufacturer's instructions) to monitor purification and identify elute fractions containing $SNAP_f$-protein ($SNAP_f$ MW = 20 kDa). Pool fractions containing $SNAP_f$-protein while avoiding contaminants (*see* **Note 5**).
8. Concentrate pooled fractions of purified $SNAP_f$-protein using centrifugal filters and dialyze into 50 mM HEPES buffer pH 7.5, 150 mM NaCl, and 1 mM DTT. The anticipated result is ~6 ml protein solution at 8 g/l (quantify using NanoDrop). Add 10% volume glycerol, freeze in liquid nitrogen, and store at −80 °C.

3.2 Immobilization of $SNAP_f$-Protein on Agarose Beads

1. Incubate 20 mg purified $SNAP_f$-protein with 0.2 g dry NHS-activated agarose beads in a total volume of 3 ml. Take sample aliquot of protein solution before incubation with agarose beads to monitor immobilization efficiency. Incubate beads according to the manufacturer's recommendations, e.g., 2 h at room temperature on a rotator (*see* **Note 6**). Take sample aliquot after incubation with agarose beads.
2. Wash beads 2× with 5 ml PBS and add 5 ml 1 M Tris pH 6.8 as quenching buffer. Incubate for 20 min at room temperature on

a rotator. Wash beads 5× with PBS, and select final buffer according the respective biological sample from which SNAP-substrate is to be depleted.

3. Immobilization efficiency of $SNAP_f$-protein by the agarose beads can be estimated from sample aliquots by SDS-PAGE analysis.
4. Store beads at 4 °C until use.

3.3 Depletion of SNAP-Substrates Using $SNAP_f$-Beads

1. Incubate biological sample containing a SNAP-tagged protein with an excess of SNAP-substrate, e.g., benzylguanine dye. Then, in order to deplete non-reacted SNAP-substrate, incubate the above reaction mix with the $SNAP_f$-beads. Take sample aliquot before incubation with beads to monitor depletion efficiency. The number of $SNAP_f$-beads needed, incubation time, and incubation temperature depend on the respective biological sample and should be optimized to minimize bead consumption and necessary incubation time. Keep beads in suspension, e.g., on a rotator during incubation.
2. After incubation, take sample aliquot and gently spin down the sample to remove beads.
3. Analyze depletion efficiency by SDS-PAGE analysis. Adjust electrophoresis time to avoid unreacted SNAP-substrate running out of the SDS-PAGE gel.

4 Notes

1. The following experimental recommendations are based on expression optimization of $SNAP_f$-protein in Rosetta2(DE3). If another *E. coli* strain is chosen for protein expression, it is recommended to first optimize expression conditions before proceeding to a 5 l scale.
2. If Rosetta2(DE3) cells are used, it is recommended to supplement the medium with chloramphenicol; see the manufacturer's recommendations.
3. Avoid EDTA throughout the purification as the $SNAP_f$-protein contains a structurally important Zn ion.
4. To enable studies of RNA-protein complexes, RNase contamination should be avoided at each step.
5. The described chromatography protocol is expected to yield highly purified $SNAP_f$-protein, which has been used successfully in several applications including single-molecule studies of the spliceosome. If a higher degree of purity is desired, additional purification steps may be added to this protocol.
6. Perform cross-linking reaction of NHS-activated resin in amine-free buffer.

Acknowledgments

Joerg E. Braun acknowledges funding from the Human Frontier Science Program (HFSP) LT000166/2013 and the European Molecular Biology Organization (EMBO) ALTF 890-2012. Joerg E. Braun is supported by funding to the laboratory of Melissa J. Moore: NIH R01 GM053007.

References

1. Keppler A, Gendreizig S, Gronemeyer T, Pick H, Vogel H, Johnsson K (2003) A general method for the covalent labeling of fusion proteins with small molecules in vivo. Nat Biotechnol 21(1):86–89. https://doi.org/10.1038/nbt765
2. Gautier A, Juillerat A, Heinis C, Correa IR Jr, Kindermann M, Beaufils F, Johnsson K (2008) An engineered protein tag for multiprotein labeling in living cells. Chem Biol 15 (2):128–136. https://doi.org/10.1016/j.chembiol.2008.01.007
3. Sun X, Zhang A, Baker B, Sun L, Howard A, Buswell J, Maurel D, Masharina A, Johnsson K, Noren CJ, Xu MQ, Correa IR Jr (2011) Development of SNAP-tag fluorogenic probes for wash-free fluorescence imaging. Chembiochem 12(14):2217–2226. https://doi.org/10.1002/cbic.201100173
4. Hoskins AA, Friedman LJ, Gallagher SS, Crawford DJ, Anderson EG, Wombacher R, Ramirez N, Cornish VW, Gelles J, Moore MJ (2011) Ordered and dynamic assembly of single spliceosomes. Science 331(6022):1289–1295. https://doi.org/10.1126/science.1198830
5. Anderson EG, Hoskins AA (2014) Single molecule approaches for studying spliceosome assembly and catalysis. Methods Mol Biol 1126:217–241. https://doi.org/10.1007/978-1-62703-980-2_17
6. Braun JE, Friedman LJ, Gelles J, Moore MJ (2018) Synergistic assembly of human pre-spliceosomes across introns and exons. Elife 7. https://doi.org/10.7554/eLife.37751
7. Braun JE, Serebrov V (2017) Single-molecule analysis of pre-mRNA splicing with Colocalization single-molecule spectroscopy (CoSMoS). Methods Mol Biol 1648:27–37. https://doi.org/10.1007/978-1-4939-7204-3_3
8. Larson JD, Rodgers ML, Hoskins AA (2014) Visualizing cellular machines with colocalization single molecule microscopy. Chem Soc Rev 43 (4):1189–1200. https://doi.org/10.1039/c3cs60208g

Chapter 3

Comparative Colocalization Single-Molecule Spectroscopy (CoSMoS) with Multiple RNA Species

Reka A. Haraszti and Joerg E. Braun

Abstract

Colocalization single-molecule spectroscopy (CoSMoS) allows studying RNA-protein complexes in the full complexity of their cellular environment at single-molecule resolution. Conventionally, the interaction between a single RNA species and multiple proteins is monitored in real time. However, comparing interactions of the same proteins with different RNA species in the same cell extract promises unique insights into RNA biology. Here, we describe an approach to monitor multiple RNA species simultaneously to enable direct comparison. This approach represents a technological development to avoid conventional inter-experiment comparisons.

Key words RNA spectroscopy, CoSMoS, Colocalization single-molecule spectroscopy, RNA-protein complex, RNP, TIRF microscopy, Single-molecule spectroscopy, Single-molecule fluorescence microscopy

1 Introduction

Multi–RNA Colocalization Single–Molecule SpectroscopySingle-molecule methods can reveal activity and interactions of individual RNAs and their complexes without being limited by statistical averaging. Colocalization single-molecule spectroscopy (CoSMoS) has enabled studying individual macromolecular complexes in highly complex biological systems [1–14]. Briefly, the molecules of interest, e.g., RNAs and their interaction partners, are labeled with fluorescent dyes and monitored in real time using total internal reflection fluorescence (TIRF) microscopy. General CoSMoS experimental procedures, data collection, and data analysis have been described elsewhere [15–19]. Here, we focus on comparing different RNAs with the same cell extract. So far, different RNAs could only be studied in different experiments, where data comparison was hampered by potential variation between experiments of this highly sensitive technique. Here, we describe an approach to overcome this challenge by measuring multiple RNA species

Véronique Arluison and Frank Wien (eds.), *RNA Spectroscopy: Methods and Protocols*, Methods in Molecular Biology, vol. 2113,
https://doi.org/10.1007/978-1-0716-0278-2_3,

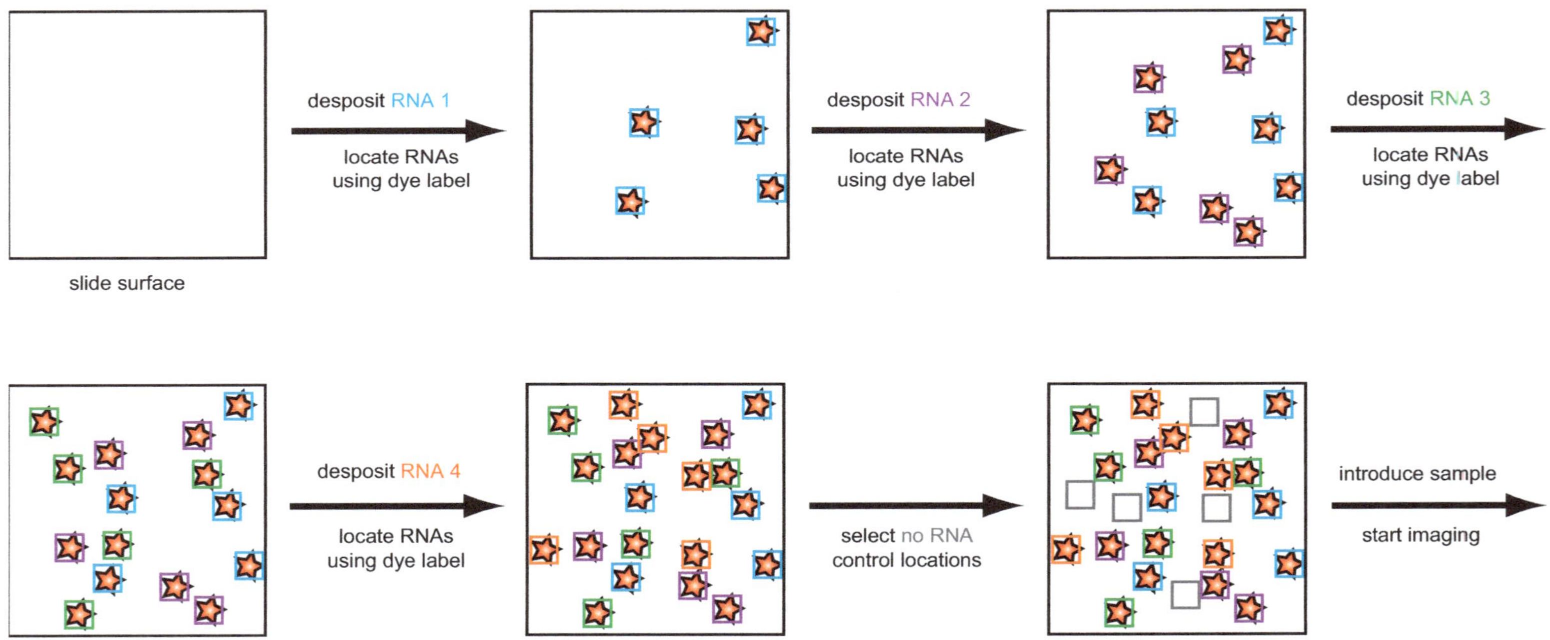

Fig. 1 Sequentially deposit and locate four different RNA species on the same CoSMoS slide surface. After RNA deposition, random "no RNA" locations are selected and serve as negative control

simultaneously in a comparative CoSMoS experiment. Analogous to analyzing multiple different samples on the same gel in a biochemical experiment, we describe depositing multiple different RNAs on the same CoSMoS slide surface to enable direct comparison (*see* Fig. 1). Sequential deposition of the RNAs allows studying, e.g., four different RNA species simultaneously while labeling all RNA species with the same dye [2].

2 Materials

2.1 Depositing and Imaging Multiple RNA Species on a Single Slide for CoSMoS

1. Total internal reflection fluorescence (TIRF) microscope setup (*see* **Note 1**).
2. Cleaned, passivated, and biotin-functionalized slides, e.g., Gold Seal No 1.5, No. 3423 (*see* **Note 2**).
3. Cleaned and passivated coverslips, e.g., Corning No. 1.5 (*see* **Note 2**).
4. RNA deposition buffer compatible with both slide passivation and biological system, e.g., 50 mM potassium phosphate buffer pH 7.4.
5. Vacuum grease, e.g., Dow Corning high-vacuum grease.
6. N_2 gas (ultrahigh-purity grade of nitrogen).
7. Filter paper, e.g., Whatman 1, 90 mm.
8. Streptavidin, e.g., Prozyme, SA10.
9. RNAs labeled with fluorescent dye and biotin (*see* **Note 3**).
10. RNase-free H_2O, filtered 0.2 μm (*see* **Note 4**).
11. Oxygen scavenging system, e.g., *Pseudomonas* protocatechuate 3,4-dioxygenase and protocatechuic acid (*see* **Note 5**).
12. Biological sample, e.g., cell extract with fluorescently labeled molecules of interest.

2.2 Data Analysis for CoSMoS Experiments with Multiple RNAs

1. Data analysis software, e.g., imscrol https://github.com/gelles-brandeis/CoSMoS_Analysis [18].

3 Methods

3.1 Depositing and Imaging Multiple RNA Species on a Single Slide for CoSMoS

The general CoSMoS experimental setup and preparations including slide and coverslip cleaning, passivation, and functionalization with biotin have been described in detail elsewhere [15–17]. Here, we focus on the step of RNA deposition and describe how to deposit multiple RNAs:

1. The slide and coverslip are thawed (if stored at −80 °C) and washed with an RNA deposition buffer compatible with both slide passivation and biological system, e.g., 50 mM potassium phosphate buffer pH 7.4, and subsequently dried using an N_2 gas stream.
2. Draw five grease lanes using a syringe filled with vacuum grease, and carefully place the coverslip onto the grease lanes to generate four flow chambers on the slide (dividing the slide into four flow chambers results in flow chambers of convenient width and enables four independent CoSMoS experiments per slide). Gently push down the coverslip to ensure complete separation of the flow chambers by the vacuum grease.
3. Choose one flow chamber. Fill and clean flow chamber with RNA deposition buffer. Draw solution through the flow chamber using folded filter paper; typically, 30 μl are sufficient per flow chamber.
4. Select a field of view on the flow chamber's slide surface, and acquire an image at the appropriate wavelength of the fluorescent dye used to label the RNAs. Confirm absence of fluorescent contamination; in case of contamination, illuminate the field of view at high laser power to bleach contaminants on the surface. Keep field of view constant throughout the complete CoSMoS experiment (*see* **Notes 6** and 7).
5. Draw 30 μl of 2.5 mg/l streptavidin in RNA deposition buffer through the flow chamber. Incubate solution for 45 s and wash flow chamber twice with 70 μl RNA deposition buffer.
6. Draw 30 μl of 10^{-11}–10^{-10} M of fluorescently labeled RNA 1 in RNA deposition buffer supplemented with appropriate oxygen scavenging system, e.g., protocatechuate-dioxygenase through the flow chamber (*see* **Note 8**). Monitor RNA deposition using the fluorescence label. Once the desired RNA density on the slide surface is reached (*see* **Note 9**), draw twice 30 μl RNA deposition buffer supplemented with oxygen scavenging system through the flow chamber to remove excess RNA. The RNA concentration to be used and incubation time depend on the biotinylation efficiency of the RNA and the desired RNA density (*see* **Note 9**).
7. Acquire an image of the field of view at the appropriate wavelength to assign RNA identity to RNA locations.
8. Repeat **steps 6** and 7 for each RNA to be deposited.
9. Draw 30 μl of the biological sample with fluorescently labeled molecules of interest, e.g., cell extract, supplemented with oxygen scavenging system through the flow chamber, and start imaging at the respective wavelengths to monitor biological process in real time.

3.2 Data Analysis for CoSMoS Experiments with Multiple RNAs

The general CoSMoS data analysis has been described in detail elsewhere [15, 17–19]. Here, we focus on the step of assigning RNA species identity, which is specific for data analysis of CoSMoS experiments with multiple RNAs. Images acquired during RNA deposition are used to identify RNA species as illustrated in Fig. 1:

1. Open images taken during RNA deposition with suitable image analysis software, e.g., imscrol https://github.com/gelles-brandeis/CoSMoS_Analysis [18].
2. Locations with fluorescent spots indicative of fluorescently labeled RNAs are identified in the image taken after deposition of RNA 1. The locations of these spots are assigned to RNA 1.
3. RNA 2 fluorescent spots are identified by removing all spot locations that are already assigned to RNA 1 from the spot location list of the image taken after the deposition of RNA 2. Locations in the image taken after the deposition of RNA 2 have to be drift corrected to the image taken after deposition of RNA 1, so that location coordinates can be compared.
4. Proceed with any additional RNA as described in **step 3** (in the example of Fig. 1 also for RNA 3 and RNA 4).
5. After determining all RNA locations, pick random locations without a fluorescent spot as "no RNA" locations. These locations are processed like RNA locations during data analysis and serve as internal negative control for nonspecific events.
6. Proceed with data analysis like for CoSMoS experiments with only one RNA species.

4 Notes

1. Detailed description on how to build a TIRF microscope suitable for CoSMoS experiments can be found here [20].
2. Cleaning and passivation of slides and coverslips, as well as functionalization of slides with biotin, can be found here [15].
3. Methods to label RNAs with fluorescent dyes for CoSMoS experiments are described here [16].
4. To enable studies of RNA-protein complexes, RNase contamination should be avoided at each step. To minimize background fluorescence, any contamination with potentially fluorescent particles should be avoided, e.g., by 0.2 μm filtration of reagents and using filter tips.
5. Establishing and benchmarking of the protocatechuate-dioxygenase oxygen scavenging system are described here [3].
6. RNAs already deposited on the surface can be used as reference for stage drift correction during the RNA deposition process.

Alternatively, streptavidin-conjugated fluorescent beads, e.g., Life Technologies, T10711, can be tethered to the slide surface as fiducial marker prior to RNA deposition.

7. Multiple fields of view enable an increase of the number of RNA locations monitored. When using multiple fields of view, acquire images after each RNA deposition step in each field of view.
8. Low binding tips and tubes are recommended due to the picomolar concentrations during RNA deposition.
9. Generally, total RNA densities of 0.2–0.5 RNAs per μm^2 are desirable to maximize monitored RNA numbers while avoiding issues arising from high RNA densities during data processing.

Acknowledgments

Joerg E. Braun acknowledges funding from the Human Frontier Science Program (HFSP) LT000166/2013 and the European Molecular Biology Organization (EMBO) ALTF 890-2012. Joerg E. Braun is supported by funding to the laboratory of Melissa J. Moore: NIH R01 GM053007.

References

1. Hoskins AA, Friedman LJ, Gallagher SS, Crawford DJ, Anderson EG, Wombacher R, Ramirez N, Cornish VW, Gelles J, Moore MJ (2011) Ordered and dynamic assembly of single spliceosomes. Science 331 (6022):1289–1295. https://doi.org/10.1126/science.1198830
2. Braun JE, Friedman LJ, Gelles J, Moore MJ (2018) Synergistic assembly of human pre-spliceosomes across introns and exons. Elife 7. https://doi.org/10.7554/eLife.37751
3. Crawford DJ, Hoskins AA, Friedman LJ, Gelles J, Moore MJ (2008) Visualizing the splicing of single pre-mRNA molecules in whole cell extract. RNA 14(1):170–179. https://doi.org/10.1261/rna.794808
4. Loveland AB, Habuchi S, Walter JC, van Oijen AM (2012) A general approach to break the concentration barrier in single-molecule imaging. Nat Methods 9(10):987–992. https://doi.org/10.1038/nmeth.2174
5. Yardimci H, Wang X, Loveland AB, Tappin I, Rudner DZ, Hurwitz J, van Oijen AM, Walter JC (2012) Bypass of a protein barrier by a replicative DNA helicase. Nature 492 (7428):205–209. https://doi.org/10.1038/nature11730
6. Crawford DJ, Hoskins AA, Friedman LJ, Gelles J, Moore MJ (2013) Single-molecule colocalization FRET evidence that spliceosome activation precedes stable approach of 5′ splice site and branch site. Proc Natl Acad Sci U S A 110(17):6783–6788. https://doi.org/10.1073/pnas.1219305110
7. Lee HW, Kyung T, Yoo J, Kim T, Chung C, Ryu JY, Lee H, Park K, Lee S, Jones WD, Lim DS, Hyeon C, Heo WD, Yoon TY (2013) Real-time single-molecule co-immunoprecipitation analyses reveal cancer-specific Ras signalling dynamics. Nat Commun 4:1505. https://doi.org/10.1038/ncomms2507
8. Lu Y, Wang W, Kirschner MW (2015) Specificity of the anaphase-promoting complex: a single-molecule study. Science 348 (6231):1248737. https://doi.org/10.1126/science.1248737
9. Yao C, Sasaki HM, Ueda T, Tomari Y, Tadakuma H (2015) Single-molecule analysis of the target cleavage reaction by the drosophila RNAi enzyme complex. Mol Cell 59

(1):125–132. https://doi.org/10.1016/j.molcel.2015.05.015

10. Graham TG, Walter JC, Loparo JJ (2016) Two-stage synapsis of DNA ends during non-homologous end joining. Mol Cell 61 (6):850–858. https://doi.org/10.1016/j.molcel.2016.02.010
11. Arauz E, Aggarwal V, Jain A, Ha T, Chen J (2016) Single-molecule analysis of lipid-protein interactions in crude cell lysates. Anal Chem 88(8):4269–4276. https://doi.org/10.1021/acs.analchem.5b04127
12. Watanabe M, Iwakawa HO, Tadakuma H, Tomari Y (2017) Biochemical and single-molecule analyses of the RNA silencing suppressing activity of CrPV-1A. Nucleic Acids Res 45(18):10837–10844. https://doi.org/10.1093/nar/gkx748
13. Hoskins AA, Rodgers ML, Friedman LJ, Gelles J, Moore MJ (2016) Single molecule analysis reveals reversible and irreversible steps during spliceosome activation. Elife 5. https://doi.org/10.7554/eLife.14166
14. Larson JD, Hoskins AA (2017) Dynamics and consequences of spliceosome E complex formation. Elife 6. https://doi.org/10.7554/eLife.27592
15. Anderson EG, Hoskins AA (2014) Single molecule approaches for studying spliceosome assembly and catalysis. Methods Mol Biol 1126:217–241. https://doi.org/10.1007/978-1-62703-980-2_17
16. Braun JE, Serebrov V (2017) Single-molecule analysis of pre-mRNA splicing with colocalization single-molecule spectroscopy (CoSMoS). Methods Mol Biol 1648:27–37. https://doi.org/10.1007/978-1-4939-7204-3_3
17. Larson JD, Rodgers ML, Hoskins AA (2014) Visualizing cellular machines with colocalization single molecule microscopy. Chem Soc Rev 43(4):1189–1200. https://doi.org/10.1039/c3cs60208g
18. Friedman LJ, Gelles J (2015) Multi-wavelength single-molecule fluorescence analysis of transcription mechanisms. Methods 86:27–36. https://doi.org/10.1016/j.ymeth.2015.05.026
19. Kaur H, Jamalidinan F, Condon SGF, Senes A, Hoskins AA (2019) Analysis of spliceosome dynamics by maximum likelihood fitting of dwell time distributions. Methods 153:13–21. https://doi.org/10.1016/j.ymeth.2018.11.014
20. Larson J, Kirk M, Drier EA, O'Brien W, MacKay JF, Friedman LJ, Hoskins AA (2014) Design and construction of a multiwavelength, micromirror total internal reflectance fluorescence microscope. Nat Protoc 9 (10):2317–2328. https://doi.org/10.1038/nprot.2014.155

Chapter 4

Dynamic Light Scattering Analysis on RNA Associated to Proteins

Serena Bernacchi

Abstract

Dynamic light scattering represents an accurate, robust, and reliable technique to analyze molecule size in solution and monitor their interactions in real time. Here, we describe how to analyze by DLS an RNA-protein interaction. In our frame, we studied complexes formed between RNA fragments derived from the genome of HIV-1 in association with the viral precursor $Pr55^{Gag}$. These interactions are crucial for the specific selection of the viral genomic RNA (gRNA) from the bulk of the viral spliced and cellular RNAs. This chapter displays how DLS allows to characterize the interactions that regulate the early steps of viral assembly.

Key words Dynamic light scattering, Protein-RNA interactions, Hydrodynamic radius, Diffusion coefficient, $Pr55^{Gag}$ precursor, Viral genomic RNA, Viral spliced RNA, Viral assembly

1 Introduction

The application of light scattering to analyze biological macromolecules was laid by Rayleigh, Mie, and Einstein [1–3]. Two branches of light scattering were then developed: the static light scattering (SLS) and the dynamic light scattering (DLS). In SLS analysis, the intensity of the scattered light is measured as a function of the scattering angle, which is related to the size and the shape of macromolecules [4], whereas in DLS experiments the intensity of the fluctuations of the scattered light is detected as a function of the time, and thus, this method provides information about the diffusion and, therefore, the hydrodynamic size of the molecules [5]. Hence, DLS is based on quasi-elastic diffusion of light (QELS) and measures the fluctuation in scattered light intensity over time of molecules diffusing in solution in a random Brownian motion [6]. The scattered light intensity is then used to generate the autocorrelation function which is related to the diffusion coefficient (D) that provides information on the hydrodynamic radius

Véronique Arluison and Frank Wien (eds.), *RNA Spectroscopy: Methods and Protocols*, Methods in Molecular Biology, vol. 2113, https://doi.org/10.1007/978-1-0716-0278-2_4,

(R_h) of the molecules via the Stokes-Einstein equation (see Subheading 3.2) [7].

This noninvasive and reliable method allows typically size measurement of molecules from 1 nm to 10 μm and whose molecular weight is below 1 kDa. Moreover, DLS is a rapid method providing data in few tens of minutes, and samples do not require labeling or any specific preparation. The analysis is possible at a vast range of solvent conditions, and one only needs knowledge of the viscosity of the solvent, its composition, and the related refractive index (RI). Compared to other techniques as analytical ultracentrifugation (AUC), the amount of sample required for DLS is quite low (as little as few μl). The measurements can be performed at various temperatures, thus providing information about thermal stability of macromolecules. Moreover, DLS is a suitable technique to monitor the complex formation, as well as to study the aggregation of macromolecules, which can be useful in preparations of samples for structural studies such as X-ray crystallography [8, 9].

Here, we used DLS to analyze the interactions occurring between RNA fragments derived from the viral genome of HIV-1 and the viral precursor $Pr55^{Gag}$. Indeed, in order to generate an infectious particle, the dimeric unspliced genomic RNA [10] must be selected from more than 100 partially or fully spliced viral RNAs and 10,000 cellular RNAs [11, 12]. This selection is a highly regulated process relying on specific interactions occurring between the 5′ untranslated region (UTR) and the beginning of the Gag coding sequences of the gRNA and the viral precursor $Pr55^{Gag}$. This last one orchestrates the viral assembly at the plasma membrane, and about 2000 copies of $Pr55^{Gag}$ and the gRNA dimer are encapsidated. The process ends with the budding of the viral particle, which typically has a diameter of about 120–150 nm. This study indicates how to compare the interaction of different kinds of viral RNAs (unspliced and spliced) with the $Pr55^{Gag}$ protein by DLS. The characterization of the mean size, the monodispersity, and the stability of the different RNA/protein complexes constitutes pieces of information leading to a better understanding of properties required for the proper process of assembly.

2 Materials

1. Prepare all the solutions using ultrapure water (using deionized water attaining a sensitivity of 18 MΩ-cm at 25 °C.
2. Buffer A: 30 ml of 30 mM Tris–HCl (pH 7.5) with 300 mM NaCl and 5 mM $MgCl_2$ (*see* **Notes 1** and **2**). This buffer is the one used for the DLS analysis of protein-RNA complexes.
3. Buffer B: 30 ml of 50 mM Tris–HCl pH 8, supplemented with 1 M NaCl, and 5 mM DTT.

This buffer corresponds to the storage buffer of $Pr55^{Gag}$ recombinant protein (i.e., the absence of aggregates) (*see* **Note 3**).

4. 0.22 μm Membrane sterile syringe filter (SARSTED) and 50 ml syringe.
5. Parafilm (Merck Millipore).
6. NanoDrop ND-1000 Spectrophotometer (Thermo Fisher Scientific, Waltham, MA, USA) or equivalent.
7. Single-use 50 μl trUView cuvette (BioRad Laboratories) (*see* **Note 4**).
8. Intensity of the scattered light and correlation times were measured here using a Zetasizer™ Nano S apparatus (4 mW He-Ne laser, $\lambda_0 = 633$ nm, scattering angle $\theta = 173°$ (Malvern, UK)).
9. DTS software (Malvern).

3 Methods

3.1 Sample Preparation

1. Buffers need to be filtered at first.
2. Recombinant proteins are stocked at −20 °C. Following their thawing, they have to be diluted into a 50 μl solution to a final concentration of 5–10 μM in the buffer B (*see* **Note 5**) and in the buffer A (*see* the Subheading 2).
3. Fill the sterile 50 μl trUView cuvettes (BioRad Laboratories), and cover carefully the cuvettes with Parafilm (Merck Millipore) (*see* **Note 6**). The protein sample is now ready for the analysis (*see* **Note 7** and Subheading 3.2.2).
4. Our RNA fragments (600 nts) need to be folded once that the aliquot is thawed. The RNA samples are stored in Milli-Q (Millipore) and then diluted to get the final ratio $Pr55^{Gag}$/RNA = 200. To properly fold the RNA, samples are denatured for 2 min at 90 °C and snap-cooled on ice for 2 min. Folding is then achieved by the addition of the binding buffer A and incubation for 15 min at 37 °C (*see* **Note 8**).

3.2 Designing the Experiment

3.2.1 Basics

In a typical DLS setup (Fig. 1, *see* **Note 9**), the sample is irradiated by a laser, and the attenuator is chosen after measuring the scattering intensity on maximum intensity; the typical values should be between 4 and 9. Recommended *count rates* for *DLS measurements* are between 100 and 500 kilo counts per second (kpcs). If the sample is too concentrated, and the solution looks white or too hazy, then the measured size of your particles will be inaccurate due to multiple scattering or viscosity effects (*see* **Note 10**). The fluctuations of the scattered light are recorded at microsecond time

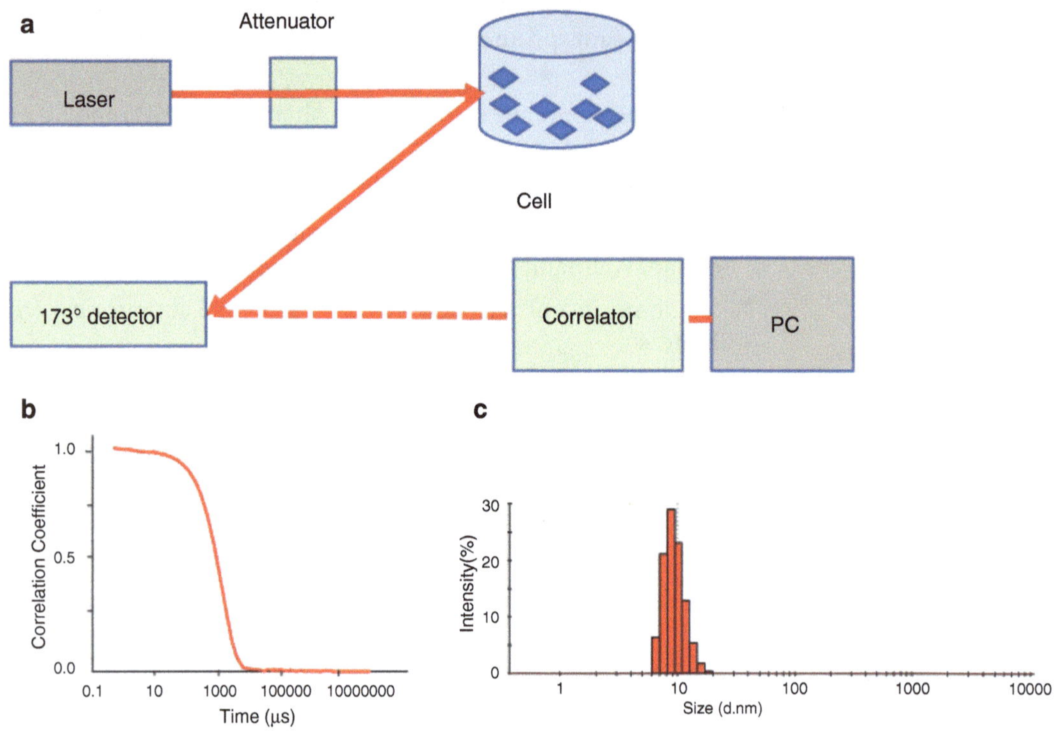

Fig. 1 (**a**) Schematic of the DLS setup. The particles in solution are irradiated by a laser beam, and the fluctuations of the scattered light intensity are recorded at microsecond time interval. (**b**) The autocorrelation function is derived, and the decay rate of the curve is proportional to the diffusion coefficient *D*. (**c**) This value is then used to determine the hydrodynamic diameter, d_h, which is represented in the intensity distribution

interval, and in a typical experiment, the software records about 10–20 files of about 10-s duration (*see* **Note 11**).

The autocorrelation function is then generated, and its intercept should be about 1, which reflects the quality of the measurements. Besides, the decay of the curve is proportional to D, and that leads to the determination of R_h which corresponds to the radius of a hard sphere diffusing in solution with the same speed as the particle being measured, or in other words, it corresponds to the size of the solvated particle that is considered as being as spherical. In practice, solvated particles can be spherical like or nonspherical; thus, R_h is only an indicator of the apparent size of the molecules or the complexes in solution that are approximated as being spherical. The R_h is relied to D by the Stokes-Einstein equation:

$$D = \frac{T k_B}{6\pi\eta R_{h.}}$$

where T is the sample temperature, η is the solvent viscosity, and k_B is Boltzmann's constant (1.380×10^{-23} kg m^2 s^{-2} K^{-1}).

The data can be represented with intensity, volume, or number-based distributions; however, the intensity distribution is usually chosen since it represents a sensitive detector for the presence of large material in the sample (i.e., aggregates) (*see* **Note 12**). Nevertheless, since DLS is based on Rayleigh theory, the scattered intensity is proportional to the sixth power of the diameter, and thus, one has to keep in mind that this analysis results heavily weighted toward larger particle size.

3.2.2 The Experiment

1. The laser should be switched on 30 min before starting the measurement to warm up.
2. The equilibration time prior to the experiment should be set at 2–5 min, and in this way, the sample should reach the exact temperature suitable for the analysis.
3. Create your own *standard operating procedure* (SOP). This last one will include the number of the measurements for each sample, the temperature, the kind of the cuvette chosen for the experiment, and also the exact composition of the medium (*see* **Note 13**).
4. Fill the cuvette with up to 50 μl of protein according the kind of cuvettes used for the DLS experiment, and check its quality. As explained before, 10–20 files of about 10 s duration would likely be enough. The distribution should be unimodal (i.e., only one peak) and monodisperse (i.e., very narrow width of distribution; *see* **Notes 14** and **15**). Finally, the autocorrelation functions were analyzed using the DTS software developed by the manufacturer, and the estimated molecular weight through the R_h should correspond to the protein molecular weight or protein multimers as it is in our example. The recombinant protein was tested in the binding buffer (buffer B), and accordingly HIV-1 $Pr55^{Gag}$ precursor in our buffer conditions corresponds to a tetramer [13] (Fig. 2).
5. Then add the previously folded RNA directly in the cuvette to get the final ratio $Pr55^{Gag}/RNA = 200$ which was chosen to analyze this interaction in conditions approaching the early steps of the viral assembly. We analyzed the interactions between $Pr55^{Gag}$ and an RNA fragment corresponding to the first 600 nts of the HIV-1 genome, called gRNA. The acquisition leaded to a corresponding mean R_h of about 170 nm (Fig. 3). Moreover, the intensity distribution of the complex displayed a unimodal and rather monodisperse character.
6. We then repeated the same analysis using two other RNA fragments of the same size and then gRNA corresponding to the spliced viral Vpr and Rev. RNA, respectively [14]. In this case, we observed that the mean R_h was about 380 and 125 nm (Fig. 4). Vpr and Rev. RNA interaction with $Pr55^{Gag}$ displayed

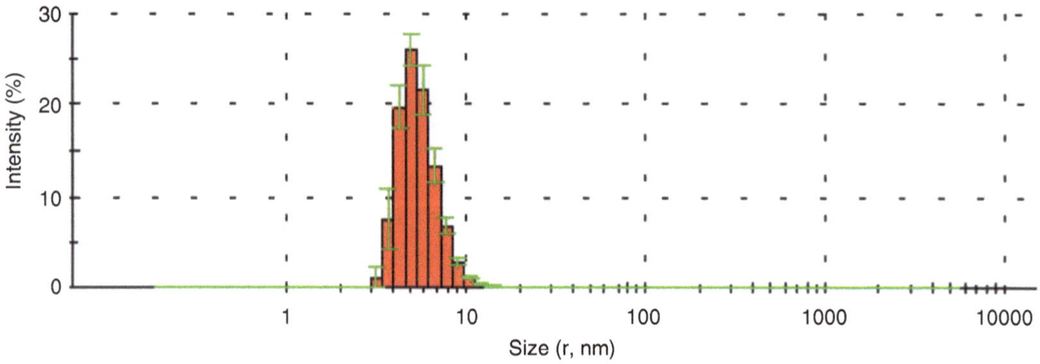

Fig. 2 Pr55^{Gag} samples were characterized by DLS in 50 mM Tris–HCl pH 8 and 1 M NaCl. The absorbance of the sample was measured and Pr55^{Gag} concentration was adjusted to 8 μM. Measurements were performed in a single 50 μl trUView cuvette (BioRad Laboratories, CA, USA), maintained at 20 °C. Intensity distribution of Pr55^{Gag} samples appeared unimodal and rather monodisperse (polydispersity index = 0.22). The mean hydrodynamic radius ($R_h = 5.1 \pm 0.7$ nm) was determined via the Stokes-Einstein equation, and this was assigned to Pr55^{Gag} trimers. Our DLS data thus indicated that our Pr55^{Gag} preparations were devoid of microaggregates, which could affect RNA binding studies

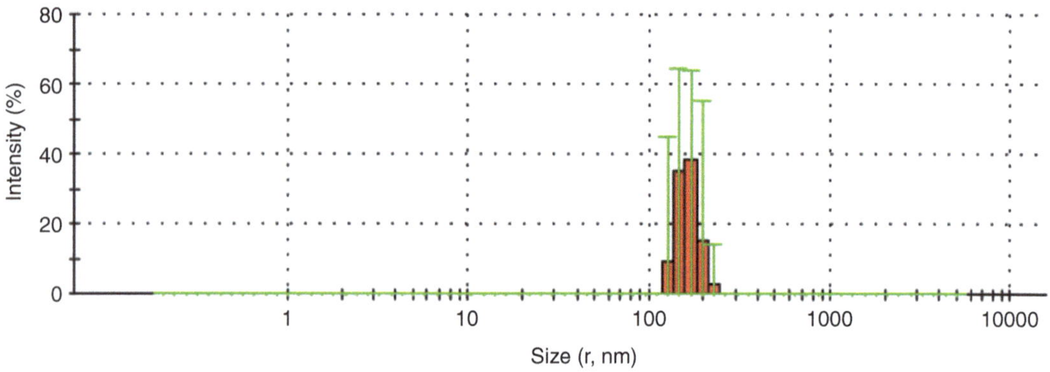

Fig. 3 The interaction between gRNA fragment and Pr55^{Gag} was characterized by DLS in 50 mM Tris–HCl pH 8 and 1 M NaCl. The ratio Pr55^{Gag}/gRNA = 200 was chosen to mimic the typical conditions of the early events of the HIV-1 viral assembly. RNA-protein complexes were achieved by adding the folded gRNA directly in the cuvette, and the complex was kept at 20 °C. Intensity distribution of Pr55^{Gag} samples appeared unimodal and monodisperse (polydispersity index = 0.17). The mean hydrodynamic radius was $R_h = 170 \pm 12$ nm

unimodal intensity distributions but rather polydisperse. This revealed that most likely several RNA-protein complexes of different sizes were present in solution; therefore, this interaction seems to be rather unspecific.

Our DLS test showed that dimeric gRNA provided more compact complexes compared to the spliced viral RNAs. In addition, DLS analysis over the time (i.e., 2–12 h after the formation of the RNA-protein complex) revealed that only the gRNA can form

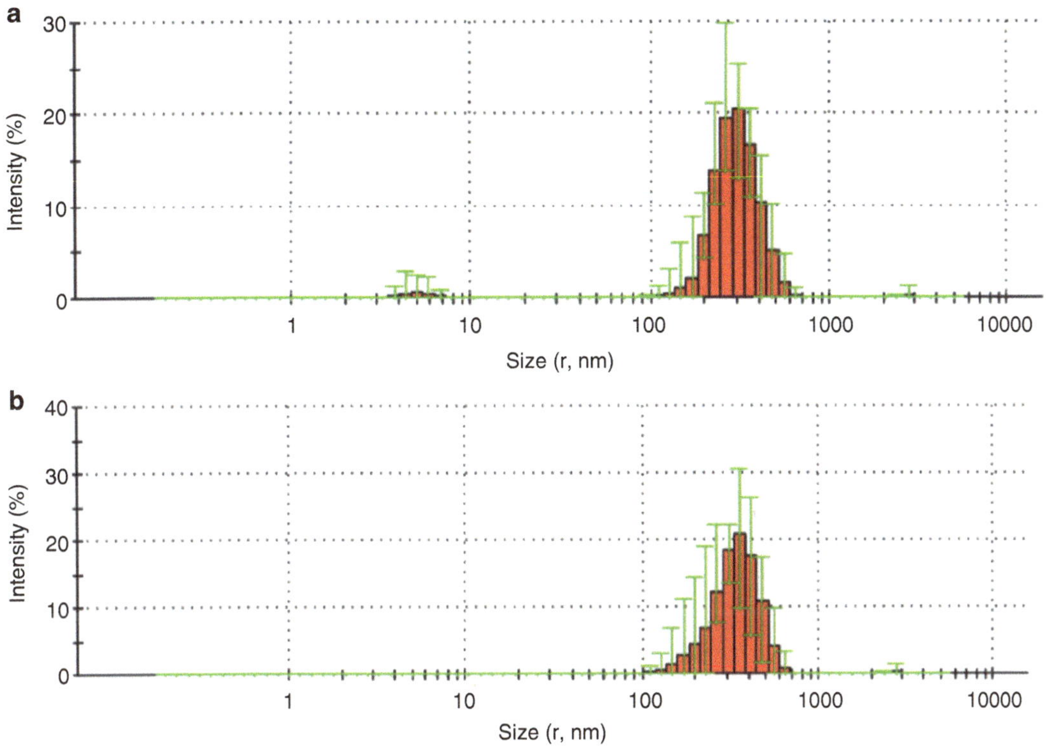

Fig. 4 The interaction between two RNA fragments and corresponding to the VPR (**a**) and REV (**b**) viral spliced RNA and Pr55^{Gag} was characterized by DLS in 50 mM Tris–HCl pH 8 and 1 M NaCl. The ratio Pr55^{Gag}/gRNA= 200 was chosen to mimic the typical conditions of the early events of the HIV-1 viral assembly. RNA-protein complexes were achieved by adding the folded gRNA directly in the cuvette, and the complex was kept at 20 °C. The mean hydrodynamic radii were $R_h = 320 \pm 11$ nm (**a**) and 425 ± 17 nm (**b**). The corresponding intensity distributions appeared unimodal but rather polydisperse (polydispersity index = 0.33 and 0.41)

stable complexes in size with the Pr55^{Gag} protein. Even if the size of gRNA-Pr55^{Gag} complexes results to be larger than the typical size of the viral particles, the capacity to stably compact RNA should play a role in efficient Pr55^{Gag} nucleation and in the viral assembly. In the natural context, other factors would contribute to stabilize protein conformation and promote their oligomerization, thus leading to proper viral particle formation.

Other techniques such as electron microscopy are necessary to provide further information about the shape of those complexes; however, in our study, DLS offered a reliable characterization of those RNA-protein interactions that regulate the early steps of the HIV-1 assembly.

4 Notes

1. All buffers used here are filtered using a 0.22 μm membrane filter (Millipore) to ensure buffer sterilization and stored at 4 °C. The buffer solutions are made fresh every week.
2. It is recommended to always rinse the filter prior to use.
3. It is recommended to check the quality of the protein in its storage buffer by a simple DLS test once that it is thawed. In this way, one can estimate easily and quickly if the recombinant aliquot of the protein is not aggregated (i.e., the corresponding intensity distribution is not unimodal) or cleaved (i.e., the observed R_h can be related to a sphere whose molecular weight (MW) approaches the expected MW of the protein or the MW of a multimer).
4. It is also possible to use quartz cuvettes instead of single-use ones. In this case, to avoid dusts that can become an issue for the analysis, it is important to clean them carefully with the alkaline Hellmanex solution (Hellma Analytics) usually 1–2 volume %. Do not leave the cuvettes in alkaline Hellmanex solution more than 1 week to avoid quartz deterioration. An increase of temperature speeds up the cleaning process; in this case, reduce the application time of the product.
5. To avoid protein aggregation, it is recommended not to thaw frozen protein aliquots more than once. Also, the DLS analysis in the storage buffer of the protein, which was chosen to preserve the protein, can be considered as a quality control of the samples.
6. Since those cuvettes are devoid of caps, Parafilm avoids dusts to enter in the cuvettes. Dust particles can be indeed a problem, and for this same reason, it is recommended to filter the samples.
7. The air bubbles in the samples should also be avoided since they can be an issue for the analysis.
8. RNA folding is a very important step in our experiences since the secondary structure elements are mandatory for efficient interaction with $Pr55^{Gag}$ protein.
9. In the given example, the intensity distribution is represented as a function of the hydrodynamic diameter, d_h, of particles diffusing in solution.
10. In our system, this value is achieved for micromolar concentrations of protein.
11. It is possible to extend the duration of the records if the molecules analyzed are particularly slow.

12. Since a small amount of larger particle species can dominate the distribution, DLS can place a strong bias on the largest particles in solution.
13. In Zetasizer™ Nano S apparatus, the medium composition can be input in the software, and accordingly, the corresponding refractive index (RI) and viscosity values for a given temperature will be associated to the experimental data set. The data analysis and the determination of the R_h require indeed the knowledge of those parameters.
14. The polydispersity index is dimensionless and is calculated from the cumulant analysis. In terms of a protein analysis, a % polydispersity less than 20% indicates that the sample is monodisperse. This % polydispersity can be roughly calculated as the half peak width of the distribution.
15. If the intensity distribution of the protein under observation looks polydisperse (i.e., most likely aggregates are present in solution), it is recommended to ultracentrifugate the samples for 1 h at 45,000 rpm or 100,000 × *g*.

References

1. Strutt JW (1871) On the scattering of the light by small particles. *Philos Mag*. Series 275:447–454
2. Einstein A (1905) über einen die Erzeugung und Verwandlung des Lichtes betreffenden heuristischen Gesichtspunkt. Ann Phys 322:132–148
3. Mie G (1908) Beitrage zur Optik trüber Medien, speziell kolloidaler Metallosungen. Ann Phys 330:337–445
4. Wyatt PJ (1993) Light scattering and the absolute characterization of macromolecules. *Anal Chim Acta* 272:1–40
5. Sutherland W (1905) A dynamical theory of diffusion for non-electrolytes and the molecular mass of albumin. Philos Mag. Series 9:781–785
6. Fujime S (1972) Quasi-elastic scattering laser light. A new tool for the dynamic study of biological macromolecules. Adv Biophys 3:1–43
7. Friksen B (2001) Revisiting the method of cumulants for the analysis of dynamic light scattering data. Appl Opt 40:4087–4091
8. Patel TR, Chojnowski G, Koul A, McKenna SA, Bujnicki JM (2017) Structural studies of RNA-protein complexes: a hybrid approach involving hydrodynamics, scattering, and computational methods. Methods 118:146–162
9. Maguire CM, Rösslein M, Wickc P, Prina-Mello A (2018) Characterisation of particles in solution – a perspective on light scattering and comparative technologies. *Sci Technol Adv Mater* 19:732–745
10. Paillart J-C, Shehu-Xhilaga M, Marquet R, Mak J (2004) Dimerization of retroviral RNA genomes: an inseparable pair. Nat Rev Microbiol 2:461–472
11. Mailler E, Bernacchi S, Marquet R, Paillart J-C, Vivet-Boudou V, Smyth RP (2016) The life cycle of the HIV-1 gag-RNA complex. Viruses 8. pii: E248. https://doi.org/10.3390/v8090248
12. Kuzembayeva M, Dilley K, Sardo L, Hu WS (2014) Life of psi: how full-length HIV-1 RNAs become packaged genomes in the viral particles. Virology 454:362–370
13. Abd El-Wahab EW, Smyth RP, Mailler E, Bernacchi S, Vivet-Boudou V, Hijnen M, Jossinet F, Mak J, Paillart J-C, Marquet R (2014) Specific recognition of the HIV-1 genomic RNA by the gag precursor. Nat Commun 5:4304. https://doi.org/10.1038/ncomms5304
14. Sinck L, Richer D, Howard J, Alexander M, Purcell DF, Marquet R, Paillart J-C (2007) In vitro dimerization of human immunodeficiency virus type 1 (HIV-1) spliced RNAs. RNA 12:2141–2150

Chapter 5

Bioinformatic Application of Fluorescence-Based In Vivo RNA Regional Accessibility Data to Identify Novel sRNA Targets

Emily K. Bowman, Mia K. Mihailovic, Bridget Li, and Lydia M. Contreras

Abstract

Data from fluorescence-based methods that measure in vivo hybridization efficacy of unique RNA regions can be used to infer regulatory activity and to identify novel RNA: RNA interactions. Here, we document the step-by-step analysis of fluorescence data collected using an in vivo regional RNA structural sensing system (iRS^3) for the purpose of identifying potential functional sites that are likely to be involved in regulatory interactions. We also detail a step-by-step protocol that couples this in vivo accessibility data with computational mRNA target predictions to inform the selection of potentially true targets from long lists of thermodynamic predictions.

Key words Regional RNA accessibility, RNA hybridization efficacy, Identification of RNA-binding sites, RNA-RNA interactions, In vivo fluorescence-based assays, Electrophoretic mobility shift assay, Target networks

1 Introduction

Regulatory noncoding RNAs are thought to harbor a vast expanse of mechanisms yet to be fully characterized in prokaryotes. Of great interest are the roles of regulatory small RNAs (sRNAs) due to their contribution to cellular homeostasis during exposures to external stresses that include starvation and oxidative stress [1]. Advances in bioinformatics and transcriptomics have enabled the prediction and validation of sRNAs in a variety of bacteria [2]; however, the known interactome of most sRNAs remains incomplete, even in model organisms such as *E. coli* [1]. Given that these sRNAs are most commonly involved in stress response as survival by tuning expression of mRNA targets, there has been considerable interest in expansion of sRNA target networks as a means of elucidating sRNA-specific cellular roles. Successful identification and subsequent manipulation of these roles have helped to define an arena of metabolic engineering that uses antisense RNAs in

Véronique Arluison and Frank Wien (eds.), *RNA Spectroscopy: Methods and Protocols*, Methods in Molecular Biology, vol. 2113, https://doi.org/10.1007/978-1-0716-0278-2_5, © Springer Science+Business Media, LLC, part of Springer Nature 2020

conjunction with other defined bioengineering methods to artificially tune gene expression for multiple purposes, including overproduction of metabolites such as amino acids and TCA cycle products [3].

Traditional methods of characterizing sRNA targets have primarily consisted of MS2-associated pulldowns or MAPS [4]. However, these methods support the profiling of only a single sRNA at a time and rely on overexpression of a tagged RNA in a way that alters the native environment and, consequently, native regulation [5]. Furthermore, as detected interactions may be indirect, each target must be individually validated by complementary approaches. Some methods have risen to the challenge of high-throughput target identification of multiple regulatory RNAs simultaneously and, importantly, in their native environment, such as cross-linking, ligation, and sequencing hybrids (CLASH) [6, 7]. However, this method is limited by low recovery of ligated RNAs and requires immense sequencing depth to obtain statistical significance of lowly abundant RNA pairs [8]. To enrich for regulatory interactions, approaches that exploit protein mediation of RNA-RNA interactions (via cross-linking and coimmunoprecipitation) have been developed [9–11]; however, these methods overlook both unchaperoned RNA regulation and regulation by proteins not identified for pulldowns. In addition, although these approaches are adept for pinpointing true interactions due to sequence specificity of scars, no current software package for analyzing ligated pairs considers sequence complementarity [8]. As such, the integration of base pair complementarity between sRNAs and their targets is missing within the analyses of interactome data [8].

Other computational approaches have exploited the specificity of Watson-Crick base pairing within RNA secondary structure and regulatory intermolecular RNA-RNA interactions to scan genomes in search of both novel sRNAs and potential mRNA targets [12]. Although accuracy of target prediction tools can be achieved, especially with homology context, true targets can often rank lowly and be overlooked, while false targets can rank highly. For example, for a subset of sRNAs shown to have direct RNA-RNA interactions with their targets, 20% of true sRNA targets ranked below the top 100 in many of the best target predictors [12]. Given the labor-intensive nature of experimental target validation, it is valuable to address limitations of target prediction programs to increase confidence of computationally-predicted targets. Target misidentification can in part be attributed to the heavy reliance of computational predictions on free energy predictions of interaction and lack of in vivo insight [12]. For example, these algorithms consider the entire sequence space of the sRNA; however, sRNAs can use as few as six nucleotides to interact with their targets [13]. The remainder of the sRNA's sequence space can either serve as binding sites for chaperones, including Hfq, ProQ, and CspC [14], structural elements, or are merely artifacts of evolution [15]. It is also well-accepted that sRNAs rely on many intracellular factors to modify their structure and enable the hybridizability of these particular regions for the target interaction [13]. Therefore, informing computational predictions with in vivo accessibility data can help elucidate potential targets that would otherwise go unstudied.

In this chapter, we report how in vivo hybridization data from iRS^3, an *E. coli* fluorescence-based assay that can survey the native regional accessibility landscape of any RNA with known sequence [16], can be coupled with computational target prediction methods, such as IntaRNA [17–19], to select in vivo-supported predicted sRNA targets. To accurately identify the accessibility propensity of all regions within any RNA using iRS^3 assay, antisense RNA (asRNA) probes are designed to hybridize to 9–16 nucleotide segments within the RNA of interest. These antisense RNAs are cloned in as the 5′ end of an inducible reporter assay that contains a hairpin loop, in which a ribosomal binding site (RBS) is sequestered by a cis-blocking region, followed by a green fluorescent protein (GFP) coding sequence (Fig. 1a). In the event the asRNA probe cannot access its target region, the ribosomal-binding site stays sequestered, inhibiting downstream translation of GFP (Fig. 1b). When the asRNA probe is able to interact with its target region, the hairpin loop structure is disrupted, revealing the RBS and allowing for translation of the downstream GFP (Fig. 1c). The accessibility of user-defined RNA regions of interest can be determined based on fluorescence shifts upon induction of the iRS^3 system, allowing a quantitative measure of hybridization favorability of any region along an RNA [16, 20]. It is important to note that the method's mimicry of true regulatory RNA interactions in vivo that occur at the regional level with high sequence specificity can tell us whether or not an RNA region will engage in interaction with its cognate. A powerful application of intracellular accessibility data of regions along an RNA of interest is the ability to gauge the likelihood that certain regions are more likely to be present in states in which they are hybridized, intra- or intermolecularly. Although this assay cannot directly pair regulatory sRNAs with targets or discriminate accessibility effects due to intra- or intermolecular hybridization, RNA hybridization data has been key in uncovering true targets of regulatory RNAs of interest [21]. A high-throughput version of this iRS^3 assay, relying on transcriptional elongation rather than fluorescent protein expression to determine hybridization (INTERFACE), has uncovered characteristic hybridization patterns of regulatory RNA regions within well-characterized *E. coli* sRNAs [21]. Specifically, by simultaneously probing accessibility in over 200 regions in 16 sRNAs, this work found that RNA regions corresponding to previously mapped mRNA-binding sites were enriched with extremely high or low

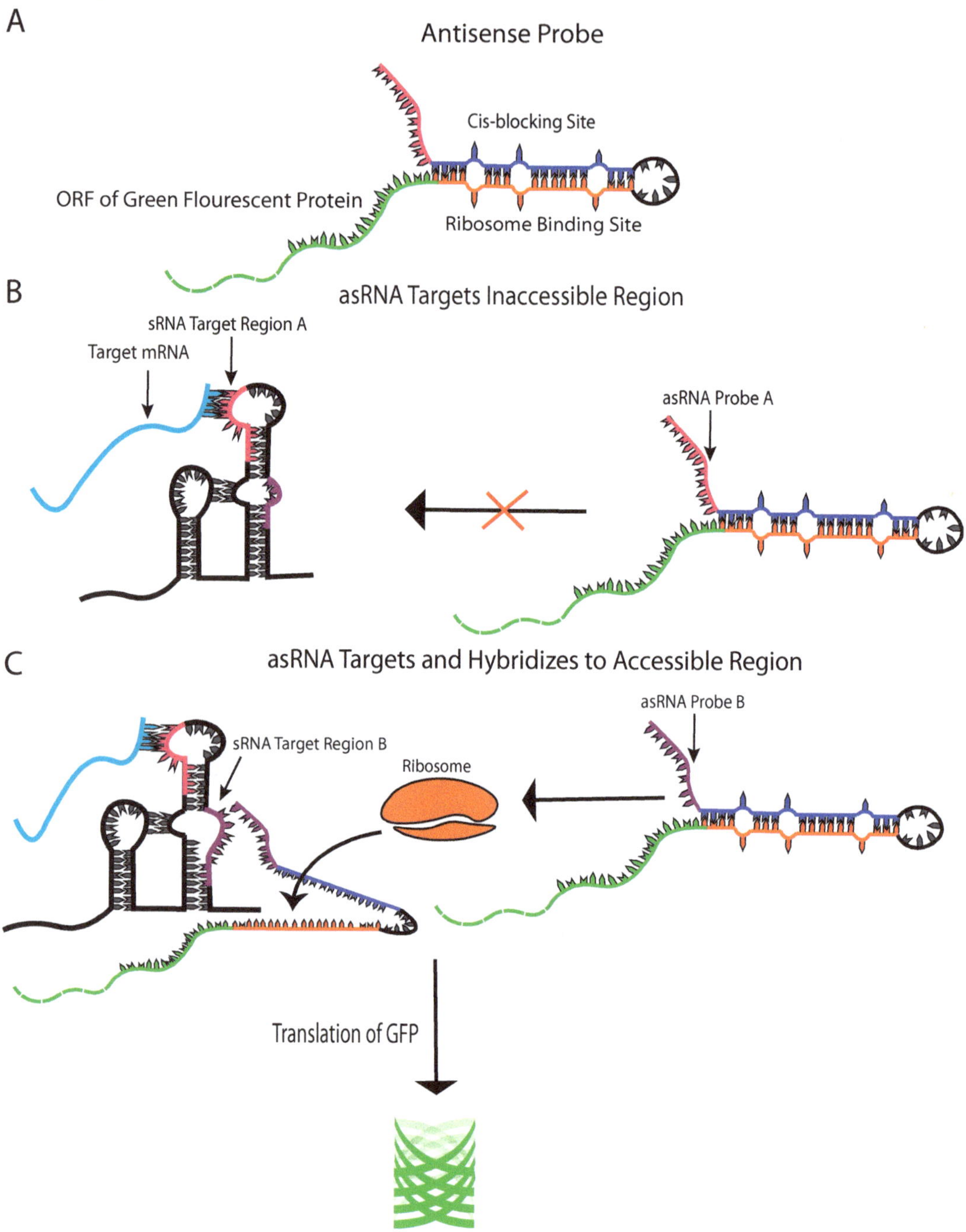

Fig. 1 Molecular basis of the iRS^3 assay. (**a**) The iRS^3 assay is capable of quantifying hybridization of any RNA region in vivo by means of its molecular design: a perfectly antisense RNA probe (gray) is followed by a hairpin in which a cis blocking site (blue) sequesters the RBS (orange) upstream of the GFP coding sequence (green). (**b**) When targeting an inaccessible region, the asRNA probe cannot hybridize to target region of interest; thus, the iRS^3 hairpin remains stable and GFP is not expressed. (**c**) When targeting an accessible region, the asRNA probe will hybridize to its target region and destabilize the hairpin to reveal the ribosome-binding site, allowing for GFP translation

accessibility in vivo. Importantly, these patterns were the basis for a hypothesis in which levels of extreme regional RNA accessibility within any RNA can serve as an indicator that the particular RNA stretch is more likely than surrounding regions to form intermolecular RNA interactions. Therefore, regions matching these extreme patterns of accessibility within an RNA of interest, but previously unassociated with known targets, were flagged. These regions were then used to enrich computationally-predicted targets that may have been previously overlooked due to low ranking. In this way, patterns of regional in vivo accessibility were fruitful filters for pinpointing true, lowly ranked, thermodynamically predicted targets in three previously uncharacterized sRNAs (SroE, SroG, Tpke70) (Table 1). Importantly, half of these confirmed targets did not overlap with mRNAs in the equivalent ranking of unfiltered IntaRNA results [21]. The establishment of binding site-indicative in vivo RNA accessibility information as successful complementation to computational predictions inspired the similar validation of accessibility-informed computationally-predicted targets in uncharacterized sRNAs (Tpke11, SroA), as well as the expansion of target networks for previously characterized sRNAs (CyaR, GlmY) (Table 1). Furthermore, the observation of extremely high accessibility of the GlmY terminator led to recognition of rare terminator-specific mRNA binding in this sRNA previously only recognized for protein sequestration activity [22] (Table 1).

This chapter details the bioinformatic application of fluorescence-based regional RNA accessibility data for the purpose of minimizing effort associated with experimental confirmation of computationally-predicted sRNA targets (Fig. 2), using the successful characterization of novel terminator mRNA-binding activity in the sRNA GlmY as a case study [21]. Specifically, the chapter covers relevant iRS^3 experimental methods (Fig. 2a), including the selection of target regions for accessibility probing, analysis, and interpretation of iRS^3 results, as well as the use of this data to identify likely functional regions. Next, we cover the computational prediction of sRNA interactions by IntaRNA to obtain a list of potential mRNA targets ranked by thermodynamic favorability (Fig. 2b). Thereafter, we detail the bioinformatic selection of computationally-predicted mRNA targets that are supported by their predicted interaction with sRNA regions exhibiting functionality as inferred from in vivo accessibility and further filtered by available transcriptomic data (Fig. 2c). Finally, this chapter details the in vitro experimental confirmation of potential targets using electrophoretic mobility shift assays (EMSAs) to identify sRNA-target binding (Fig. 2d). Importantly, the approach detailed in this chapter has the potential to expand the known targetome of any RNA anticipated to engage in RNA-RNA interactions in vivo. We furthermore anticipate that adapting this approach for investigation of environment-

Table 1

Comprehensive list of acessibility-informed computationally-predicted sRNA targets that were confirmed via in vitro binding assays in Mihailovic and Vazquez-Anderson 2018 [21]. Encoded functions of mRNA targets were obtained from UniProt [36]. Column 3 indicates whether the mRNA target was selected from predictions corresponding to the entire sRNA sequence ("Full Sequence") or only the proposed binding site sequence inferred from regional accessibility data ("Binding Site"). Grey font indicates targets that were also flagged via un-informed computational predictions. For SroE, SroG, SroH, and Tpke70, this is defined by an equivalent rank-based cutoff to that of the re-ranked INTERFACE-informed predictions; for all other sNRAs, it is defined as un-informed prediction containment within the top 15

sRNA	Confirmed mRNA target	Identified via Full Sequence or Binding Site-Specific Prediction?
CyaR	*zapC* cell division protein	Binding Site
GcvB	*thrL* thr operon leader peptide	Binding Site
	gadX HTH-type transcriptional regulator	Full Sequence
GlmY	*yphC* zinc-binding protein	Binding Site
	add adenosine deaminase	Full Sequence
SroA	*rhtA* threonine/homoserine exporter	Full Sequence
	rplK 50s ribosomal protein	Full Sequence
SroE	*ligB* DNA ligase B	Full Sequence
SroG	*cdsA* phosphatidate cytidylyltransferase	Full Sequence
	cysJ sulfite reductase (NADPH) flavoprotein alpha-component	Full Sequence
	yfbR 5'deoxynucleotidase	Full Sequence
	yncD probable tonB dependent receptor	Full Sequence
	mepM murein DD-endopeptidase	Full Sequence
	potA spermidine/putrescine import ATP binding protein	Full Sequence
	pgI glucose-6-phosphate isomerase	Full Sequence
	fecE iron dicitrate transport ATP binding protein	Full Sequence

(continued)

Table 1 (continued)

SroH	*kbl* AKB ligase	Full Sequence
Tpke11	*pgsA* PGP synthase	Full Sequence
	prlC oligopeptidase A	Full Sequence
	yehU sensor histidine kinase	Full Sequence
Tpke70	*mcrA* 5-mehtylcytosine-specific restriction enzyme A	Full Sequence
	torT periplasmic protein TorT	Full Sequence
	glnK nitrogen regulatory protein P-II 2	Full Sequence

specific regional accessibility patterns will boost its utility in characterization and subsequent modification of RNAs for metabolic engineering applications.

2 Materials

2.1 iRS³ General Materials and Reagents

1. Microcentrifuge.
2. 2.5 μL–1000 μL Pipette set.
3. Disposable pipette tips.
4. 1.7 mL Polypropylene microtubes.
5. PCR tubes.
6. Nuclease-free water.
7. Thermocycler.
8. Luria broth.
9. Agar.
10. Petri dishes.
11. 25 mL Culture tubes.
12. Kanamycin stock solution: 10 mg/mL or 100 mg/mL in nuclease-free water.
13. Incubator set to 37 °C with shaking capability.
14. Spectrophotometer.

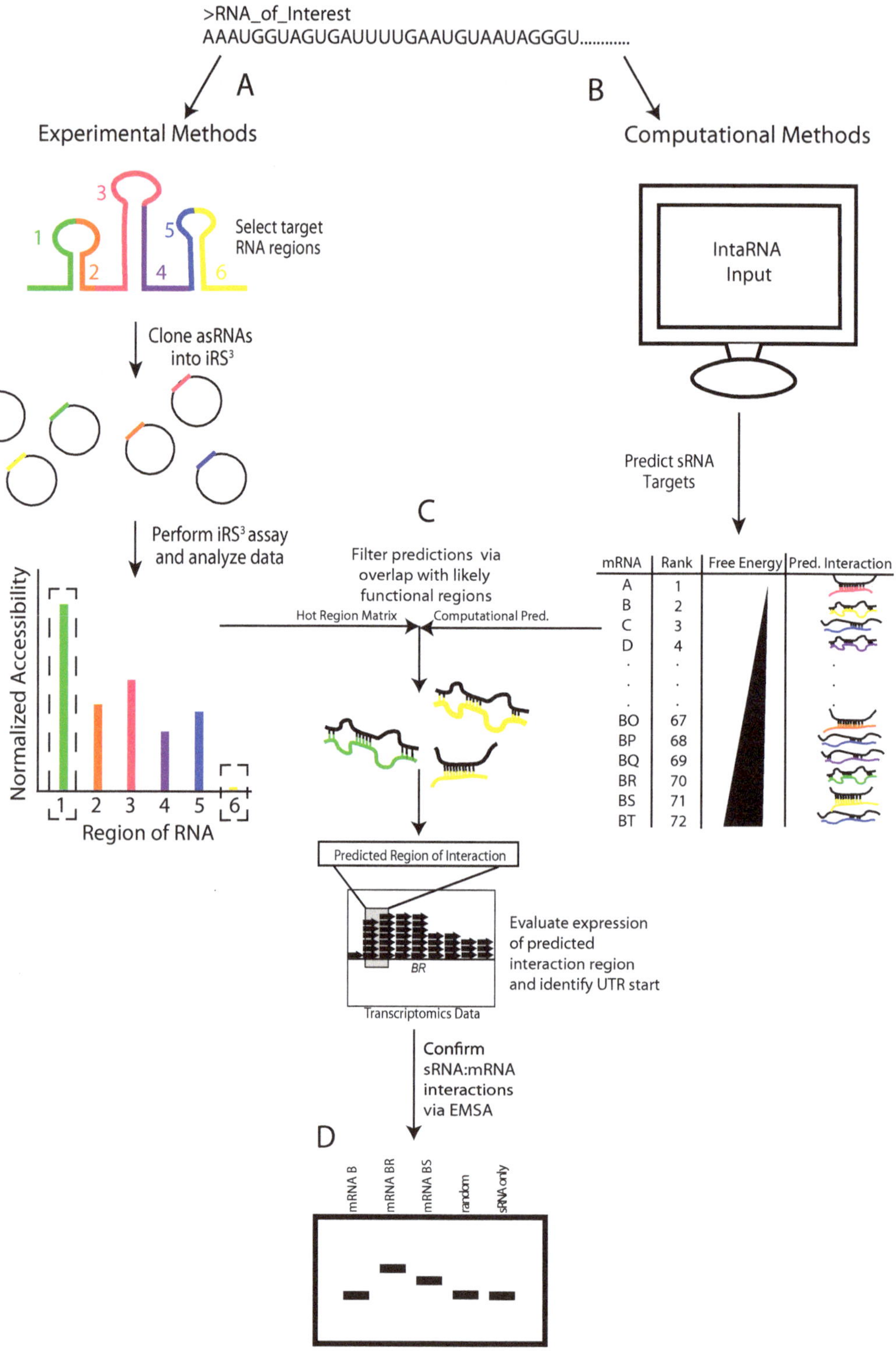

Fig. 2 Overview of the bioinformatic application of iRS^3 data for the purpose of suggesting in vivo-supported computationally-predicted mRNA targets for experimental validation. Experimental (**a**) and computational (**b**) methods can be performed concurrently for a given RNA sequence of interest (top). (**a**) Steps of the iRS^3

2.2 iRS³ asRNA Probe Insertion

1. 50 ng pN-iRS³GG vector (Addgene plasmid #98858).
2. Linker buffer: 50 mM Tris–HCl pH 8.0, 100 mM NaCl, 1 mM EDTA.
3. Primers diluted in linker buffer to 100 μM.
4. 10X T4 DNA ligase buffer (NEB).
5. T4 DNA ligase (400,000 U/mL, NEB).
6. Bsmb1/Esp3I (10 U/μL, NEB).
7. Nuclease-free water..
8. 0.025 μm Nitrocellulose membrane filters.
9. Electrocompetent *E. coli* cells for cloning.
10. 4-Chloro-DL-phenylalanine.

2.3 iRS³ Experimentation

1. Transformation-competent *E. coli* K-12 MG1655 or another experimental strain of choice.
2. Sterile 250 mL Erlenmeyer flasks.
3. Sterile 200 μL 96-well black clear-bottom plates.
4. 20 mL Syringes with Luer lock tip.
5. Syringe-tip filters (0.22 μm PVDF 30 mm diameter).
6. Arabinose stock solution: 20% w/v, filter sterilized through 0.22 μm filter.
7. 1X Phosphate-buffered saline (PBS): 137 mM NaCl, 2.7 mM KCl, 8 mM Na_2HPO_4, and 2 mM KH_2PO_4, sterilized through 0.22 μm filter.
8. 5 mL Polystyrene round-bottomed tubes.
9. BD FACSFlow sheath fluid.
10. BD FACSCalibur flow cytometer and BD CellQuest Pro software.

Fig. 2 (continued) experimental design and method. First, target regions within the RNA must be selected. Next, asRNA probes corresponding to the selected target regions are cloned into the iRS³ plasmid. Upon experimentation, accessibility data is normalized, and potential functional regions (green region 1 and yellow region 6 due to their extreme accessibility and inaccessibility in vivo, respectively) are identified and compiled into a "hot region matrix." (**b**). Computational predictions of sRNA targets are obtained via IntaRNA, or another preferred program, and ranked by thermodynamic favorability. Color of sRNA in predicted sRNA:mRNA interactions correspond to the sRNA region (panel **a**) predicted to interact with the given mRNA target. (**c**) Experimental and computational results are bioinformatically coupled to select for computationally-predicted targets whose predicted interaction relies on potential functional regions identified via iRS³ experimentation. This pool is further filtered by available transcriptomic data. (**d**) By selecting predicted targets that are supported by in vivo accessibility data, experimental effort associated with EMSAs can be reduced

2.4 Computational Prediction of mRNA Targets of sRNAs

1. Access to IntaRNA (Web server or local installation).
2. RNA of interest (i.e., sRNA) sequence (and/or individual sequence of functional regions if applicable; *see* **Note 1**) in FASTA format.
3. GFF and FASTA files for potential targets of RNA of interest.
4. Rstudio [23] including dplyr [24], biostrings [25], and stringr [26] packages.

2.5 Bioinformatic Filtration of Computationally-Predicted mRNA Targets

1. Anaconda (Python launcher) [27] including Spyder (https://github.com/spyder-ide) and Pandas [28].
2. Microsoft Excel.
3. Integrative Genomics Viewer (IGV) [29].

2.6 General Materials and Reagents for Electrophoretic Mobility Shift Assays (EMSAs)

1. Microcentrifuge (*see* **Note 2**).
2. 2.5–1000 μL pipette set (*see* **Note 2**)
3. Disposable pipette tips (*see* **Note 3**).
4. Nuclease-free water (*see* **Note 3**).
5. RNase-free 0.6 mL/1.7 mL polypropylene microtubes.
6. 50 mL Conical centrifuge tubes (Falcon)
7. RNase decontamination solution (RNase*Zap*).
8. UV spectrophotometer (*see* **Note 2**).
9. 10X Tris/borate/EDTA (TBE) buffer.
15. 70°C Heat block compatible with 0.6 mL microtubes.

2.7 In Vitro Transcription (IVT)-Compatible PCR Amplification of sRNAs and mRNAs

1. Access to genomic DNA of selected experimental strain.
2. PCR tubes.
3. 1.7 mL Polypropylene microtubes.
4. Deoxynucleotide solution mix (NEB).
5. 5x Phusion high-fidelity (HF) buffer (NEB).
6. Phusion high-fidelity (HF) DNA polymerase (2000 U/mL, NEB).
7. Thermocycler.
8. Illustra GFX PCR DNA and Gel Band Purification Kit (GE Healthcare).

2.8 IVT of sRNAs and mRNAs

1. MEGAscript T7 transcription kit (Invitrogen).
2. [P32]-UTP, 10 mCi/mL (Perkin Elmer) (*see* **Note 2**).
3. 200 Proof ethanol (OmniPur).
4. RNA clean and concentrator RCC kit 5 (Zymo Research).
5. Mini-PROTEAN Tetra Cell (Bio-Rad).

6. Power Pac HV Power Supply (Bio-Rad).
7. Urea gel concentrate (National Diagnostics Ultra-Pure SequaGel).
8. Urea gel diluent (National Diagnostics Ultra-Pure SequaGel).
9. Urea gel buffer (National Diagnostics Ultra-Pure SequaGel).
10. Urea gel loading dye (gel loading buffer II, Invitrogen).
11. 10% Ammonium persulfate (APS) solution, sterilized through 0.22 μm filter.
12. Tetramethylethylenediamine (TEMED).
13. Low-range ssRNA ladder (NEB).
14. ChemiDoc XRS + gel imager (Bio-Rad).

2.9 EMSA

1. Access to a fume hood.
2. 40% wt/vol Acrylamide:bisacrylamide, 19:1.
3. Glycerol.
4. APS.
5. TEMED.
6. PROTEAN II xi cell with 16 × 20 cm plates (Bio-Rad).
7. Power Pac HV Power Supply (Bio-Rad)..
8. 5X EMSA binding buffer: 20 mM Tris–HCl pH 8.0, 1 mM DTT, 1 mM $MgCl_2$, 20 mM KCl, 10 mM $NaHPO_4$ adjusted to pH 8.0.
9. 37 °C Water bath or incubator designated for radiation work.
10. 70 °C Heat block.
11. EMSA loading buffer: 10 mM Tris, 50% glycerol, 0.0001% wt/vol bromophenol blue.
12. Gel dryer (Bio-Rad).
13. Phosphorimaging cassette (bioWORLD).
14. Access to PhosphorImager (GE Health Life Science Typhoon).

3 Methods

3.1 iRS^3 Experimental Design and Execution

This section summarizes a previously published chapter that details cloning, experimental procedure, and data analysis for using the iRS^3 assay to probe regional accessibility in natively expressed RNAs [20] (*see* **Note 4**). Here, we briefly review the iRS^3 protocol and highlight the important design consideration of target region selection for the RNA of interest as well as data analysis and normalization. Importantly, this section emphasizes the identification of regions that show functional characteristics previously defined, specifically, those in the 5′ and 3′ ends that show extremely high or low accessibility [21].

3.1.1 Selection of Regions Within RNA of Interest for Accessibility Probing

As previously described [20], target regions identified for iRS^3 probing should be between 9 and 16 nucleotides to maintain targeting specificity and minimize asRNA probe secondary structure [30]. Furthermore, we recommend confirming the sequence specificity of any proposed asRNA probe via sequence alignment, i.e., BLASTn (NCBI) [31].

Aside from these guidelines, selection of targeting regions within an RNA of interest is entirely dependent on the downstream utility of information. For example, known binding sites of a well-characterized RNA could be targeted to infer site- or environment-specific activity [32]. Alternatively, if the user intent is to use in vivo accessibility to infer previously uncharacterized functionality of an RNA, it is important to consider that target region adjustments implicating only a few base pairs have exhibited drastic hybridization changes [21], likely due to dependence on toeholds [33]. Thus, in this case, full, overlapping coverage of the RNA of interest is recommended to increase the likelihood that sites with strong binding activity are not overlooked.

To reduce experimental effort, estimation of hybridization efficiency via a biophysical model that considers thermodynamic and structural ensemble predictions can be utilized. Importantly, this model was optimized with previously collected hybridization data for the purposes of capturing influences of intracellular factors [30]. Reducing iRS^3 experimental effort within full-coverage constraints (by considering predicted hybridization efficacy, as estimated using the aforementioned biophysical model) is the basis for the second design technique discussed in this section; notably, this strategy was used to design target regions for sRNA of interest, GlmY (Table 2, Fig. 3a, b).

1. Rational Design Based on Prior Functional Knowledge: This approach is recommended for conducting targeted regional hybridization evaluation to inform sRNA redesign. Given the focused nature of this approach, it is necessary to assay RNA regions whose functionality is known, such as highly suspected (based on sequence and structural features) or experimentally confirmed binding sites. Once accessibilities have been measured, site-specific activity contributions can be inferred based on previously characterized functional attributes of the molecule to advise redesign.

   This target region selection approach was previously used for the successful modular redesign of the CsrA sponge sRNA, CsrB, to construct a library of CsrB mutants with differential binding activity [32]. Specifically, the accessibilities of 9–16 nucleotide stretches within CsrB that contain the canonical protein target-recognition sequences (ANGGA) were probed. Site-specific activity of these regions was inferred based on the hypothesis that active target-binding sites on the RNA will be

Table 2
Accessibility table of GlmY regions of interest, including the region number, region position(s), normalized accessibility data with standard error, and antisense RNA probe sequence. Note the extremely high accessibility in the terminator region of the sRNA, region 18, with small error. The full sequence of the GlmY sRNA is found at the bottom of the table

GlmY region of interest	Start NT	End NT	Normalized accessibility	asRNA probe sequence
1	1	9	0.12 ± 0.019	UGAGCCACU
2	4	16	0.38 ± 0.041	CGGUGAAUGAGCC
3	16	29	0.69 ± 0.053	GGCUGACAUAAGUC
4	30	45	0.00 ± 0.009	UAGCACGUCCCGAAGG
5	45	53	0.37 ± 0.189	AUUUUAUGU
6	49	64	0.68 ± 0.122	GCGUCAUUCGUAUUUU
7	65	80	0.58 ± 0.090	GCAGGCACCUUGUUGU
8	75	83	0.66 ± 0.175	ACGGCAGGC
9	84	99	0.62 ± 0.078	UGAUAUCAGAAGUUGG
10	92	100	0.41 ± 0.095	CUGAUAUCA
11	101	116	0.47 ± 0.040	GGUUGAUAUAGCUACG
12	114	127	0.51 ± 0.023	UUUCGCCCGAUGGU
13	124	133	0.24 ± 0.045	UCGACGUUUC
14	132	147	0.59 ± 0.113	AGGCGGUGCCUUAACUC
15	141	150	0.58 ± 0.180	AUAAGGCGGU
16	150	165	0.53 ± 0.150	CGGCUUUGUUAUGGAA
17	166	181	0.75 ± 0.099	AAAGCCGGGAAUUACC
18	174	184	1.00 ± 0.012	AACAAAGCCGG

Full GlmY sequence: AGUGGCUCAUUCACCGACUUAUGUCAGCCCCUUCGGGACGUGCUACA UAAAAUACGAAUGACGCACAACAAGGUGCCUGCCGUCCAACUUCUGAUAUCAGCGUAGCUAU AUCAACCAUCGGGCGAAACGUCGAGUUAGGCACCGCCUUAUUCCAUAACAAAGCCGGGUAAU UCCCGGCUUUGUU

occluded to the asRNA probe. Based on these inferences, target recognition sites were redistributed on CsrB mutants to affect overall protein-binding activity of CsrB [32]. This stepwise approach to rational design via probing for accessibility could be applied to any RNA that has known functional sites. We further anticipate that comparison of site-specific activity under varying environmental conditions will be valuable to designing condition-specific RNA regulation.

2. Generation of Full Accessibility Profiles with Minimized Experimental Effort: Previously, a sparse knowledge gradient (SpKG) algorithm was used to help inform identification of target

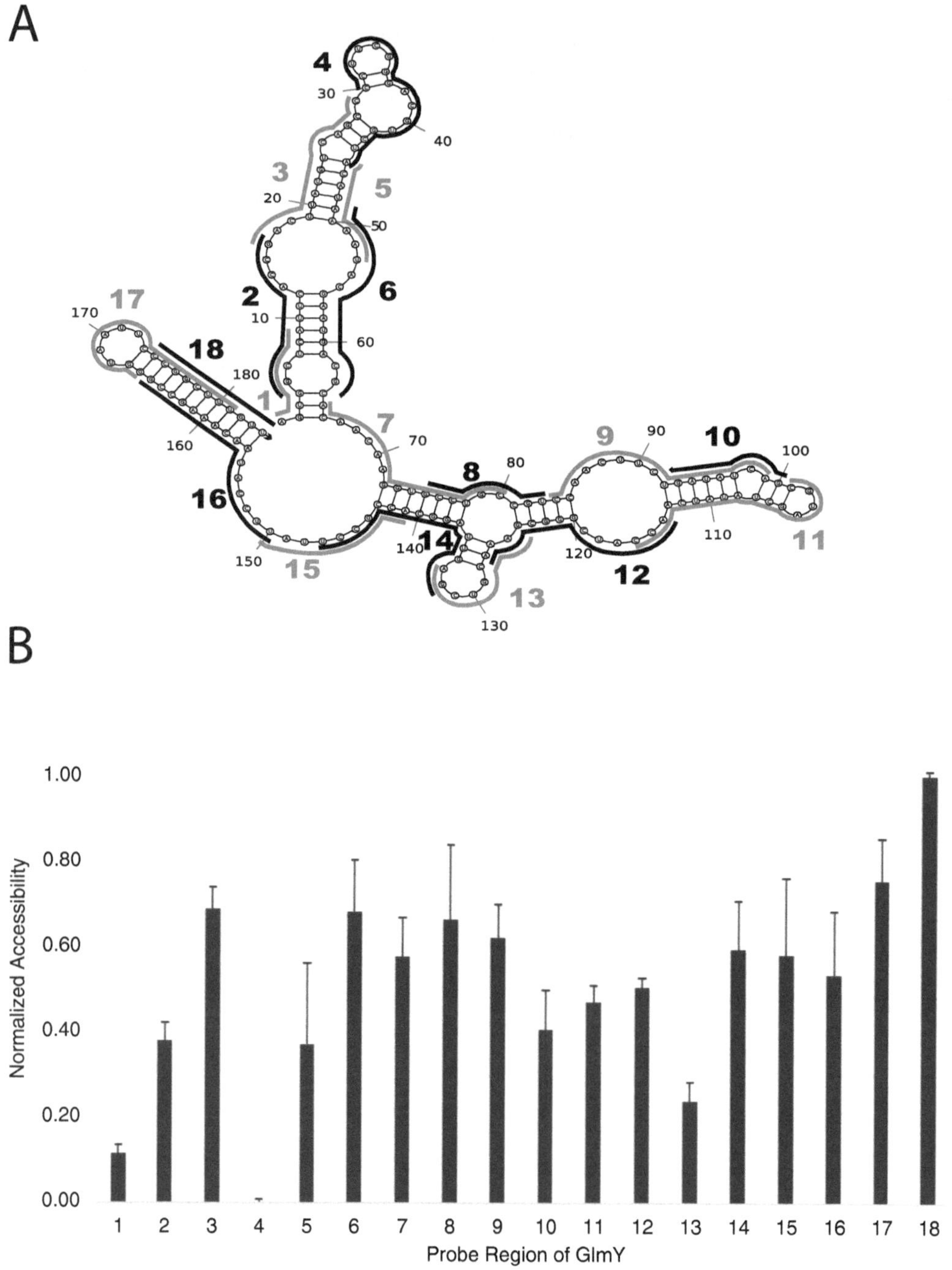

Fig. 3 Target map (**a**) and in vivo accessibility profile (**b**) of GlmY regions. (**a**) Nupack-predicted [34] minimum free energy secondary structure of RNA of interest, GlmY. Mapped onto the GlmY structure are all individual regions whose accessibility was interrogated by iRS[3]. Note the strong secondary structure of the terminator loop (fully containing regions 17 and 18). (**b**) Graph of normalized accessibility of GlmY with standard error bars. It is interesting to note that the terminator loop region exhibits high accessibility **in vivo** despite its strong predicted secondary structure

regions for accessibility probing to make full accessibility profiles of multiple sRNAs, including GlmY (Fig. 3b) [21]. Briefly, a machine learning algorithm was adapted to a weighted set cover problem to get full coverage of an RNA of interest within asRNA probe length and overlap constraints. Importantly, this algorithm minimizes experimental effort by simultaneously enriching the likelihood that extremely accessible regions are assayed and by minimizing the total number of regions that are targeted.

Furthermore, this target region design strategy has built-in overlap tolerance to increase likelihood that functional regions are assayed. For example, considerable target region overlap can be seen in the case of GlmY (Fig. 3a). Due to the comprehensive nature of this approach, this targeting strategy offers a complete and information-rich accessibility profile that can support identification of novel functional regions and selection of corresponding predicted mRNA targets for experimental validation.

3.1.2 asRNA Probe Insertion

Once target regions are selected, unique iRS^3 constructs containing asRNA probes corresponding to each region must be cloned (from parent plasmid pN-iRS^{3}GG). In plasmid pN-iRS^{3}GG, a PheS cassette serves as an asRNA placeholder that confers strain sensitivity to 4-chloro-DL-phenylalanine. Importantly, the PheS cassette is flanked by BsmbI sites, which enable golden gate cloning-based replacement of PheS by asRNA probes. Thus, successful insertion of asRNA probe sequences allows for selection on agar plates containing 4-chloro-DL-phenylalanine. An in-depth detail of this cloning process can be found in Leistra and Mihailovic et al. 2018 [20].

3.1.3 iRS^3 Experimentation

iRS^3 experiments should be performed as previously described [20]. Briefly, iRS^3 plasmids are transformed into the desired experimental strain (*see* **Note 5**) and shake-incubated at 37 °C (40 mL in flasks or 200 μL in plates) until early exponential phase. Each culture is then split into two volumes, and 20% arabinose is added to corresponding induced samples for a final concentration of 0.8%. At desired sampling OD, all samples are resuspended in 1X PBS for green fluorescence measurement via flow cytometry. To ensure proper function of the assay, we recommend including iRS^3 control constructs that represent the widest range of fluorescence shift expected—a scramble iRS^3 construct (small fluorescence shift upon induction), in which the asRNA probe is a randomized sequence, and the open RBS (large fluorescence shift upon induction), a mutated version of the scramble construct in which the CB sequence is modified to support exposure of the RBS independent of asRNA hybridization. Previously used random sequences for use in *E. coli* K-12 MG1655 and BW25113 are outlined elsewhere [20].

3.1.4 Data Analysis

1. Calculation of Accessibility: Fluorescence results are expected to have a non-normal distribution; therefore, median fluorescence is assumed representative of the population. Accessibility, or hybridization efficacy, of an RNA region is calculated as the average logarithm of the ratio of the fluorescence signal obtained from induced iRS^3 plasmid conditions to uninduced (background) fluorescence.
2. Intra-RNA Accessibility Normalization: When evaluating accessibility profiles of multiple RNAs of interest for the purpose of identifying likely functional regions, it may be valuable to intramolecularly normalize accessibility results from 0 to 1. In this manner, any abundancy-attributed absolute accessibility differences between RNAs of interest (sRNAs), which are expected due to sRNA tendency to differentially express between conditions associated with their function, are eliminated. Normalized GlmY accessibility results can be seen in Table 2 and Fig. 3b. Importantly, such normalization allows for ease of recognition of extremely high/low accessibility regions within a library of RNAs of interest.

3.1.5 Identification of Likely Functional Regions to Make a Hot Region Matrix

In light of previous successes in using regional position and in vivo RNA accessibility results to ultimately validate computational target predictions outside of the typically tested, top-ranked pool [21], we suggest selecting likely functional regions based on their (i) extreme normalized accessibility and (ii) position in the 5′ or 3′ ends of the RNA of interest (i.e., sRNA) (*see* **Note 6**). After intra-RNA normalization, regions of extreme accessibility on the RNA of interest should be apparent, particularly with the visual aid of an accessibility chart, as seen with example sRNA of interest, GlmY (Fig. 3b). For example, high average normalized accessibility (> 0.75 on 0–1 scale) in the GlmY terminator region (regions 17 and 18, with particularly low standard deviation in region 18) prompted the labeling of nucleotides 166–184 as likely functional (Table 2) [21].

3.2 IntaRNA Prediction of sRNA Targets

Here, the use of IntaRNA to predict mRNA targets of RNAs of interest is detailed. We describe two approaches to using IntaRNA for target prediction—using the IntaRNA Web server (http://rna.informatik.uni-freiburg.de/IntaRNA/Input.jsp) or a locally installed version of the software (*see* **Note 7**). In the featured case of sRNAs as RNAs of interest, we suggest the restriction of potential target mRNA sequences as [likely] 5' UTRs (*see* **Note 8**) and the first 100 nucleotides of the coding sequence to account for the high frequency of sRNA interaction at 5′ ends of mRNAs [1]. The output of IntaRNA predictions is a list of potential mRNA targets ranked by predicted thermodynamic favorability. A representative set of IntaRNA predictions for RNA of interest GlmY is shown in Fig. 4a.

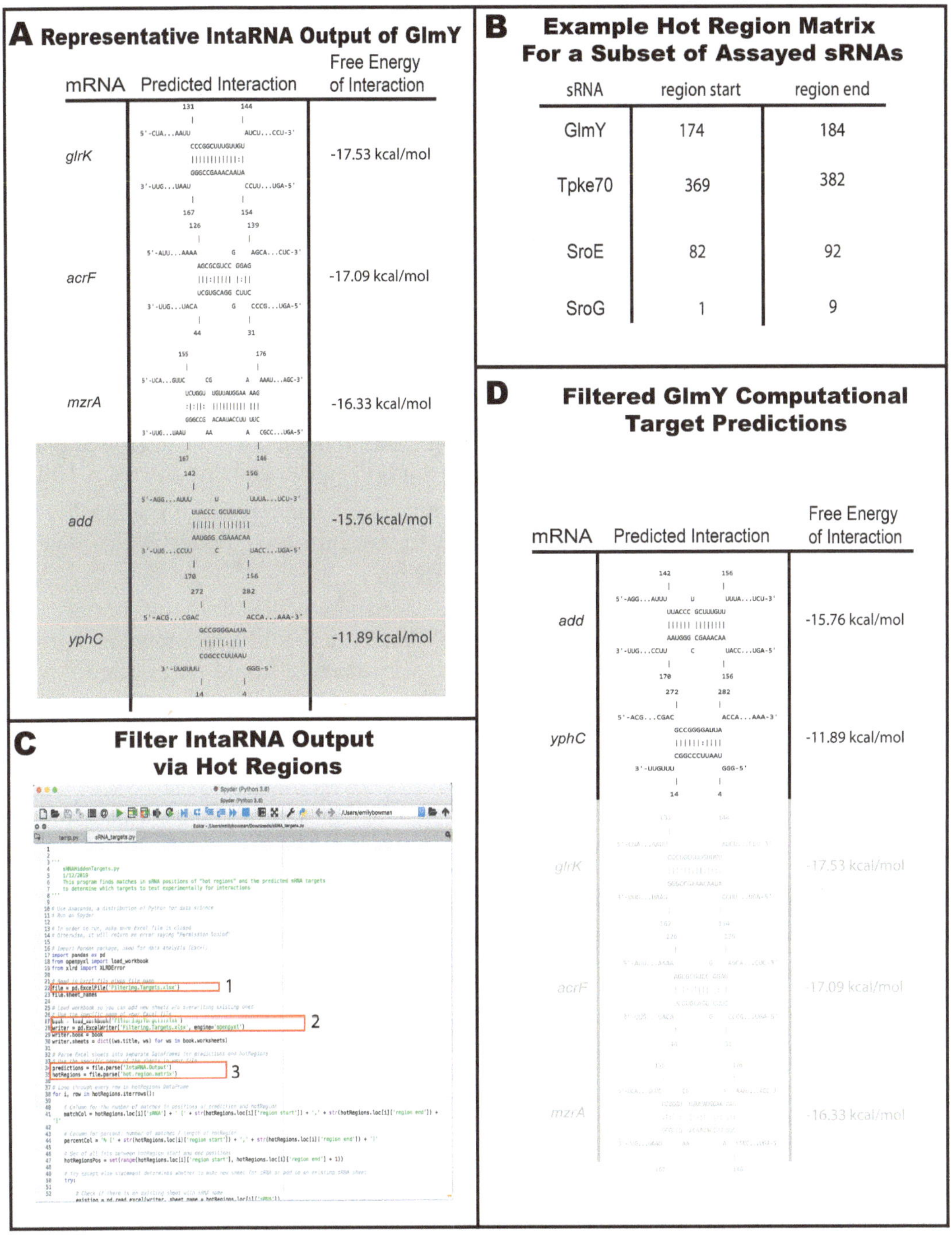

Fig. 4 Example schematic of filtering computational target predictions via hot regions. (**a**) Representative output from IntaRNA of GlmY predicted mRNA target interactions, ranked via predicted free energy of hybridization. Note that predicted target *yphC* corresponds to IntaRNA predictions of only the likely functional region as the query RNA. Targets listed in gray font fall below the top five predicted targets (pre-bioinformatic filtering) that would likely be chosen for further study. (**b**) "Hot region matrices" corresponding to GlmY and a subset of other assayed sRNAs from Mihailovic and Vazquez-Anderson 2018 [21]. (**c**) Screen capture of

3.2.1 Computational Target Prediction Using IntaRNA Web Server

1. Obtain sequence of query RNA of interest (i.e., the entire sRNA sequence of interest or the sequence of a functional region of an sRNA of interest; *see* **Note 1**), and copy and paste into the query ncRNA box. The full sequence of the example RNA of interest, GlmY, is seen in Table 2. The sequence used for predictions corresponding to only GlmY's accessibility-identified functional site (highly accessible terminator regions 17 and 18 in Fig. 2a, Table 2) is as follows: GGTAATTCCCGGCTTTGTT.
2. Select "Get target RNA sequences from *prokaryotic* NCBI Genome," and input relevant Target NCBI RefSeq ID (NC_000913 for *E. coli* K-12 MG1655 or, in the case of GlmY, NZ_CP009273 for BW25113). Specify possible target mRNA regions as −200 to +100 around all start codons in the genome by selecting "start codon" for "Extract sequences around" and setting "nt up" as 200 and "nt down" as 100.
3. Alternatively, copy and paste a curated list of potential target sequences in FASTA format into the "Target RNA (long) in FASTA" box (*see* **Note 8**).
4. Leave the remainder of settings as default ("Number of (sub) optimal interactions" should be 1; for "Suboptimal interaction overlap," "can overlap in query" should be selected; "Min. number of base pairs" in seed should be 7).
5. Enter a title of the run in the description box if desired.
6. Provide email for updates on the status of the job.
7. Once complete, export "IntaRNA_result" tabular data for further analysis.

3.2.2 Computational Target Prediction Using Locally Installed IntaRNA

1. Format the sequence of RNA of interest (e.g., sRNA) and/or the sequence of the corresponding likely functional region into a FASTA file (*see* **Note 1**) ("sRNAs.fasta").
2. Create file containing gene names, orientation, and start codon coordinates corresponding to all mRNAs in relevant genome. First, download both the GFF and FASTA file of the genome of interest from NCBI. Download "getGeneLocs" function (https://github.com/emkbowman/Bioinformatic-

Fig. 4 (continued) Spyder Python program sRNA-targets.py re-ranks IntaRNA output (panel A) based on whether the predicted interaction involves corresponding region(s) listed in the hot region matrix (panel B). Locations marked require user inputs: (1) and (2) require the name of the excel file from which to read hot matrix and IntaRNA inputs as well as to which to write outputs; (3) the names of the sheets that correspond to the hot region matrix and the IntaRNA predictions. (**d**) Representative output of the filtered GlmY computational target predictions. Note that the bottom two targets from panel **a**, previously ranked below *glrK*, *acrF*, and *mzrA*, are now highly considered for experimental validation due to their predicted interaction with GlmY's likely functional region

Application-of-Fluorescence-based-in-vivo-RNA-Regional-Accessibility-Data-/blob/master/getGeneLocs.R), and open in Rstudio. Direct script to desired GFF file by replacing "directoryoffile" with your file name (line 6) (directory/filename) (e.g., "GFF = read.delim(desktop/folderofgff/nameoffile.gff")). Click "Run" in the upper right-hand corner of the display to extract mRNA genomic start and end coordinates and gene orientation from GFF file; the output will be a .csv file named "clean_gff.csv" with five columns, containing, from left to right, an index (column 1), the position of the first nucleotide of the start codon (column 2), the name of the gene (column 3), the position of the last nucleotide in the coding sequence (column 4), and the directional orientation of the gene in the genome (binary output in which "+" corresponds to forward, and "-" corresponds to reverse) (column 5).

3. From clean_gff.csv, the "left-" and "right-hand" coordinates of the −200 to +100 sequence of each mRNA must be calculated in two additional columns for subsequent sequence extraction. We recommend using an IF statement for both coordinates to account for gene orientation (i.e., for "left-hand" coordinate, if the gene orientation = "+," then "left-hand" coordinate = start − 200; otherwise, "left-hand" coordinate = start +200).
4. Create a reference "clean GFF" file from information obtained/calculated in **steps 2** and **3**. Specifically, open and save a new spreadsheet as a tab delimited file named "UTRs_list.txt." Move full columns (including column headers) containing the gene name, gene orientation, left-hand coordinate, and right-hand coordinate into the first, second, third, and fourth columns, respectively.
5. Create relevant "Intatargets.fasta" file by using the clean GFF: download UTR_SequenceExtract.py (https://github.com/emkbowman/Bioinformatic-Application-of-Fluorescence-based-in-vivo-RNA-Regional-Accessibility-Data-/blob/master/UTR_SequenceExtract.py), and open using Python IDLE. Direct code to newly made "UTRs_list.csv" and run. Output should be −200 to +100 sequences of all genes in genome of interest in FASTA format with sequence names corresponding to the unique gene names.
6. Install IntaRNA locally (http://www.bioinf.uni-freiburg.de/Software/). Run the following Linux command to predict energetics of interaction between all possible mRNA targets (IntaTargets.fasta) and all query RNAs (i.e., full sequences of RNAs of interest and/or likely functional region sequences only) (sRNAs.fasta): "IntaRNA -q sRNAs.fasta -t IntaTargets.fasta --outmode=C –outCsvCols='id1,start1,end1,id2,start2,

end2,E'>IntaRNAOut.csv." This command will run thermodynamic predictions for each query RNA (−q) against each target (−t).

7. IntaRNA output will be formatted as a .csv file containing the name of the target (column 1, titled "id1"), the start and end positions of the predicted interaction within each given target mRNA sequence (columns 2 and 3 titled "start1" and "end1," respectively), the name of the query RNA (column 4, titled "id2"), the start and end positions of the predicted interaction within the query RNA sequence (columns 5 and 6 titled "start2" and "end2," respectively), and the predicted free energy of interaction (column 7 titled "E").

3.3 Bioinformatic Filtering of mRNA Targets Based on Hot Regions Identified via iRS^3

Once the experimental iRS^3 and computational IntaRNA prediction portions (Fig. 2a, b, respectively) of the method workflow are complete for a given RNA of interest, mRNA targets can be filtered based on insights obtained from iRS^3 (*see* **Note 9**). Specifically, this section first describes the development of a hot region matrix, which consists of in vivo-identified likely functional regions of the RNA of interest. For the case of GlmY, a representative hot region matrix can be seen in Fig. 4b. Upon formatting of IntaRNA results, we describe the use of a Python script for the purposes of filtering computational target predictions by overlap with likely functional regions (listed in the hot region matrix) (Fig. 4c). In this way, predicted targets are re-ranked prior to selection for experimental confirmation to account for observed regional in vivo behaviors. Representative re-ranked targets for GlmY are pictured in Fig. 4d. Lastly, we describe an optional step in which mRNA targets that rank highly in the re-ranked computational target predictions are evaluated for expression in the predicted sRNA interaction regions (*see* **Note 8**).

3.3.1 Creating a "Hot Region Matrix"

Based on the regions of interest identified by iRS^3 (Subheading 3.5), create an Excel file containing the specifications (i.e., position) of likely functional regions, or "hot regions," inferred from iRS^3 data, for each relevant RNA of interest (Fig. 4b). The program used in Subheading 5.3 requires the following information for each likely functional region, or "hot region": a column labeled "sRNA" containing the name of the sRNA(s) as entered in the IntaRNA "sRNAs.fasta" input; a column labeled "region_start," containing the start nucleotide position of the proposed functional region; and a column labeled "region_end," containing the end nucleotide position of the end of the proposed functional region.

3.3.2 Formatting IntaRNA Target Predictions

To support the recognition of IntaRNA results by the program used in Subheading 5.3 for IntaRNA prediction filtering, IntaRNA results must be exported into an Excel file and columns renamed. Name the spreadsheet tab "IntaRNA.output." If reformatting Web server-based IntaRNA predictions, columns do not need to be renamed. If reformatting locally performed IntaRNA predictions, relabel columns from "id1," "start1," "end1," "id2," "start2," "end2," and "E" labels to "target_name," "target_start_pos," "target_end_pos," "sRNA_name," "sRNA_start_pos," "sRNA_end_pos," and "energy," respectively.

3.3.3 Filtering Target Predictions by Overlap with Hot Regions to Obtain mRNA Candidates for Experimental Confirmation

This subsection details a program to find positional overlap between likely functional regions ("hot regions") and computationally-predicted interactions between the RNA of interest and potential mRNA targets. The program requires properly formatted spreadsheets containing the "hot region" matrix and IntaRNA predictions (*see* Subheading 5.2) corresponding to RNA (s) of interest. The program will create an Excel tab corresponding to each "hot region" in the matrix, in which mRNA target predictions corresponding to the RNA of interest are ordered by the number of nucleotide positions (within the RNA of interest) that overlap between the "hot region" and the predicted interaction with each potential target, and then by energetic favorability. Importantly, the steps outlined in this subsection are not necessary when the query RNA inputted to IntaRNA is the sequence of the likely functional region only (*see* **Note 9**).

1. Move Excel spreadsheet containing the hot region matrix to the same Excel file as that of the formatted IntaRNA predictions. Label the sheet "hot.region.matrix."
2. Download sRNA_targets.py from https://github.com/bl37/sRNA-targets.
3. Run sRNA_targets.py on Spyder, directing the program to the appropriate Excel file containing both the "hot region" matrix (hot.region.matrix) and the predicted targets (IntaRNA.output). Ensure that names of the individual sheets and columns match the required format of the code (Fig. 4c).
4. Perform IntaRNA rank exclusion (*see* **Note 10**).
5. Filter again by a positional overlap cutoff between the "hot region" and the predicted interaction with each potential target (*see* **Note 11**) (Fig. 4d).
6. Choose predicted mRNA targets for which to continue with experimental confirmation from filtered list. Specifically, select a pool of filtered mRNA targets [for which experimentation is feasible] with the most energetically favorable predicted interactions.

3.3.4 Filtering mRNA Candidates and Select Representative Sequences for Experimental Confirmation by Transcriptomics

If relevant transcriptomics are available, ensure that predicted target-binding site (i.e., on an mRNA) exhibits expression using IGV [29] (Fig. 2d) (*see* **Note 8**).

1. Download and open IGV (https://software.broadinstitute.org/software/igv/download). In the top left corner, load the relevant genome by selecting from the drop-down "Genomes" menu. If the relevant genome is not hosted on the IGV server, follow instructions for creating a GENOME file from a corresponding indexed FASTA file (*see* https://software.broadinstitute.org/software/igv/LoadGenome for instructions).
2. Click "Files > load from file" to upload relevant expression data in .bam.bai format. Search for relevant gene names associated with top-ranked filtered targets using the search bar in the top of the window.
3. Keep only mRNA targets that show transcriptional expression of both the gene coding sequence and the binding site predicted to interact with the RNA of interest/sRNA (*see* **Note 12**). If a predicted mRNA [−200 to +100]-binding site is expressed but is contained within the upstream gene, do not keep (*see* **Note 13**).
4. Select start of representative mRNA sequence for experimental interaction validation to match transcriptional sites as observed in transcription data (i.e., choose the first nucleotide of the sequence to be the first nucleotide in the mRNA's 5′ UTR with observed expression) (*see* **Note 12**). For genes that appear co-transcribed, if predicted mRNA [−200 to +100]-binding site is not contained within upstream gene, choose a pseudo 5′ UTR start as the first nucleotide after end of the preceding gene.
5. Choose the end of the representative mRNA sequence for experimental interaction validation to be at least 30–45 nucleotides downstream of the predicted sRNA interaction in an attempt to maintain any native structural features of the mRNA target, ideally making the experimental mRNA sequence 120 nucleotides or longer (Fig. 5a) (*see* **Note 13**).

3.4 Experimental Validation of Predicted and Filtered RNA: RNA Interactions via EMSA

In order to identify direct targets of the RNA of interest (i.e., GlmY), in vitro binding assays between the [radiolabeled] RNA of interest and representative sequences of mRNA candidates (Fig. 5a) (*see* **Note 14**) chosen from Subheading 5 are conducted. Previously, in this way, direct binding between GlmY and targets *add* and *yphC* (along with all other sRNA:mRNA pairs listed in Table 1) was confirmed. Importantly, EMSAs also offer the opportunity to validate the contribution of specific sites to binding. For example, by running binding assays between confirmed GlmY targets (*add* and *yphC*) and a truncated version of GlmY (in which the terminator hairpin was excluded, Fig. 5b), we confirmed the dependence of the terminator for interaction [21].

3.4.1 IVT-Compatible Amplification of RNA of Interests and mRNA Target Candidates

1. Design primers to amplify RNAs of interest/sRNAs and representative sequences of candidate mRNA targets (*see* **Note 15**). Importantly, corresponding forward primers must contain an upstream overhang consisting of the T7 promoter sequence (specifically, four random nucleotides + TAATACGACTCACTATAGGGAGA) (*see* **Note 16**). See Fig. 5b for primers used to amplify RNA of interest GlmY for IVT (*see* **Note 17**).
2. Assemble IVT-compatible PCR amplification for DNA encoding RNA of interest and mRNA candidates (*see* **Note 14**) from *E. coli* genomic DNA. Genomic DNA can be obtained by boiling a single colony, diluted in 50 μL nuclease-free water at 96 °C for 5 min. On ice, add genomic DNA (4 μL), 10 μM forward and reverse insert primers (2.5 μL each, diluted in nuclease-free water), 10 mM dNTPs (1 μL), 5X Phusion high-fidelity (HF) buffer (10 μL), Phusion HF DNA polymerase (0.5 μL), and nuclease-free water to 50 μL.
3. Cycle all PCR reactions as follows: (i) 98 °C for 30 s; (ii) 25 cycles of 98 °C for 10 s, 3° above lowest melting temperature of primer pairs (excluding overhang region for melting temperature calculations) for 30 s, and 72 °C for 2.5 min (vector backbone) or 30 s/kb; and (iii) final extension 72 °C for 10 min (iv) hold at 4 °C.
4. Clean up PCR product using GE PCR cleanup kit per manufacturer instructions. Measure DNA concentrations via spectrophotometry.

3.4.2 IVT of Potential mRNA Targets for EMSA Analysis

1. Follow MEGAscript T7 kit instruction manual to in vitro transcribe representative mRNA sequences. Briefly, at room temperature, mix 2 μL ATP solution, 2 μL CTP solution, 2 μL GTP solution, 2 μL UTP solution, 2 μL 10X reaction buffer, 1–1.5 pmol of DNA template (*see* **Note 18**), 2uL enzyme mix, and nuclease-free water to 20 μL. Mix thoroughly and incubate for 6 h at 37 °C (*see* **Note 19**). After the full 6 h, add 1 μL Turbo DNase, mix well, and incubate for 15 min at 37 °C to remove DNA template.
2. Recover transcribed RNA by following manufacturer instructions of RNA Clean & Concentrator-5 (Zymo Research) (*see* **Notes 20** and **21**). Briefly, bring the volume of the RNA sample to 50 μL using nuclease-free water. Add 100 μL RNA binding buffer to reaction and mix; then add 150 μL 100% OmniPur ethanol. Mix well and transfer to assembled Zymo-Spin Column (in a collection tube), and centrifuge for 30 s at 16,000 rcf. Discard eluent, add 400 μL RNA prep buffer to column, and centrifuge for 30 s. Discard eluent, add 700 μL of RNA wash buffer, and centrifuge 30 s. Discard eluent, add 400 μL of RNA wash buffer, and centrifuge for 2 min. Discard eluent and centrifuge 30 s to ensure that column filter is dry. Place column in new RNase-free tube and elute RNA with 12–15 μL nuclease-free water.

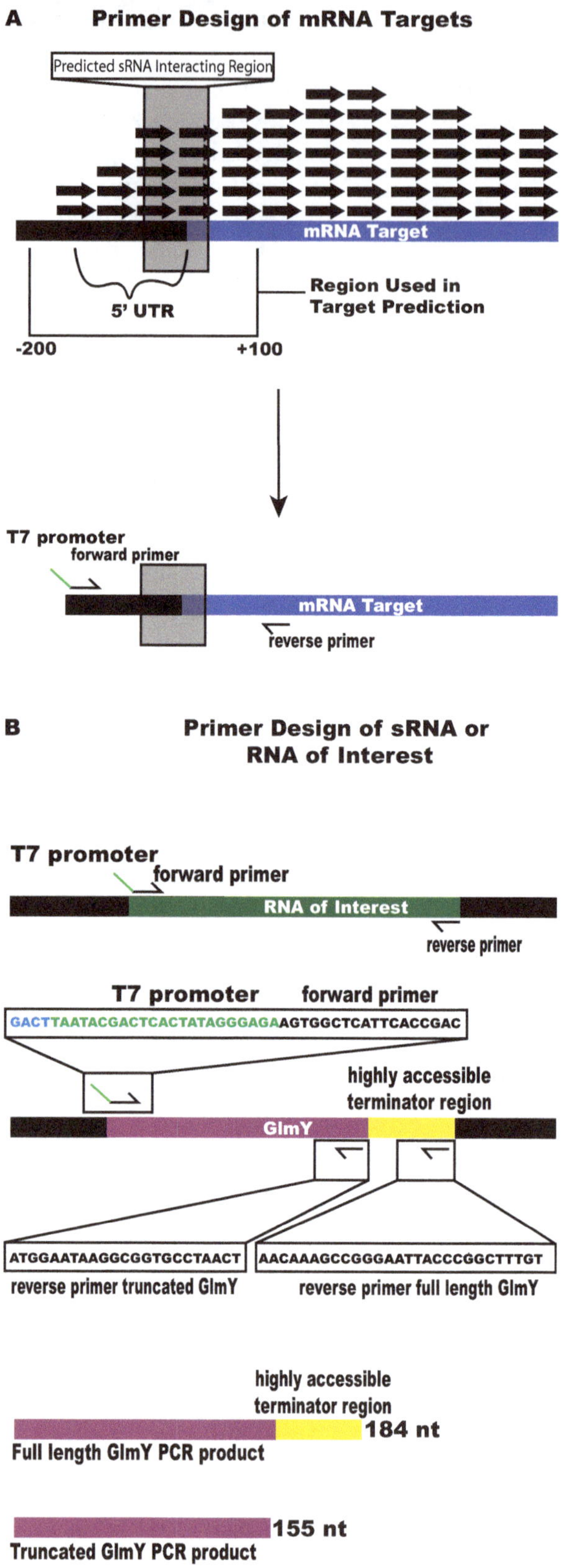

Fig. 5 Block DNA representation of amplification of representative mRNA sequences (**a**) and RNA of interest (**b**). (**a**) Identification of representative mRNA target sequence and subsequent primer design for IVT-compatible

3. Alternatively, RNA recovery can be performed via chloroform purification. Add 115 μL of nuclease-free water and 115 μL of Ammonium Acetate Stop Solution (from MEGAscript T7 kit) and mix. Add two volumes of 25:24:1 phenol/chloroform/isoamyl alcohol, and invert for 1 min. Centrifuge at 13,000 rcf at 4 °C for 3 min. Transfer aqueous phase (top) to two equivalent volumes of 24:1 chloroform/isoamyl alcohol; then centrifuge at 13,000 rcf at 4 °C for 3 min. Transfer top phase to tubes containing 1 mL isopropyl alcohol and 1.5 μL GlycoBlue each, and leave to precipitate overnight at −20 °C. On the next day, centrifuge samples at 13,000 rcf at 4 °C for 15 min. Remove all liquid carefully so as not to disturb the formed RNA pellet; then add 1 mL of 95% ethanol to each sample. Centrifuge sample at 13,000 rcf at 4 °C for 1 min. Again, remove liquid carefully so as not to disturb the RNA pellet. Add 1 mL of 75% ethanol to each sample and centrifuge at 13,000 rcf at 4 °C for 1 min. Remove liquid carefully, and leave samples to dry in a solvent hood for up to 2 h. Once dry, add 30 μL of nuclease-free water and resuspend RNA.
4. Measure RNA concentration via spectrophotometry (*see* **Note 22**).
5. It is important to verify transcript size and quality via denaturing gel electrophoresis prior to binding assays. Clean all components of Mini-PROTEAN casting module with RNase*Zap* and nanopure water, and assemble while maintaining RNase-free conditions. To make an 8% urea gel, mix 4 mL urea concentrate, 4 mL urea diluent, and 1 mL urea buffer. Immediately prior to casting gel in Mini-PROTEAN casting module, add 80 mL 10% APS and 4 uL TEMED. Allow the gel to polymerize; then pre-run gel at 100 V while preparing samples. Mix 2–4 μL of RNA (normalizing for varying RNA concentrations) and RNA ladder with 5 μL of RNA loading buffer II (NEB) and nuclease-free water to 15 μL. Denature RNA mixtures for 5 min at 70 °C. Before loading samples, displace urea precipitates in wells by forcing buffer into the well using a

Fig. 5 (continued) amplification via PCR. In an attempt to maintain any native structural features of the mRNA target, the observed 5′ UTR start (based on transcriptomics) is chosen as the sequence start, and the predicted binding region is flanked by at least 30–45 nucleotides downstream. (**b**) Primer design to amplify RNA of interest for in vitro transcription. In the case of GlmY, two different reverse primers were designed in order to amplify both the full-length sequence, containing the highly accessible terminator region, and a truncated version that omits this region. Importantly, by showing that the sRNA no longer bound its newly identified mRNA targets when lacking the terminator region, the contribution of the GlmY terminator to RNA binding was confirmed [21] syringe and needle. Load samples into gel and run at 100 V for 3 h. Disassemble gel and stain in 100 mL Tris/acetate/EDTA buffer with 10 μL Sybr Green II. Let incubate in the dark for 30 min prior to imaging under Sybr Green setting.

3.4.3 IVT of sRNAs (or RNAs of Interest)

1. Follow instructions for IVT of target mRNA candidates (Subheading 6.2), with the following changes: replace up to 1.5 μL UTP for an equivalent volume of [alpha-P_{32}] UTP (3000 Ci/mmol 10 mCi/ml, 250 μCi, PerkinElmer) (*see* **Note 2**).
2. Recover RNA via preferred method from Subheading 6.1. Collect all eluent in a chemically compatible, sealable container; quarantine until radioactivity is below background and dispose.
3. Remove free dNTPs using DTR gel filtration cartridges (Edge-Bio) per manufacturer's instructions.
4. Measure RNA concentration via spectrophotometry using designated radioactive cuvette and appropriate precautions.

3.4.4 EMSAs to Confirm Direct RNA of Interest: mRNA Interactions

1. The day before the experiment, cast a 5% nondenaturing (native) polyacrylamide gel in an RNase-free PROTEAN II xi casting module. Begin by cleaning the gel plates with RNase-*Zap* and drying completely. Mix 3 mL 10× TBE electrophoresis buffer, 7.5 mL 40% wt/vol acrylamide-bisacrylamide stock solution, 3 mL glycerol, and 46.5 mL H_2O. Add 0.15 g APS and 36 μL TEMED and immediately pour into casting module. Let polymerize for 2 h at room temperature and then overnight at 4 °C.
2. In 0.6 mL conical tubes, mix internally labeled sRNAs and respective mRNA (*see* **Note 14**) in consistent molar ratios with mRNA in excess (*see* **Note 23**) in 12–15 μL reactions containing 1X EMSA binding buffer (*see* **Note 2**) and 10% glycerol, normalizing for radioactivity of sRNAs in the case that multiple sRNAs will be loaded into one gel. Denature for 5 min at 70 °C; then incubate at 37 °C for 1.5 h.
3. During the incubation period, assemble the PROTEAN II xi cell in preparation for running. Load 10 μL of EMSA loading buffer into each well and pre-run the gel at 25 mA in 0.5X TBE running buffer to determine visually whether the gel is running consistently.
4. Once binding reactions are done incubating, load the total volume of each reaction into corresponding wells (*see* **Note 2**). Run for at approximately 3–4 h at 150 V in 0.5X Tris/borate/EDTA running buffer.

5. Disassemble the gel by gently separating the plates and transferring gel to blotting paper slightly larger than the gel itself. Using plastic wrap, wrap the gel so that the excess wrap is on the side of the exposed gel (not that of the blotting paper). With the gel side up, dry the gel at 80 °C for 1.5 h and expose overnight in a phosphorimaging cassette.
6. Phosphorimage the next day using the following settings: 1000 V, 50 μm resolution.

4 Notes

1. Some RNAs of interest may contain iRS^3-identified likely functional regions within strong minimum free energy (MFE)-predicted secondary structures, for example, the terminator loop within GlmY has a predicted free energy of −16.3 kcal/mol [34] (Fig. 3b). Because computational target predictions predict most energetically favorable interactions between the query and target, it is common that regions predicted as highly structured are not represented in computational interaction predictions, despite extreme accessibility in vivo. In the case that no/few targets are predicted to interact with a likely functional region that is both highly accessible in vivo and predicted to be within a strong structure, we recommend additionally compiling predictions that correspond to the sequence of the likely functional region only. By accounting for measured in vivo accessibility in the computational predictions in this manner, *yphC*, confirmed GlmY target which was not captured in the top 100 predictions corresponding to the full-length sRNA, was selected for experimental validation.
2. When working with radioactive materials, use a microcentrifuge, pipette set, and UV spectrophotometer designated for contact and potential contamination with radioactive materials. Furthermore, use proper safety precautions, including always working behind a shield and opening concentrated samples only in solvent hoods.
3. Maintain an RNase-free environment when handling RNA. For example, use designated RNA-only nuclease-free pipette tips and water.
4. Although this chapter details the use of iRS^3 for evaluating regional accessibility with an RNA of interest of type sRNA, this fluorescence-based method can be adapted to test in vivo accessibility of any RNA class of interest (tRNAs, mRNAs, etc.).
5. iRS^3 was originally optimized for use in *E. coli* K-12 MG1655; however, no hindrances to using other experimental strains of *E. coli* have been predetermined.

6. Previously, regions whose center nucleotide was within 20% of either 5′ or 3′ end of the sRNA were considered [21]. For non-sRNA RNAs of interest, we anticipate it will be worthwhile to consider the entire sequence space for selection of likely functional regions based on accessibility information.
7. The IntaRNA Web server supports the input of only a single query RNA (RNA of interest or sRNA). Performing IntaRNA predictions locally supports high-throughput prediction of multiple query RNAs simultaneously.
8. Previously, 5′ UTRs were estimated as the sequence space 200 nucleotides upstream of all start codons; however, a curated list of 5′ UTRs based on previously compiled expression data, e.g., from RegulonDB [35], could also be used. If using a curated list of 5′ UTR starts, filtering mRNA candidates and selecting representative sequences for experimental confirmation via transcriptomics may be redundant.
9. The filtering steps described in Subheading 5 are not necessary for any predictions corresponding to the sequence of the likely functional region only because all predictions will already correspond to the said region.
10. We have previously used a rank cutoff of 100 or 300, chosen based on known rankings of true targets [12].
11. Predicted interacts were previously excluded if fewer than five nucleotides of the likely functional region were implicated in the predicted interaction with its target [21]. However, we anticipate that a percentage-based cutoff would be valuable for normalizing overlap differences due to varying lengths of likely functional regions.
12. These guidelines are suggested in order to imitate the native in vivo mRNA transcript in in vitro binding assays.
13. mRNAs with no appreciable expression data were also kept as potential targets. In this case, the start of the 5′ UTR sequence (for use in binding assays) was chosen to align with that of the sequence inputted to computational target predictions (i.e., 200 nucleotides upstream of the start codon) or as the nucleotide following the preceding gene, whichever was shorter.
14. For every sRNA tested for interaction with candidate mRNA targets, two negative control EMSA binding reactions should be included: an sRNA-only control, in which no candidate target is added, and a nonspecific interaction control, in which a random RNA sequence is used instead of that of a candidate target. Previously, the random sequence [in DNA] "AAACCCCGTGGCCTGTGAGCGGCGAAGGCCCCAAC GGGAGATACGCGCCCACGACTCGGCGCGAGTACGG GCCGTGGCGGTGTTCGTCTGGGTGTGGTCTACACAA TGCGGGCGGTGCGTCTGCTCCGGCCGGCCTCTGGT GGCTCGCAT" was used [21].

15. Alternatively, the equivalent DNA block sequence can be ordered from a DNA synthesis company.
16. The sequence "GATC" was previously used [21].
17. Upon target confirmation via in vitro binding assays, it may be valuable to evaluate the contribution of the likely functional region to binding. Previously, in light of likely functional regions positioned at far ends of the sRNA, binding assays with truncated sRNAs (excluding the likely functional region) were performed. Alternatively, binding assays with sRNAs containing point mutations in the likely functional region can be executed.
18. Molarity of DNA template or RNA can be easily estimated using the NEBioCalculator (http://nebiocalculator.neb.com).
19. Previously, per manufacturer's instructions, due to the relatively short sequence length used for in vitro binding assays (<500 nucleotides), a 6-h incubation time was used for all RNAs of interest and candidate mRNA sequences.
20. RNA recovery via the kit or chloroform purification can be performed with similar results.
21. Properly collect and dispose of denaturants (*e.g.,* guanidinium thiocyanate) in compliance with official regulations.
22. Typically, a well-transcribed reaction will contain over 1000 ng/μL of RNA.
23. 1:10 Molar sRNA/mRNA ratios were previously used for EMSA; however, smaller ratios may better mimic the native in vivo environment.

Acknowledgments

This work is supported by the Welch Foundation (Grant F-1756 to LMC) and the National Science Foundation (Grant MCB 1716777 to LMC and DGE-1610403 to MKM).

References

1. Dutta T, Srivastava S (2018) Small RNA-mediated regulation in bacteria: a growing palette of diverse mechanisms. Gene 656:60–72. https://doi.org/10.1016/j.gene.2018.02.068
2. Gelderman G, Contreras LM (2013) Discovery of posttranscriptional regulatory RNAs using next generation sequencing technologies. In: Alper HS (ed) Systems metabolic engineering: methods and protocols. Humana Press, Totowa, NJ, pp 269–295. https://doi.org/10.1007/978-1-62703-299-5_14
3. Leistra AN, Curtis NC, Contreras LM (2019) Regulatory non-coding sRNAs in bacterial metabolic pathway engineering. Metab Eng 52:190–214. https://doi.org/10.1016/j.ymben.2018.11.013
4. Carrier M-C, Lalaouna D, Massé E (2016) A game of tag: MAPS catches up on RNA interactomes. RNA Biol 13(5):473–476. https://doi.org/10.1080/15476286.2016.1156830
5. Figueroa-Bossi N, Bossi L (2018) Sponges and predators in the small RNA world. Microbiol

Spectr 6(4). https://doi.org/10.1128/microbiolspec.RWR-0021-2018
6. Kudla G, Granneman S, Hahn D, Beggs JD, Tollervey D (2011) Cross-linking, ligation, and sequencing of hybrids reveals RNA-RNA interactions in yeast. Proc Natl Acad Sci U S A 108(24):10010–10015. https://doi.org/10.1073/pnas.1017386108
7. Liu T, Zhang K, Xu S, Wang Z, Fu H, Tian B, Zheng X, Li W (2017) Detecting RNA-RNA interactions in E. coli using a modified CLASH method. BMC Genomics 18(1):343. https://doi.org/10.1186/s12864-017-3725-3
8. Schönberger B, Schaal C, Schäfer R, Voß B (2018) RNA interactomics: recent advances and remaining challenges [version 1; referees: 2 approved]. F1000Res 7(1824). https://doi.org/10.12688/f1000research.16146.1
9. Melamed S, Peer A, Faigenbaum-Romm R, Gatt Yair E, Reiss N, Bar A, Altuvia Y, Argaman L, Margalit H (2016) Global mapping of small RNA-target interactions in Bacteria. Mol Cell 63(5):884–897. https://doi.org/10.1016/j.molcel.2016.07.026
10. Melamed S, Faigenbaum-Romm R, Peer A, Reiss N, Shechter O, Bar A, Altuvia Y, Argaman L, Margalit H (2017) Mapping the small RNA interactome in bacteria using RIL-seq. Nat Protoc 13:1. https://doi.org/10.1038/nprot.2017.115. https://www.nature.com/articles/nprot.2017.115#supplementary-information
11. Waters SA, McAteer SP, Kudla G, Pang I, Deshpande NP, Amos TG, Leong KW, Wilkins MR, Strugnell R, Gally DL, Tollervey D, Tree JJ (2017) Small RNA interactome of pathogenic E. coli revealed through crosslinking of RNase E. EMBO J 36(3):374–387. https://doi.org/10.15252/embj.201694639
12. Pain A, Ott A, Amine H, Rochat T, Bouloc P, Gautheret D (2015) An assessment of bacterial small RNA target prediction programs. RNA Biol 12(5):509–513. https://doi.org/10.1080/15476286.2015.1020269
13. Beisel CL, Storz G (2010) Base pairing small RNAs and their roles in global regulatory networks. FEMS Microbiol Rev 34(5):866–882. https://doi.org/10.1111/j.1574-6976.2010.00241.x
14. Holmqvist E, Vogel J (2018) RNA-binding proteins in bacteria. Nat Rev Microbiol. https://doi.org/10.1038/s41579-018-0049-5
15. Peer A, Margalit H (2014) Evolutionary patterns of Escherichia coli small RNAs and their regulatory interactions. RNA (New York, NY) 20(7):994–1003. https://doi.org/10.1261/rna.043133.113
16. Sowa SW, Vazquez-Anderson J, Clark CA, De La Peña R, Dunn K, Fung EK, Khoury MJ, Contreras LM (2015) Exploiting post-transcriptional regulation to probe RNA structures in vivo via fluorescence. Nucleic Acids Res 43(2):e13–e13. https://doi.org/10.1093/nar/gku1191
17. Busch A, Richter AS, Backofen R (2008) IntaRNA: efficient prediction of bacterial sRNA targets incorporating target site accessibility and seed regions. Bioinformatics 24. https://doi.org/10.1093/bioinformatics/btn544
18. Wright PR, Georg J, Mann M, Sorescu DA, Richter AS, Lott S, Kleinkauf R, Hess WR, Backofen R (2014) CopraRNA and IntaRNA: predicting small RNA targets, networks and interaction domains. Nucleic Acids Res 42 (Web Server issue):W119–W123. https://doi.org/10.1093/nar/gku359
19. Mann M, Wright PR, Backofen R (2017) IntaRNA 2.0: enhanced and customizable prediction of RNA–RNA interactions. Nucleic Acids Res 45(Web Server issue):W435–W439. https://doi.org/10.1093/nar/gkx279
20. Leistra AN, Mihailovic MK, Contreras LM (2018) Fluorescence-based methods for characterizing RNA interactions in vivo. Methods in Mol Biol 1737:129–164. https://doi.org/10.1007/978-1-4939-7634-8_9
21. Mihailovic MK, Vazquez-Anderson J, Li Y, Fry V, Vimalathas P, Herrera D, Lease RA, Powell WB, Contreras LM (2018) High-throughput in vivo mapping of RNA accessible interfaces to identify functional sRNA binding sites. Nat Commun 9(1):4084. https://doi.org/10.1038/s41467-018-06207-z
22. Gruber CC, Sperandio V (2015) Global analysis of posttranscriptional regulation by GlmY and GlmZ in Enterohemorrhagic Escherichia coli O157:H7. Infect Immun 83 (4):1286–1295. https://doi.org/10.1128/IAI.02918-14
23. R Core Team (2018) R: a language and environment for statistical computing. R Foundation for Statistical Computing, Vienna, Austria
24. Müller K, Wickham H, François R, Henry L (2018) dplyr: A Grammar of Data Manipulation
25. H. Pagès P Aboyoun, R. Gentleman and S. DebRoy (2018) Biostrings: Efficient manipulation of biological strings
26. Wickham H (2018) stringr: Simple, Consistent Wrappers for Common String Operations

27. Anaconda Software Distribution (2017) Computer software. Vers. 2–2.4.0. Anaconda, Inc.
28. McKinney W (2011) pandas: a Foundational Python Library for Data Analysis and Statistics
29. Thorvaldsdóttir H, Robinson JT, Mesirov JP (2013) Integrative genomics viewer (IGV): high-performance genomics data visualization and exploration. Brief Bioinform 14 (2):178–192. https://doi.org/10.1093/bib/bbs017
30. Vazquez-Anderson J, Mihailovic MK, Baldridge KC, Reyes KG, Haning K, Cho SH, Amador P, Powell WB, Contreras LM (2017) Optimization of a novel biophysical model using large scale in vivo antisense hybridization data displays improved prediction capabilities of structurally accessible RNA regions. Nucleic Acids Res 45(9):5523–5538. https://doi.org/10.1093/nar/gkx115
31. Altschul SF, Gish W, Miller W, Myers EW, Lipman DJ (1990) Basic local alignment search tool. J Mol Biol 215(3):403–410. https://doi.org/10.1016/S0022-2836(05)80360-2
32. Leistra AN, Amador P, Buvanendiran A, Moon-Walker A, Contreras LM (2017) Rational modular RNA engineering based on in vivo profiling of structural accessibility. ACS Synth Biol. https://doi.org/10.1021/acssynbio.7b00185
33. Green AA, Silver PA, Collins JJ, Yin P (2014) Toehold switches: De-novo-designed regulators of gene expression. Cell 159(4):925–939. https://doi.org/10.1016/j.cell.2014.10.002
34. Zadeh JN, Steenberg CD, Bois JS, Wolfe BR, Pierce MB, Khan AR, Dirks RM, Pierce NA (2011) NUPACK: analysis and design of nucleic acid systems. J Comput Chem 32 (1):170–173. https://doi.org/10.1002/jcc.21596
35. Gama-Castro S, Salgado H, Santos-Zavaleta A, Ledezma-Tejeida D, Muñiz-Rascado L, García-Sotelo JS, Alquicira-Hernández K, Martínez-Flores I, Pannier L, Castro-Mondragón JA, Medina-Rivera A, Solano-Lira H, Bonavides-Martínez C, Pérez-Rueda E, Alquicira-Hernández S, Porrón-Sotelo L, López-Fuentes A, Hernández-Koutoucheva A, Del Moral-Chávez V, Rinaldi F, Collado-Vides J (2016) RegulonDB version 9.0: high-level integration of gene regulation, coexpression, motif clustering and beyond. Nucleic Acids Res 44(D1):D133–D143. https://doi.org/10.1093/nar/gkv1156
36. (2017) UniProt: the universal protein knowledgebase. Nucleic Acids Res 45(D1): D158–D169. https://doi.org/10.1093/nar/gkw1099

Chapter 6

Surface Plasmon Resonance for Investigating Molecular Interactions with RNA

Carmelo Di Primo

Abstract

Surface plasmon resonance (SPR)-based instruments have become gold-standard tools for investigating molecular interactions involving macromolecules. The major advantage is that the measured signal is sensitive to changes in mass. Therefore, all kinds of complexes can be analyzed including those with compounds as small as cations. SPR is mainly used to determine the dissociation equilibrium constant and the binding rates of a reaction if slow enough. SPR is well suited for analysis molecular interactions with nucleic acids because these negatively charged macromolecules do not have a tendency to stick to the sensor chip surface as some proteins can do. To illustrate the use of SPR with RNA molecules, we describe methods that we used for monitoring the interaction between the protein Rop from *E. coli* and two RNA-RNA loop-loop complexes. One is derived from the natural target of Rop, RNAI-RNAII. The other one is an RNA-RNA complex formed between a shortened version of the TAR element of HIV-1 and a structured RNA, TAR* rationally designed to interact with TAR through loop-loop interactions. These methods can be easily adapted to other complexes involving RNA molecules and to other SPR instruments.

Key words RNA, Hairpins, Loop-loop interactions, RNA-binding protein, Ternary complexes, Surface plasmon resonance

1 Introduction

Surface plasmon resonance (SPR)-based biosensors are instruments that allow real-time monitoring of molecular interactions between a partner immobilized onto a sensor chip surface and another one injected in a continuous flow of buffer. This biophysical technology was first commercialized in the early 1990s by Biacore (GE Healthcare, Uppsala, Sweden) [1–3]. SPR became rapidly popular, and instruments developed by different companies are now widely spread in academic and private laboratories. SPR has become a gold standard for measuring rate and equilibrium constants [4]. SPR can also be used for measuring active concentrations of proteins, i.e., the fraction of native protein that specifically recognizes a partner and not the total concentration, as generally

Véronique Arluison and Frank Wien (eds.), *RNA Spectroscopy: Methods and Protocols*, Methods in Molecular Biology, vol. 2113, https://doi.org/10.1007/978-1-0716-0278-2_6,

determined by UV-visible spectroscopy, which may include non-active proteins and impurities [5–9]. This very useful application of SPR is surprisingly barely used, which is an error because wrong concentrations used to analyze the SPR data will lead to wrong rate and equilibrium constants.

SPR is always introduced as a label-free technology because the measured signal depends only on mass changes that result from complex formation. Therefore, all kinds of complexes can be investigated (with cations, nucleic acids, proteins, lipids, cells, viruses). Some studies still controversially have suggested that conformational changes during complex formation would also contribute to the SPR signal. Compared to ITC, a true label-free technique, SPR requires to immobilize one partner of the complex onto a surface, either covalently by chemical coupling (amine, thiol), by capturing it with an anchor (antibody, complementary oligonucleotide strand), or by metal chelation (for His-tag proteins, for instance). This must be taken into account when a SPR experiment is designed as this could have an impact on the binding properties of the complexes. A major drawback of the method is that most of the sensor chip surfaces commercially available are made of a layer of carboxymethylated dextran matrix that is negatively charged at physiological pH. This can generate nonspecific interaction with the solid support when positively charged molecules are injected. We showed recently that nonspecific binding can be overcome provided that the SPR experiment is carefully designed, in particular the control experiment [9]. When nonspecific binding occurs, the SPR signal from a control channel left blank cannot be simply subtracted from the signal from the channel functionalized with the target because the amount of molecules that stick to the surface is actually not similar on both channels: nonspecific binding is actually lower when part on the surface is already occupied by a target.

Historically, SPR was developed to characterize protein-protein interactions, in particular antigen-antibody complexes, but the use of SPR is now extended to all kinds of molecules. Some instruments are now so sensitive that there is no limit to the mass that can be detected, according to the manufacturers. In the laboratory, we have been using SPR for years for analyzing complexes with structured RNAs from viruses [10–16]. We also used SPR to select RNA aptamers against one RNA target from mammalian cells by the SELEX method directly from an RNA target immobilized onto a sensor chip surface [17]. The combination of SPR and SELEX speeded up the selection process because selection and evaluation were performed simultaneously. Over the years, we observed that SPR was a very efficient tool for nucleic acids. Being polyanions, they do not stick to the negatively charged sensor chip surface at physiological pH as some other compounds can do. They can be easily immobilized by streptavidin-biotin coupling or captured by an oligonucleotide anchor complementary to its 5′ or 3′ end. With nucleic acids, regeneration of the functionalized surface, i.e., instantaneous dissociation of the resulting complex, can be easily achieved.

This chapter presents protocols for investigating proteins binding to RNA molecules with a Biacore™ 3000 instrument (Biacore, GE Healthcare, Uppsala, Sweden), an instrument that has been on the market for over 15 years. Although old, this very robust biosensor is still used in many laboratories including ours since 2003. These protocols can be used with more recent instruments from Biacore (T100/T200 generation) and SPR instruments with equivalent specifications from other suppliers. To illustrate the use of SPR for investigating interactions with RNA molecules, we describe the interaction between two RNA loop-loop complexes recognized by the protein Rop from *E. coli* [18, 19]. The first one is formed between a shortened version of the trans-activating responsive element (TAR) RNA of the human immunodeficiency virus type 1 (HIV-1) and a stem-loop RNA, TAR*, that was rationally designed to form a "kissing" complex with the viral target (Fig. 1) [20, 21]. The second RNA-RNA complex is formed between two RNA hairpins, RNAIi and RNAIIi (Fig. 1), that display inverted loops compared with the RNAI-RNAII complex from *E. coli*, which is the natural target of the protein Rop [18, 19, 22].

The results show that SPR is a very efficient tool for investigating a ternary complex formed between two RNA hairpins and a protein that recognizes specifically loop-loop complexes. These experiments were carried out at low temperature (5 °C), which allowed the stabilization of the nucleic acid-protein complex that would have been difficult or even impossible to analyze at higher temperatures due to the weak affinity of the protein for the RNA-RNA complex. The results also show that, although the protein Rop is able to interact with an artificial RNA-RNA loop-loop complex, its affinity is lower than that obtained with the RNA-RNA complex derived from its natural target.

2 Materials

2.1 Preparation of RNA Molecules and ColE1 Rop Protein

1. A nucleic acid synthesizer: We used an Expedite 8908 synthesizer (Millipore).
2. RNA molecules can be synthesized in house chemically, by transcription, or purchased from commercial suppliers (*see* **Note 1**).
3. Sephadex™ G-25 fine (GE Healthcare).
4. Centrifugal concentrators with molecular weight (MW) cutoff below that of the protein.

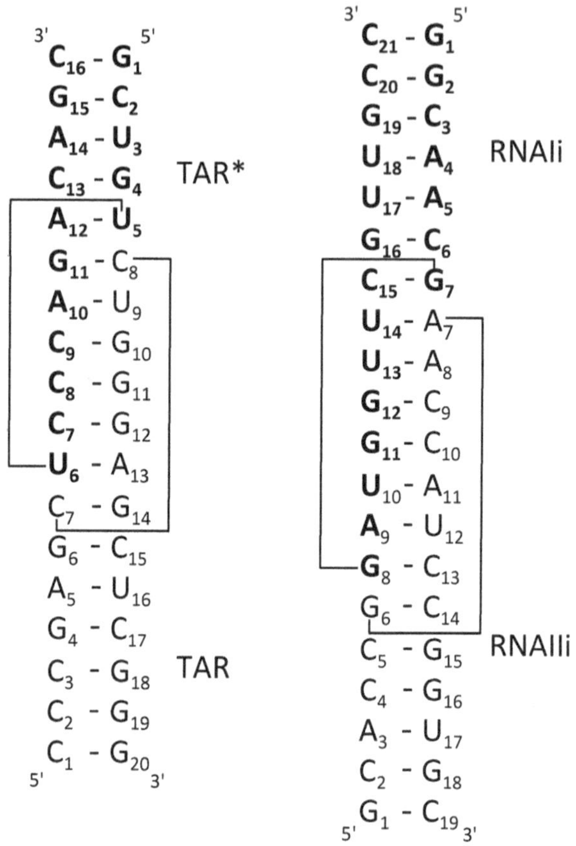

Fig. 1 Sequence and secondary structure of TAR-TAR* (left) and RNAIi-RNAIIi complexes (right). TAR* (in bold) was rationally designed to form a loop-loop complex with TAR, a stem-loop derived for the full-length TAR hairpin of HIV-1. RNAIi (in bold) and RNAIIi are hairpins that display inverted loops compared with the wild-type hairpins from the *E. coli* plasmid ColE1. These loops inversions were shown to stabilize the complex. TAR and RNAIIi were biotinylated after synthesis in house (see **Note 1**) at their 5′ end with the 5′EndTag labeling DNA/RNA kit and the biotin (long arm) maleimide reagent from Vector Labs (Peterborough, United Kingdom) following the instructions of the manufacturer

2.2 Interaction Studies

1. Biacore™ 3000 instrument (Biacore, GE Healthcare) or equivalent instrument (*see* **Note 2**).
2. Streptavidin-coated sensor chips either purchased from Biacore (GE Healthcare, Uppsala, Sweden) or XanTec Bioanalytics (Düsseldorf, Germany) (*see* **Note 3**).
3. Solution 1: 0.5% (w/v) SDS in MilliQ water.
4. Solution 2: 50 mM glycine pH 9.5 (adjust pH with NaOH).
5. Solution 3: 1% sodium hypochlorite.
6. Solution 4: mixture of NaCl 1 M and NaOH 50 mM prepared in MilliQ water.

7. Solution 5: mixture of EDTA 30 mM, urea 3.6 M, and formamide 40% prepared in MilliQ water.
8. 1.5 mL Eppendorf LoBind DNA and protein tubes.
9. Running buffer that will flow in the instrument: 10 mM sodium phosphate pH 7.2 at 21 °C containing 140 mM potassium chloride, 20 mM sodium chloride, 5 mM magnesium chloride, and 0.005% P20 (Biacore) (*see* **Note 4**).

3 Methods

3.1 Preparation of the Instrument

RNA molecules are known to be less stable than DNA. Even short RNA oligonucleotides (<30–50 nt), produced chemically or enzymatically, can be degraded within seconds by RNases if the buffers used to prepare them and flowed in the instrument are contaminated. As a general rule, prepare systematically fresh buffers and never use solutions more than 1 week old. Then clean carefully the instrument before running the experiments.

1. Set the temperature at 23 °C (at least).
2. Dock a "Maintenance" sensor chip (*see* **Note 5**).
3. Check and clean with MilliQ water the connector block connected to the microfluidic cartridge (IFC) and the injection needle if crystallized salts are present (*see* **Note 6**).
4. "Prime" at least twice the instrument with 0.22 μm filtered fresh MilliQ water.
5. Run "Desorb" with solutions 1 (0.5% (w/v) SDS in MilliQ water) and 2 (50 mM glycine-NaOH, pH 9.5) following the instructions of the Biacore control software displayed on screen.
6. "Prime" again the instrument with MilliQ water.
7. "Prime" twice with solution 3 (1% sodium hypochlorite).
8. "Prime" twice with MilliQ water to remove traces of cleaning solutions.
9. To finish, "Prime" with running buffer.

3.2 Preparation of the Sensor Chip

1. Use a streptavidin-coated sensor chip (*see* **Note 7**) to immobilize the biotinylated RNA target.
2. Dock the sensor chip into the instrument after inspecting its surface (*see* **Note 5**).
3. Set the temperature instrument at 23 °C (*see* **Note 8**).
4. "Prime" again the instrument with running buffer.
5. Run a sensorgram in manual mode on all flow cells at 5 μL/min.

6. Remove non-covalently attached streptavidin by injecting three 1-min pulses of solution.
7. Wash the surface with three 1-min pulses of running buffer.

3.3 Preparation and Immobilization of the Biotinylated Targets

1. Prepare 200 μL of biotinylated TAR and RNAIIi in running buffer at 50 nM in 1.5 mL Eppendorf LoBind tubes.
2. Denature the RNA samples 90 s at >95 °C, put the tubes on ice for 10 min and then at room temperature for 10 min, spin the tubes to remove condensation, and put them on the autosampler.
3. Run a sensorgram on one flow cell at 5 μL/min.
4. Inject the RNA sample for 30 s to check how fast the immobilization is. Using 50 nM solution, the slope of binding to the streptavidin-coated surface should be slow enough so that the immobilization level of the target can be easily controlled.
5. Inject again the RNA sample to reach about 500 resonance units (RU) of biotinylated target onto the surface (*see* **Note 9**).
6. Inject running buffer for 2 min. This will help in removing any traces of biotinylated RNA in the tubings of the instrument and in the channel of the IFC that was flowed with the biotinylated target (*see* **Note 10**).
7. Set the temperature of the instrument to 5 °C (*see* **Note 11**).

3.4 Preparation of the RNA and Protein Samples

1. Prepare 200 μL of TAR* and RNAIi at 1.5 μM in running buffer, in 1.5 mL Eppendorf tubes.
2. Heat the RNA samples at >95 °C for 2 min, and put them on ice for at least 5 min and then at room temperature for at least 15 min (*see* **Note 12**).
3. Put the RNA tubes in the sample compartment.
4. Prepare the protein Rop by serial dilution between 0 and 14 μM and 0 and 5.6 μM for the TAR-TAR* and RNAIi-RNAIIi complexes, respectively, in running buffer containing 1.5 μM of TAR* or RNAIi (*see* **Note 13**).
5. Put the RNA-protein mixtures in the sample compartment.

3.5 Binding of the Protein Rop to the RNA-RNA Complexes

The Biacore 3000 can be controlled with its control software that allows also writing methods to operate the instrument automatically. The sequence of operations used for monitoring the interaction between Rop and the RNA-RNA complexes contains three blocks of sequences: the first one, DEFINE APROG, defines the sequence of injections; the second one, DEFINE LOOP, defines the positions of the samples that will be injected; and the third one, MAIN, defines the sequences of operations in the whole method and includes the two first blocks with given names.

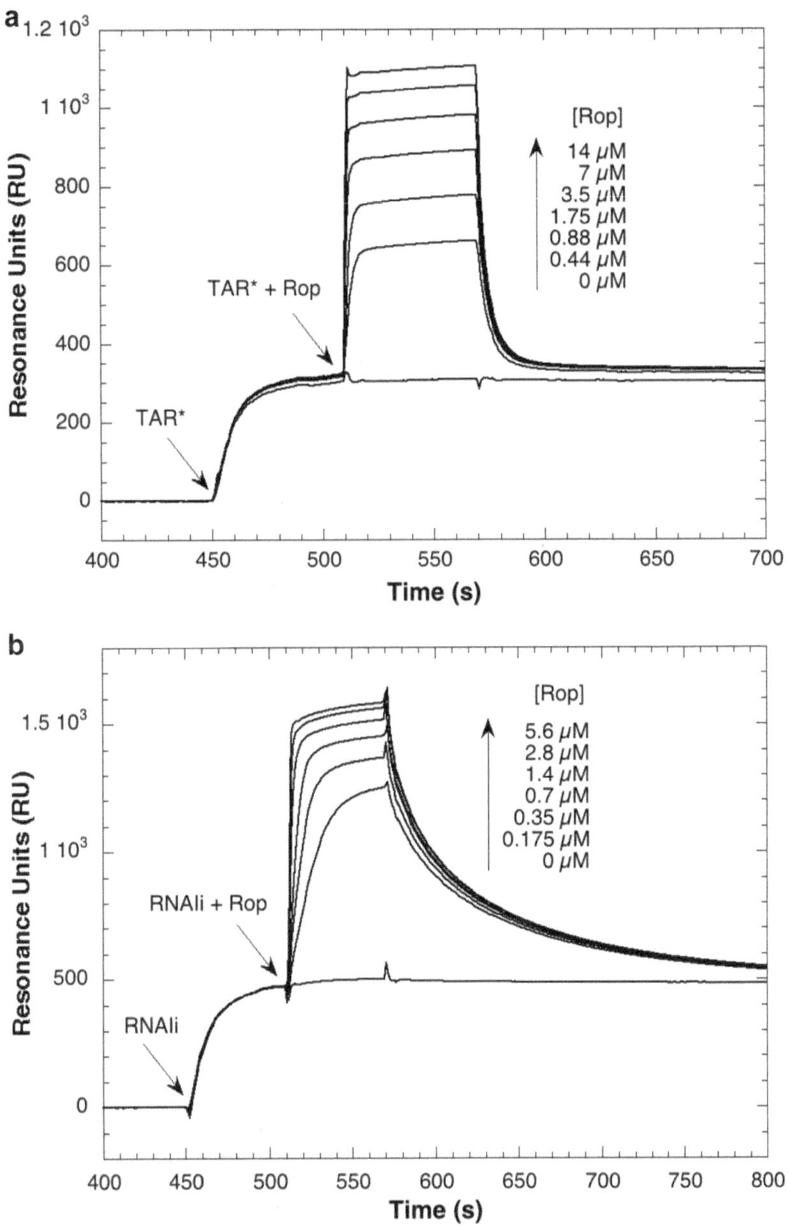

Fig. 2 Binding of the protein Rop to RNA-RNA loop-loop complexes. TAR* (**a**) and RNAli (**b**) were injected during 1 min (first arrow) at 1.5 μM in the running buffer over the TAR or RNAIIi functionalized surfaces . Then a mixture of TAR* or RNAli at constant concentration (1.5 μM) and Rop at increasing concentration (vertical arrow) was injected during 1 min (second arrow) immediately after the first injection ("COINJECT" command). After each binding cycle the surface was regenerated with a 1-min pulse of Solution 5 ("INJECT" command) followed by a 1-min pulse of running buffer to wash the surface

The sequence of operations is as follows:

1. First block:
   DEFINE APROG Binding (Binding is the name of the block).

   PARAM %RNA %Rop (defines the samples that will injected sequentially in the loop; first the RNA and then the protein).

   FLOW 20 (sets the flow rate of the instrument to 20 μL/min).

   FLOWPATH 3 and 4 (buffer is flowed in flow cells 3 and 4 of the IFC; biotinylated TAR and RNAIIi were immobilized in flow cells 3 and 4, respectively; each flow cell was used as a control channel for the other target).

   WAIT 04:00 (buffer is flowed through the IFC for 4 min before injecting the samples).

   COINJECT %RNA 20%Rop 20 (the RNA partner is injected first for 1 min, and immediately after, the protein samples containing the RNA partner at saturating concentration (*see* **Note 13**) are injected also for 1 min).

   INJECT r2f3 20 (20 μL of solution 5 is injected from position r2f3 of the rack to regenerate the sensor chip surface; *see* **Note 14**).

   INJECT r2f4 20 (20 μL of running buffer is injected from position r2f4 of the rack to wash the flow cells).

   WASH n (washes the needle of the autosampler).

   WASH l (washes the loop of the IFC).

   WASH s (washes the sample line of the IFC).

   WAIT 10:00 (waits for 10 min before ending the block).

   END

2. Second block (binding of Rop to TAR-TAR*):
   DEFINE LOOP Again (Again is the name of the block).

   LPARAMETER %RNA %Rop (the RNA partner and the protein samples as defined above in block named Binding will be injected from indicated positions of the racks).

   TIMES 1 (only one injection is performed per sample).

   r2f4 r2f4 (buffer is injected for double-referencing of the sensorgrams).

   r1a1 r1a4 (the RNA partner TAR* prepared at 1.5 μM is injected first from position r1a1 and then Rop mixed with TAR* from position r1a4).

   r1a1 r1a5 (same as above with Rop at increasing concentration, also for the next positions).

   r1a1 r1a6.

r1a1 r1b1.

r1a1 r1b2.

r1a1 r1b3.

r1a1 r1b4.

r2f4 r2f4 (running buffer is injected again to control stability of the surface).

An additional block can be written similarly to monitor the interaction between Rop and RNAIi-RNAIIi loop-loop complex.

3. Third block:
MAIN.

TEMP 5 (sets the temperature of the instrument to 5 °C).

FLOWCELL 3–4 (buffer and samples are flowed through flow cells 3 and 4).

LOOP Again ORDER (injects the samples in the order of the positions listed in block 2).

APROG Binding %RNA %Rop (operates the instrument according to the sequence of operations listed in blocks 1 and 2).

ENDLOOP (stops the sequence of operations).

APPEND continue (leaves the instrument on standby flowing running buffer once the program is finished).

END

Figure 2 shows an example of sensorgrams obtained when Rop was injected at increasing concentrations over the TAR-TAR* (Fig. 2a) and RNAIi-RNAIIi (Fig. 2b) loop-loop complexes. Similar sequences of commands can be performed on other SPR instruments for monitoring interactions involving nucleic acids. Last-generation instruments are user-friendlier to operate them automatically with control software that are wizard based.

3.6 Analysis of the Sensorgrams

The dissociation equilibrium constant, K_D, of Rop binding to the RNA-RNA loop-loop complexes can be determined by steady-state analysis by simply plotting the SPR signal at the end of the injection of the protein as a function of its concentration. The resulting curves are reported in Fig. 3 for Rop binding to TAR-TAR* (Fig. 3a) and RNAIi-RNAIIi (Fig. 3b). The dissociation equilibrium constant, K_D, is equal to 687 ± 35 nM and 62 ± 3 nM for the former and latter complexes, respectively. Not surprisingly, Rop displays a higher affinity for the RNA-RNA complex derived from its natural target RNAI-RNAII than for another complex formed between a structured RNA derived from the TAR element of HIV-1 and a hairpin RNA, TAR*, rationally designed to recognize TAR by loop-loop interactions.

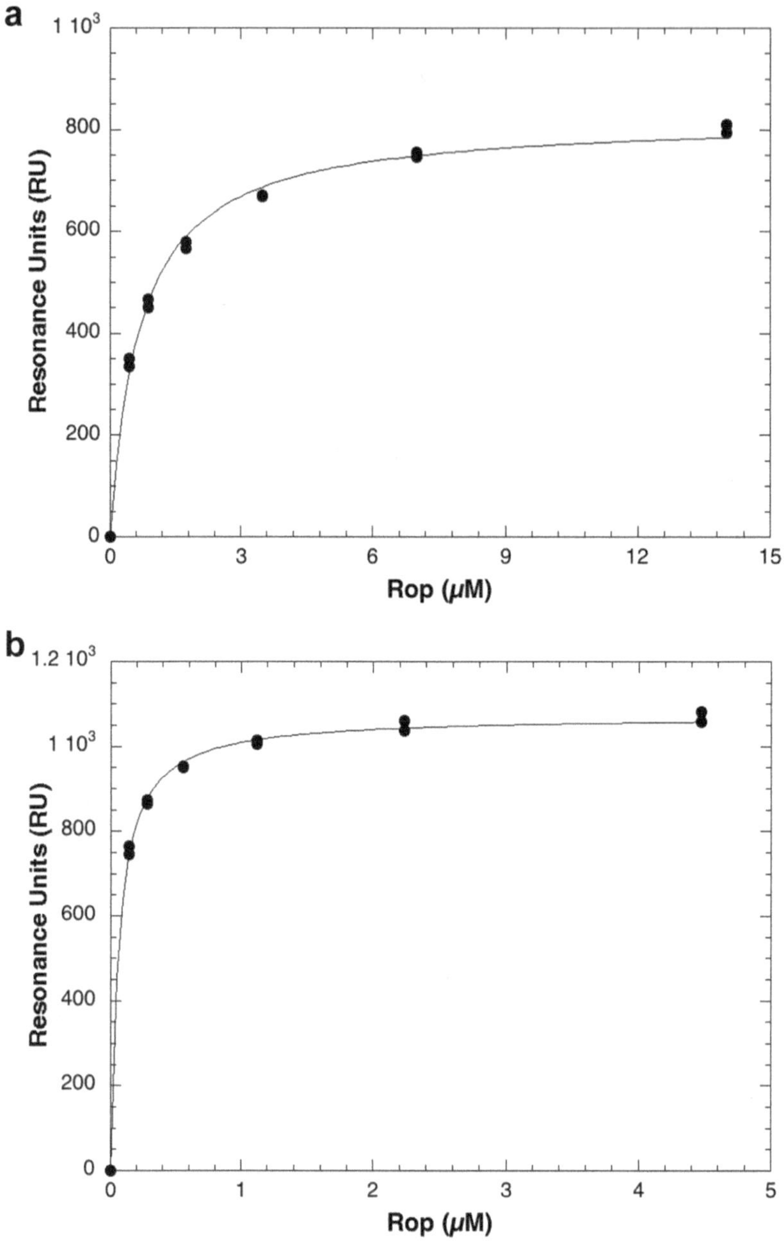

Fig. 3 Steady-state analysis of Rop binding to TAR-TAR* (**a**) and RNAIi-RNAIIi (**b**) RNA-RNA loop-loop complexes. The SPR signal obtained at the end of the injection of the mixture of Rop + TAR* (Fig. 2a) or Rop + RNAIi (Fig. 2b) was plotted as a function of Rop concentration to determine the dissociation equilibrium constant, K_D, of the nucleic acid-protein complex. K_D corresponds to the concentration of Rop at half saturation of the SPR signal

4 Notes

1. RNA molecules can be purchased from several commercial suppliers. RNA oligonucleotides are about 10 times more expensive to buy than DNA. Over years working with RNA molecules, we observed differences in quality and quantity provided by the suppliers. Very few are good for synthesizing RNA molecules. Dharmacon (GE Healthcare) is one of them. As a general rule, never trust suppliers regarding quality of the samples. Quality of the synthesis should be checked systematically either by mass spectrometry or by running a 7 M urea 10–20% polyacrylamide (depending of the size of the RNA molecule) denaturing gel loaded with at least 2 μg of RNA per well and then colored by stain-all solution, as described previously [23]. Further purification will be required if n-1 RNA bands are present. We also strongly recommend "standardizing" all RNA samples, whether chemically synthesized, produced by transcription in house or purchased from suppliers, by passing them through a G-25 spin column (GE Healthcare) or running a dialysis with Slide-A-Lyzer™ MINI dialysis devices with a cuplike design available in 0.1, 0.5, and 2 mL capacities (ThermoFisher Scientific). This will remove impurities from the synthesis so the oligonucleotides will be in similar buffer conditions no matter how they were produced.
2. Biacore is the world leader in the SPR market. The company offers instruments from 2 to 16 flow cells, thermostated between 4 and 40–45 °C for most of them, and with sample racks that can be also temperature controlled. This is useful for fragile samples that must be kept below 10 °C to avoid denaturation. A large temperature range is also of interest for determining thermodynamic parameters by measuring equilibrium constants as a function of temperature (van't Hoff analysis). Other companies such as Reichert Technologies, FortéBio, Horiba, and IBIS Technologies have commercialized SPR instruments. When time comes to buy an instrument, the purchase price is obviously an important element to consider, but it is not the only one. Sensitivity, reproducibility, number of complexes that can be analyzed simultaneously, consistency of the signals, and how the surface reacts with sticky molecules (positively charged) or with chemical compounds for regenerating the functionalized surface should be also considered. Suppliers always sell instruments tested with molecules that behaved ideally. Take time to test the instrument with your

preferred molecules or test them with someone who owns the instrument and is trained with it.

3. Streptavidin-coated sensor chips can be purchased directly from Biacore and XanTec Bioanalytics that sell sensor chips compatible with Biacore instruments. Alternatively, the surface can be prepared in house with streptavidin purchased from Sigma-Aldrich or IBA Lifesciences (Göttingen, Germany), or other suppliers, and carboxymethylated dextran sensor chips, typically CM5 or CMD200m sensor chips, from Biacore and XanTec Bioanalytics, respectively. Follow the instructions of the suppliers for long-term storage of streptavidin (we store it at 2.5 mg/mL at −80 °C). To activate the surface, follow the instructions of the manufacturers that come with the sensor chips. In our hands, with an instrument set at 25 °C at a flow rate of 5 μl/min, 7-min activation of the surface with a mixture of NHS/EDC from Biacore followed by 7-min injection of streptavidin at 0.25 mg/mL in 10 mM sodium acetate buffer, pH 5, covalently attached enough streptavidin onto the surface to immobilize biotinylated oligonucleotide for most experiments. To increase biotinylated RNA target level, if necessary, simply increase streptavidin density by increasing the duration of activation of the surface and injection of the protein.
4. P20 sold by Biacore can be replaced by Tween-20. We have been using it for years without noting any adverse effects on the experiments with nucleic acids and proteins. Biacore recommends using this nonionic detergent to avoid adsorption of molecules on tubings and plastics of the microfluidic cartridge and to help molecules flow. By default, we use Tween-20 at the percentages recommended by the manufacturer, 0.005% and 0.05% for Biacore 3000 and Biacore T200, respectively. If there is any doubt about possible effects, prepare two buffers, with and without surfactant, and check reproducibility and quality of the SPR curves.
5. Before inserting any sensor chip in the instrument, remove it from its cassette and check for dust and suspicious traces on the glass side of the support. Blow compressed air to remove dust. Use a repositionable piece of tape carefully placed on the glass surface to remove any traces that resist air. Do not touch the sensor chip on the dextran side. Free dextran surfaces can be cleaned with milliQ water and dried with compressed air without altering the binding capacity of the sensor chips.
6. Crystallized salts present on the injection needle and the connector block means that the instrument requires cleaning. No matter who used the instrument before, extensively clean the instrument when working with RNA molecules. Last-generation instruments from Biacore are less prone to salt

deposition on the injection needle because it is systematically washed with milliQ water. Even with this improvement, the instrument must be cleaned to remove RNases that will degrade RNA in seconds.

7. Streptavidin-coated sensor chips can be purchased directly from Biacore (GE Healthcare). They can also be purchased at a lower cost from XanTec Bioanalytics (Düsseldorf, Germany). We did not observed differences between these sensor chips for investigating interactions between nucleic acids, but we observed that surfaces from Biacore and XanTec could behave differently with respect to nonspecific binding when proteins or small ligands were injected. Check this by injecting the partners on nonfunctionalized sensor chips. Increasing the ionic strength of the running buffer can reduce nonspecific binding, but this can also destabilize the interaction of interest. Biacore and XanTec sell polyethylene glycol-based sensor chips to reduce nonspecific binding. We indeed observed that nonspecific binding decreased with some of our molecules but was not totally removed. Nonspecific binding can be removed from the signal of interest using appropriate control molecules in particular when the control molecule and the target can be captured [9].

8. In some SPR instruments, the sample compartment and/or the sample racks can also be thermostated. This can be very useful to keep protein samples at low temperature to avoid denaturation. We usually keep the sample compartment at 10 °C with proteins.

9. The amount of biotinylated RNA immobilized onto a streptavidin sensor chip can be easy controlled by changing the concentration of the RNA sample or the contact time (injection duration). 25–50 nM is a good range of concentration. Concentrations above 100 nM or higher will make it difficult to control the immobilization level because of a too fast rate of immobilization. For kinetic studies, it is usually recommended to use low densities to avoid artifacts such as rebinding of the analyte to ligand during the dissociation phase. With RNA and nucleic acids in general that can be immobilized onto the surface with a unique orientation, we observed that immobilization level was less critical than when proteins were immobilized by amine coupling. Always check how density levels of the target affect the binding curves.

10. The IFC can be contaminated if too high concentrations (>250–500 nM) of biotinylated compounds are injected. Every time we could not immobilized a biotinylated target, it was because of a contamination either with free biotin or with a biotinylated sample used in a previous experiment, never

because of the streptavidin sensor chip. A fast drop of the slope of the immobilization curve is a sign of contamination. If this occurs, the best way to decontaminate the IFC and the instrument is to inject free streptavidin prepared in running buffer at 1–2 mg/mL. If you suspect your samples to be contaminated with free biotin, pass them through a G-25 Sephadex column. Keep and use as a control a biotinylated nucleic acid, RNA or DNA, which worked in previous experiments. This will help to distinguish between contamination and a problem during the biotinylation of the RNA samples either chemically synthesized or biotinylated after in vitro transcription.

11. Decreasing the temperature of the instrument helped to increase the affinity of the protein for the RNA-RNA complex. At room temperature, the affinity was insufficient to measure it. Not all instruments allow to reach temperatures around 5 °C. This is also a specification to consider before buying an instrument. Large range of temperature can also be useful for thermodynamic studies by van't Hoff analysis, although it is more time-consuming than isothermal titration calorimetry.
12. The denaturation and refolding of structured RNAs is a crucial step that will ensure that these molecules adopted the most thermodynamically favorable conformation. Doing that will not guarantee that alternative foldings do not exit. Even with short RNA molecules (less than 30 nucleotides), we observed conformations stable enough, in equilibrium, that made the sensorgrams impossible to analyze (Palau et al., 2013). The advantage of nucleic acids is that structural heterogeneity can be easily assessed running, for instance, a native polyacrylamide gel. We always denature/refold RNAs in the buffer used for the interaction rather than preparing the samples first in MilliQ water to denature/refold them and then adding buffer. The reason is that we observed that the SPR signal was always higher when the RNA samples were denatured/refolded in buffer rather than in milliQ water.
13. Rop is a protein that stabilizes loop-loop RNA complexes. If the protein is injected alone after the injection of the RNA partner, then part of the RNA-RNA formed that did not bind to the protein will dissociate simultaneously with the binding of Rop to the RNA-RNA complex still present on the sensor chip surface. The only way to avoid these simultaneous reactions that will make this analysis difficult is to inject a mixture of RNA at constant and saturating concentration and protein at variable concentrations.
14. Regeneration is a crucial step during SPR experiments. This step is used to dissociate instantaneously the partners that bound to the immobilized or captured target in order to

perform new cycles of binding. If this step is not properly controlled, it greatly compromises the use of SPR for investigating molecular interactions. Compounds used for regenerating the functionalized surface have to be efficient to dissociate the complexes but without damaging the target and the surface. NaOH injected for 1 min at room temperature is very efficient for regenerating complexes involving DNA. It can actually also work with RNA molecules, but stability of the target has to be checked by performing several cycles of binding/regeneration to ensure that the RNA is not hydrolyzed. EDTA is efficient for regenerating complexes with structured RNA stabilized by magnesium ions. Whatever the regenerant used, the SPR will decrease with time. How fast this occurs will depend on the target, the regenerant, the number of binding/regeneration cycles, and, last but not least, how fresh the buffers flowed in the instrument are. With RNA molecules, we never use buffers more than 1 week old, even if there were filtered.

Acknowledgments

We thank the structural biophysico-chemistry facility (UMS 3030/US001 CNRS/Inserm) of the IECB (Pessac, France) for access to the Biacore 3000 instrument and Lætitia Minder, assistant engineer in charge of the SPR facility of the institute, for technical assistance. The SPR instrument was acquired with the support of the "Conseil Régional Nouvelle Aquitaine" (former Région Aquitaine).

References

1. Jönsson U, Fägerstam L, Ivarsson B, Johnsson B, Karlsson R, Lundh K, Löfås S, Persson B, Roos H, Rönnberg I (1991) Real-time biospecific interaction analysis using surface plasmon resonance and a sensor chip technology. BioTechniques 11:620–627
2. Karlsson R, Michaelsson A, Mattsson L (1991) Kinetic analysis of monoclonal antibody-antigen interactions with a new biosensor based analytical system. J Immunol Methods 145:229–240
3. Fägerstam LG, Frostell-Karlsson A, Karlsson R, Persson B, Rönnberg I (1992) Biospecific interaction analysis using surface plasmon resonance detection applied to kinetic, binding site and concentration analysis. J Chromatogr 597:397–410
4. Morton TA, Myszka DG (1998) Kinetic analysis of macromolecular interactions using surface plasmon resonance biosensors. Meth Enzymol 295:268–294
5. Pol E (2010) The importance of correct protein concentration for kinetics and affinity determination in structure-function analysis. J Vis Exp. https://doi.org/10.3791/1746
6. Karlsson R (2016) Biosensor binding data and its applicability to the determination of active concentration. Biophys Rev 8:347–358
7. Visentin J, Minder L, Lee J-H, Taupin J-L, Di Primo C (2016) Calibration free concentration analysis by surface plasmon resonance in a capture mode. Talanta 148:478–485
8. Di Primo C, Taupin J-L, Visentin J (2018) Comments on 'direct quantitative measurement of the kinetics of HLA-specific antibody interactions with isolated HLA proteins. Hum Immunol 79:129
9. Visentin J, Couzi L, Dromer C, Neau-Cransac-M, Guidicelli G, Veniard V, Coniat KN,

Merville P, Di Primo C, Taupin J-L (2018) Overcoming non-specific binding to measure the active concentration and kinetics of serum anti-HLA antibodies by surface plasmon resonance. Biosens Bioelectron 117:191–200

10. Boucard D, Toulmé J-J, Di Primo C (2006) Bimodal loop-loop interactions increase the affinity of RNA aptamers for HIV-1 RNA structures. Biochemistry 45:1518–1524

11. Palau W, Masante C, Ventura M, Di Primo C (2013) Direct evidence for RNA-RNA interactions at the 3′ end of the hepatitis C virus genome using surface plasmon resonance. RNA 19:982–991

12. Lebars I, Legrand P, Aimé A, Pinaud N, Fribourg S, Di Primo C (2008) Exploring TAR-RNA aptamer loop-loop interaction by X-ray crystallography, UV spectroscopy and surface plasmon resonance. Nucleic Acids Res 36:7146–7156

13. Ducongé F, Di Primo C, Toulmé JJ (2000) Is a closing 'GA pair' a rule for stable loop-loop RNA complexes? J Biol Chem 275:21287–21294

14. Darfeuille F, Arzumanov A, Gryaznov S, Gait MJ, Di Primo C, Toulmé J-J (2002) Loop-loop interaction of HIV-1 TAR RNA with N3'-->P5' deoxyphosphoramidate aptamers inhibits in vitro tat-mediated transcription. Proc Natl Acad Sci U S A 99:9709–9714

15. Masante C, Jaubert C, Palau W, Plissonneau J, Besnard L, Ventura M, Di Primo C (2015) Mutations of the SL2 dimerization sequence of the hepatitis C genome abrogate viral replication. Cell Mol Life Sci 72:3375–3385

16. Di Primo C, Rudloff I, Reigadas S, Arzumanov AA, Gait MJ, Toulmé J-J (2007) Systematic screening of LNA/2'-O-methyl chimeric derivatives of a TAR RNA aptamer. FEBS Lett 581:771–774

17. Dausse E, Barré A, Aimé A, Groppi A, Rico A, Ainali C, Salgado G, Palau W, Daguerre E, Nikolski M et al (2016) Aptamer selection by direct microfluidic recovery and surface plasmon resonance evaluation. Biosens Bioelectron 80:418–425

18. Eguchi Y, Tomizawa J (1990) Complex formed by complementary RNA stem-loops and its stabilization by a protein: function of ColE1 rom protein. Cell 60:199–209

19. Eguchi Y, Tomizawa J (1991) Complexes formed by complementary RNA stem-loops. Their formations, structures and interaction with ColE1 rom protein. J Mol Biol 220:831–842

20. Chang KY, Tinoco I (1994) Characterization of a 'kissing' hairpin complex derived from the human immunodeficiency virus genome. Proc Natl Acad Sci U S A 91:8705–8709

21. Chang KY, Tinoco I (1997) The structure of an RNA 'kissing' hairpin complex of the HIV TAR hairpin loop and its complement. J Mol Biol 269:52–66

22. Gregorian RS, Crothers DM (1995) Determinants of RNA hairpin loop-loop complex stability. J Mol Biol 248:968–984

23. Di Primo C, Dausse E, Toulmé J-J (2011) Surface plasmon resonance investigation of RNA aptamer-RNA ligand interactions. Methods Mol Biol 764:279–300

Chapter 7

RNA Folding and Unfolding Under Force: Single-Molecule Experiments and Their Analysis

Laurent Geffroy, Thierry Bizebard, and Ulrich Bockelmann

Abstract

We have previously described (Geffroy et al. Methods Mol Biol 1665:25–40, 2018) how to unfold (or fold) a single RNA molecule under force using a dual-beam optical trap setup. In this chapter, we complementarily describe how to analyze the corresponding data and how to interpret it in terms of RNA three-dimensional structure. As with all single-molecule methods, single RNA molecule force data often exhibit several discrete states where state-to-state transitions are blurred in a noisy signal. In order to cope with this limitation, we have implemented a novel strategy to analyze the data, which uses a hidden Markov modeling procedure. A representative example of such an analysis is presented.

Key words RNA, Single molecule, Optical trap, Force, Unzipping

1 Introduction

Numerous forces act on RNA molecules in the biological cells. These forces are typically associated with fundamental processes involving RNA (e.g., transcription, translation, protein binding, unwinding, and annealing of double-stranded RNA), and they have important consequences on the three-dimensional (3D) structures of the RNA molecules and thus on their functions.

Relatively few techniques exist to study the influence of mechanical forces on the 3D structure of RNA molecules. Among these techniques, single-molecule methodologies—such as the one presented in this chapter—offer important advantages compared to ensemble measurements. The most noteworthy of these advantages is the absence of ensemble averaging—thus allowing to directly probe temporal fluctuations and molecule-to-molecule variability. This aspect is particularly highlighted in nucleic-acid (including RNA) single-molecule force measurements: signals observed by these techniques depend not only on structure and base sequence of the nucleic acid but also on case-by-case singularity of the single molecule itself.

Véronique Arluison and Frank Wien (eds.), *RNA Spectroscopy: Methods and Protocols*, Methods in Molecular Biology, vol. 2113, https://doi.org/10.1007/978-1-0716-0278-2_7,

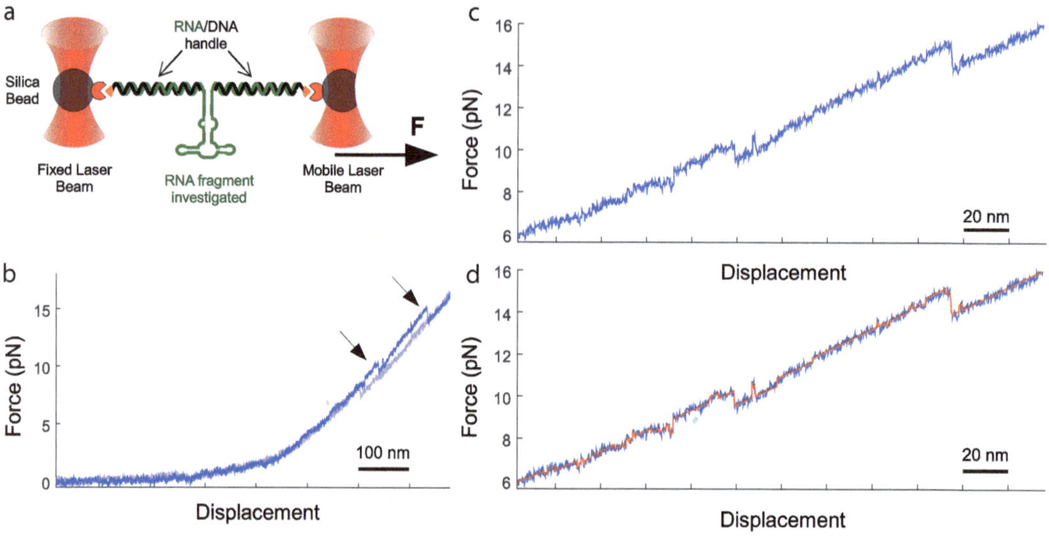

Fig. 1 Principle of the experimental setup. (**a**) A dual optical trap is used to manipulate a DNA/RNA molecular construction in a dumbbell configuration. In this configuration, the investigated RNA fragment is hybridized to two RNA/DNA handles containing both a biotin moiety (orange) at one of their extremities. Biotin moieties are linked to streptavidin (red)-coated silica beads. One optical tweezers is moved at constant velocity to stretch (or unstretch) the nucleic acid construction; the other trap is used to measure the force (for complete details, *see* ref. 3). (**b**) A typical RNA force vs. displacement curve. As the molecular construction is stretched (dark blue curve), one observes a sawtooth signal region between 10 pN and 15 pN (marked by black arrows) corresponding to the progressive unfolding of the RNA fragment of interest. When the movement of the mobile trap is reversed (light blue curve), one can observe refolding of the RNA fragment: in this case, the two (folding and unfolding) curves are very similar – this minimal hysteresis is indeed a very important condition to be fulfilled for the single-molecule experiments to be rigorously analyzed by the methodology described in this paper which assumes the system to be in "close-to-equilibrium" conditions (*see* **Note 1**). (**c**) A close-up view of Fig. **b** showing the region where the abrupt transitions corresponding to (progressive) RNA unfolding are observed. (**d**) Same as **c** with the HMM-modelized curve superimposed in red

Single-molecule force experiments on RNA molecules can be achieved in several slightly different configurations [1, 2]. In the one described in this paper, a double optical tweezers setup is used which allows to manipulate single RNA molecules in an "unzipping" configuration. In the configuration that we use, forces pull the two extremities of one single RNA/DNA hybrid molecule containing the RNA sequence under investigation (*see* Fig. 1a and its legend for more details). When high enough forces are applied onto the molecular construction, one can observe the mechanical unwinding of the 3D structure of the investigated RNA fragment. Typical forces exerted to observe these unwinding events are in the range 10–20 pN (depending on RNA sequence and structure), and importantly, this range of forces corresponds to the forces exerted on RNA (and DNA) by many essential molecular motors acting on them [4]—highlighting the biological relevance of the methodology.

The double optical tweezers setup that we routinely use to study the influence of mechanical forces on single RNA molecules was thoroughly described in a chapter of the Methods in Molecular Biology (MiMB) series [3]: in this paper, we had focused on the description of the setup of the dual optical trap, the preparation of the molecular constructions, and the description on how to perform a single RNA molecule force measurement in the unzipping configuration. In this chapter, we complementarily describe how we perform data analysis in the same experimental configuration. We will also shortly describe how these measurements can be related to RNA two-dimensional (2D) and 3D structural features. This last point will hopefully convince other scientists that our described methodology can be a very useful tool to characterize RNA 2D/3D structural features, adding complementary information to other well-known techniques.

2 Materials

2.1 Software

1. MATLAB (MathWorks).
2. vbFRET: freely available software [5].

3 Methods

As aforementioned, our previous paper in the Methods in Molecular Biology (MiMB) series [3] has thoroughly described the dual optical trap setup that we use to manipulate the single molecules; in the same publication, we also described how to design and prepare the molecular constructions necessary to mechanically probe the RNA fragment of interest – a schematic view of the setup is presented in Fig. 1a, and the legend of this figure shortly summarizes the important points of our methodology. Finally, the last paragraphs of this precedent MiMB paper described how to collect force (exerted on the single molecule) vs. displacement (of the mobile trap) curves on single RNA/DNA molecules. An example of such a curve is shown in Fig. 1b.

In this chapter, we thus complementarily describe the procedure we routinely use to analyze and interpret the corresponding data.

3.1 Data Analysis

To be able to make some meaningful analysis of the mechanical properties of the RNA sequence of interest, one has to collect many (at least several tens and preferably more than a hundred) force vs. displacement curves of the molecule of interest. Of course, the same molecule can be repeatedly stretched (the optical tweezers are moved away from each other) and then unstretched (the optical tweezers are moved back closer to each other)—this several times (usually 1–5 times), until the molecular construction breaks. At this point, additional data can be collected on another "fresh" molecular construction and so on until sufficient data is accumulated for accurate analysis.

A typical force vs. displacement experimental curve—such as the one shown in Fig. 1b, c—will display, in an otherwise smoothly varying signal, a small number of abrupt rips—and these indeed correspond to the progressive unfolding of the RNA fragment of interest. In principle, these features can be analyzed by simple eye inspection. However, this procedure is very cumbersome and also, more importantly, can be more prone to user bias and misinterpretations than an automatic procedure. That's why we have devised to implement an improved strategy using Hidden Markov Modeling (HMM) analysis of the data. We have chosen this strategy because, fundamentally, HMM analysis is well suited to find out discrete states in noisy data and find the most probable path through these states [6]—thus well adapted to our experimental data.

We detail below the overall procedure that we use.

3.1.1 Data Analysis Procedure (See **Note 2**)

A simplified flowchart of the whole procedure is presented in Fig. 2.

1. In our case, the data acquisition (described in our previous paper; *see* ref. 3) is performed with a custom-built LabVIEW script. The data file for one single molecular construction consists of a list containing the measured forces and the associated (mobile trap) displacement values. The first step of the analysis is to import these data in the MATLAB environment—this procedure of course depends on the specific data acquisition software that you have used.
2. From this point, we will use two well-known force vs. displacement theoretical models to analyze and fit our data, succinctly:
   - At low forces ($F < 2$ pN) for double-stranded nucleic acids segments and at all forces used in our described experiments (forces are always less than ~25 pN) for single-stranded nucleic acid segments, the best theoretical force (F) vs. end-to-end distance x (*see* **Note 3**) model is the classical worm-like chain (WLC) formulation of Marko and Siggia [7]:

$$F = \frac{k_B T}{4A}\left(\frac{1}{(1 - x/L_c)^2} - 1 + 4\frac{x}{L_c}\right) \quad (1)$$

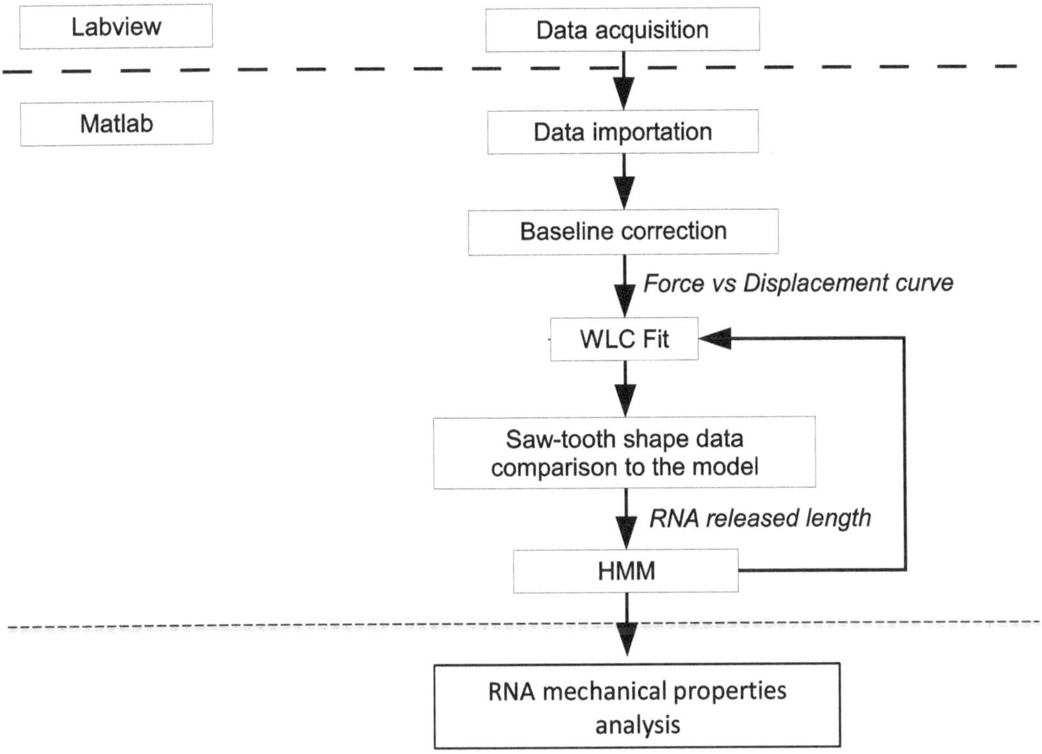

Fig. 2 Flowchart of the analysis procedure. The force vs. displacement data acquisition is performed in the LabVIEW environment. The subsequent steps of the analysis are done in the MATLAB environment. After importing the data in this environment and making a baseline correction, a first WLC theoretical model is applied to modelize the behavior of the double-stranded nucleic acid (handles); then a second WLC model is used to account for the progressive unfolding of the investigated RNA fragment (corresponding to the "sawtooth shape" data in the original force vs. extension curves). This latter analysis allows to extract from the experimental curves a new parameter: the single-stranded RNA released length (i.e., length of investigated RNA unfolded at a certain force). Following this step, the single-stranded RNA released length vs. force relationship is analyzed with a HMM software to identify the characteristic intermediate states visited during RNA unfolding by force. The last three steps of the procedure can be reiterated to optimize data processing (*see* main text for details). Finally, the intermediate states visited during RNA unfolding by force are characterized and analyzed

(where k_B is the Boltzmann constant, T is the absolute temperature, and A and L_c are, respectively, the persistence length and the contour length of the molecule).

- For forces F 2 pN < F < 25 pN in the case of double-stranded nucleic acid segments, the best theoretical model for our experiments is the so-called extensible WLC model [8] which uses the following equation:

$$x = L_0\left(1 - \sqrt{\frac{k_B T}{4A_0 F}} + \frac{F}{K_0}\right) \tag{2}$$

(where A_0, L_0, and K_0 are, respectively, the persistence length, the contour length, and the elastic modulus of the molecule).

3. In our experimental setup, the force is measured indirectly (*see* ref. 3 for complete details): we actually measure a relative force exerted on the RNA/DNA construction (i.e., zero force is not defined). To correct this, we use the property that, at low distance between the two traps, the force exerted on the molecular construction is almost equal to zero – and thus constant (rigorously, according to WLC model Eq. 1, for distances such that $F < < 0.1$ pN, therefore, negligible compared to the experimental noise of our force measurements, which is usually around ~0.2 pN). In practice, for our own experimental setup where the diameter of the two beads are of 0.96 μm and the contour length of our molecular construction is of 1.64 μm, to fulfill the above condition, we ordinarily use the data corresponding to a distance d between the two traps 1.7 μm < d < 2.2 μm, to define the baseline. The mean measured apparent force corresponding to this range of distance is calculated and then subtracted from all raw (apparent) force values to obtain absolute force values (*see* **Note 4**).
4. For an intermediate distance between the two traps, the force exerted on the molecule increases nonlinearly but smoothly with the displacement. This part of the curve corresponds to the deformation of the RNA/DNA hybrid handles, while the RNA fragment investigated remains fully folded – so, this part of the curve contains none of the abrupt rips discussed in the next step. In our experimental conditions, this intermediate distance part of our curves typically corresponds to conditions where force $F < 5$ pN (*see* Fig. 1b). Of course, for other experimental conditions, adjust this cutoff as necessary. When this selection has been performed, one has to fit the corresponding data to Eq. 2. As can be seen from the expression of this equation, the model is adjusted using a set of three free parameters: the contour length L_0, the persistence length A_0, and the elastic modulus K_0 of the molecule (*see* **Note 5**).
5. For an even higher distance between the two traps (typically for force range 5 pN < F < 15–20 pN; *see* Fig. 1b), the force vs.-displacement curves of our molecules exhibit characteristic sawtooth shapes—i.e., with a few of the aforementioned abrupt rips (Fig. 1c). It is the signature of the progressive unfolding of the investigated RNA fragment under the applied force. As there does not exist any model which predicts this sawtooth shape, we therefore use an indirect strategy: we extrapolate each experimental point of the original force vs. displacement curve as corresponding to part of the investigated RNA fragment being progressively unfolded. For a fully unfolded single-

stranded RNA fragment of length n nucleotides, we can use the WLC model of Eq. 1 with parameters $L_c = n * 0.59$ nm and $A = 1$ nm [8, 9] to calculate its extension at a certain force. This calculation is then used to compute the force vs. displacement curves of a fully unfolded RNA fragment for a set of different lengths of RNA fragment (*see* **Note 6**). This contribution can be added to the previously fitted WLC model for the double-stranded RNA/DNA part of the construction. We then use a least-square minimization to find the most likely unfolded RNA contour length corresponding to each experimental point. This quantity is called the released single-stranded RNA length (*see* **Note 7**).

6. This released single-stranded RNA length (abbreviated as "released RNA length," thereafter) is then put in the right format to be used by the HMM software vbFRET. This format consists of an array with two rows. The first row contains the recorded displacements. The second row (the "acceptor signal" of vbFRET) contains the released RNA lengths normalized between 0 and 1 (*see* **Note 8**). The HMM software then finds the most likely number of states and the transitions between them through an idealized path of the data.
7. The results of the vbFRET software analysis are saved, and the idealized path is plotted with the data. The path must at least exhibit a low state and a high state which both correspond to the expected RNA released length – i.e., 0 for the low state and the contour length corresponding to the fully unfolded RNA fragment for the upper state (*see* **Note 9**). This requirement is a control that the sawtooth shape observed corresponds to the opening of the RNA fragment of interest.
8. When the HMM fit shows that the force range used for the WLC fit of **step 4** does include some partially unfolded RNA intermediate states (they can sometimes occur below 5 pN), this means that the **step 4** fit was not fully correct. Hence, the results of the HMM analysis must be used to assess the range of low forces where no intermediate state is present, and use this new range to perform again **step 4** and the following steps.
9. Finally, to make a quality control of the whole procedure, the original force vs. displacement curve is plotted together with its idealized HMM-predicted path. An example of this is shown in Fig. 1d – in this particular case, it is clear that the HMM procedure was able to rightfully predict several RNA unfolding intermediate states.

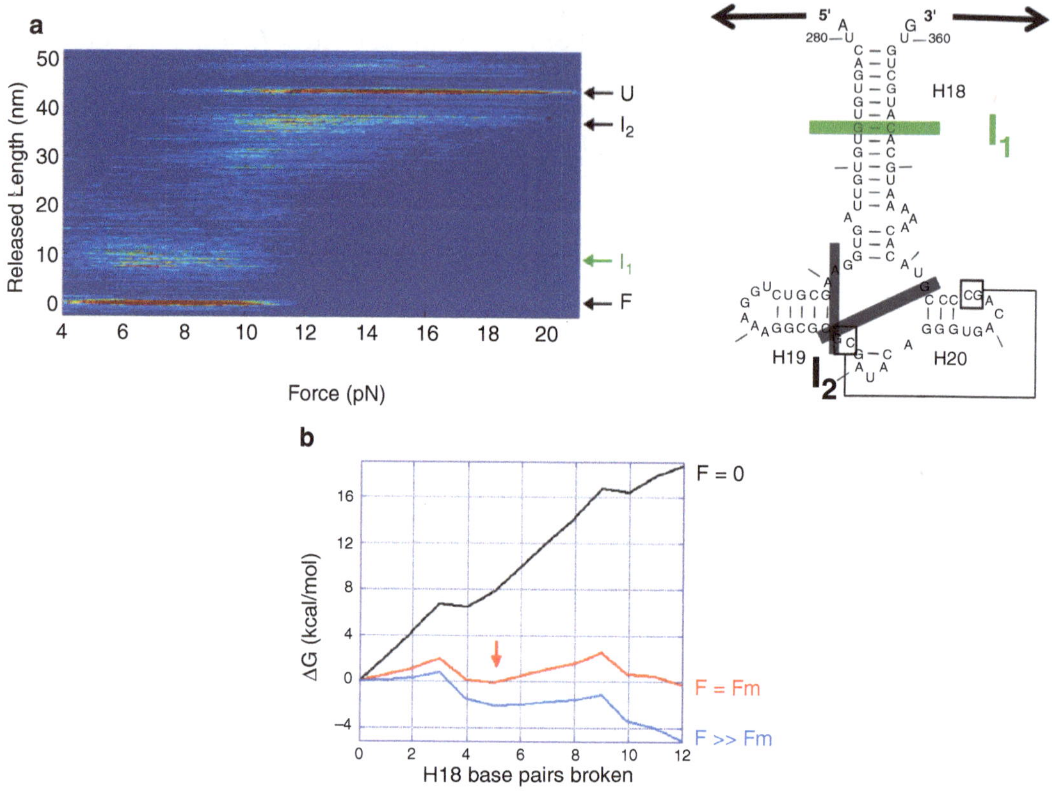

Fig. 3 (**a**) Left: superposition of the HMM-modelized RNA released length vs. force curves from 254 individual measurements on single molecules of the RNA fragment under investigation. All the curves have been superimposed in a heat map histogram where red (dark blue) indicates frequently (rarely) observed states. Positions in the histogram of intermediate states discussed in the text (I_1 and I_2) are indicated, alongside the positions of the fully folded (F) and fully unfolded (U) RNA fragment. Right: secondary structure of the rRNA fragment. The colored bars indicate the positions of the two RNA unfolding intermediate states I_1 and I_2 (there are two possibilities for I_2: *see* text). The two black arrows indicate the directions of the forces which pull the two extremities of the RNA fragment. (**b**) Intermediate I_1—its location can be rationalized as follows: on this graph, the calculated free energy landscape of helix H18 in our experimental conditions is shown. At zero force, free energies are calculated with software Mfold [11]. At forces $F > 0$, a RNA-stretching model is used to obtain the effect of force on each partially unfolded species (calculations as in ref. [12]). At zero force, the helix is obviously fully stable (black curve). At a critical force $F = Fm$ ($\sim$10 pN in our experimental conditions; red curve), a species with a partially unfolded H18 ($\sim$5 base pairs broken; its location is shown with a red arrow) is predicted to be significantly populated; this species most likely corresponds to our observed intermediate I_1 (which we experimentally map at 7 ± 2 base pairs): this local minimum corresponds to the end of the region that contains a sequence of consecutive GU-UG-AU base pairs and thus exhibits weak local stability. Finally, at forces significantly higher than Fm (blue curve), the whole H18 helix is destabilized and unfolded, as observed in our experiments

3.2 Interpretation of HMM Analysis Using Known or Predicted RNA 3D Structure

The analysis of RNA force vs. displacement data detailed in the previous paragraphs can be used to assess at the molecular level the influence of force on the 3D structure of the investigated RNA fragment (*see* **Note 10**). In this paragraph, we will use one example we have thoroughly examined to illustrate this point.

The RNA fragment we have studied consists of a small fragment of *E. coli* 23S rRNA (from nucleotides 281 to 359 of this rRNA). The three-dimensional structure of this RNA fragment is only known in the context of the fully assembled ribosome, but chemical probing and structure prediction strongly suggest that the 3D structure of the free RNA fragment is very similar [10]. The secondary structure of this RNA fragment (shown in Fig. 3a, right panel) shows that it consists of the three 23S rRNA helices H18, H19, and H20 and their three-way junction.

We have analyzed many individual force vs. displacement curves for this RNA fragment. Following the analysis presented in the previous paragraphs, we have superposed all the corresponding, HMM-analyzed, released single-stranded RNA length vs. force curves, and this superposition is shown in a heat map histogram (Fig. 3a, left panel). From this histogram, it is readily apparent that the RNA fragment unfolds under force via two main unfolding intermediates (and some other minor unfolding intermediates). The two main intermediates are indicated on Fig. 3a and have been named I_1 and I_2. Knowing the structure of the RNA fragment, we can rationalize the location of these intermediates as follows:

- Intermediate I_1 corresponds to a released length of single-stranded RNA of 8 ± 2 nm—which, using the aforementioned characteristic contour length of single-stranded RNA (i.e., 0.59 nm per nucleotide), corresponds to the unfolding of 14 ± 4 nucleotides. From the 3D structure of the RNA fragment, we can assign that this intermediate I_1 corresponds to the partial unfolding of helix H18 and the release of its first seven base pairs (Fig. 3a, right panel). The location of this intermediate might seem surprising at first sight, but careful analysis shows that it is not. This analysis is detailed in Fig. 3b and its legend.
- Intermediate I_2 corresponds to a released length of single-stranded RNA of 37 ± 2 nm—corresponding to the unfolding of 60 ± 4 nucleotides of the RNA fragment. Examination of the 3D structure of this fragment indicates that there are two possible locations of this intermediate—which are shown in Fig. 3a, right panel. The first possibility corresponds to unfolding of helices H18 and H19 (58 nucleotides unfolded)—letting only folded helix H20 plus the entangled pseudoknot (pictured in Fig. 3a as two joined boxes). The second possibility corresponds to unfolding of helices H18 and H20 (plus pseudoknot), letting only folded helix H19 and corresponding to 63 nucleotides unfolded. In our measurements, I_2 corresponds to 60 ± 4 nucleotides unfolded—thus in good agreement with the two figures deduced from the RNA structure.

4 Notes

1. Low hysteresis between unfolding and refolding curve: this condition has to be fulfilled such as to be in "close-to-equilibrium" conditions—which fully validate the analysis of force vs. extension data as described in this paper (for a thorough discussion on this point, *see* the Supplementary Informations of ref. [13]). If this is not the case, one has to reduce the speed of the mobile trap: we ordinarily use a speed of 50 nm/s. Some of us showed that velocities up to 200 nm/s are generally adequate for accurate force measurements [13]. Anyway, a slow velocity is often preferable, and 50 nm/s is a good starting point. If this is not sufficient—i.e., one still observes high hysteresis between unfolding and refolding curves—consider reducing this velocity.
2. All custom software programs referenced in the manuscript are available for academic use upon request to the authors.
3. The raw data distance parameter d is indeed trap-to-trap distance; in the equations, the end-to-end distance of the molecule (x) is used instead. To convert from one to the other, use the formula $d = \phi + F/k_F + F/k_M + x$, with ϕ being the diameter of the beads and k_F and k_M being, respectively, the fixed trap and the mobile trap stiffness.
4. The baseline is even better defined for an experiment where the single molecule is stretched and then unstretched—i.e., where the molecule does not break during the stretching phase—because both the points at the beginning of the measurement and those at the end can be used for fitting the baseline.
5. In our case, we ordinarily perform the fit using the trap-to-trap distance instead of the end-to-end distance of the molecule: as such, instead of fitting the elastic modulus of the molecule K_0, we use an apparent stiffness parameter K which is a combination of K_0 and of the two trap stiffness k_F (fixed trap stiffness) and k_M (mobile trap stiffness).
6. We ordinarily make the calculations by varying this length by 0.1 nm increments: in our hands, this resolution is usually sufficient to obtain released length vs. force curves with good enough signal/noise ratio.
7. The force vs. displacement curve is analyzed between typically 5 and 20 pN to fully capture all RNA unfolding events. With this set of values, and given our trap stiffness, the WLC model is close to a linear relationship—which ensures a constant noise over the span of the analysis.
8. This normalization of released RNA length is performed using, for each curve, its minimal and maximal values; after this

conversion, the released RNA length vs. distance curve looks like a standard FRET vs. time trace. There are several software which implement a HMM modeling of these FRET traces (*see* ref. 14 for a review of these software).

9. In our configuration, the maximum expected released length corresponds to the length of the fully unfolded RNA fragment of interest minus the diameter of the type A RNA double helix—i.e., 2.2 nm [13].
10. For the analysis proposed in this paragraph to be feasible, the 3D structure of the RNA fragment should be known (or predicted). But there are also limitations on the size of the investigated RNA fragment for the results of force vs. displacement experiments to be interpretable in terms of RNA 3D structure: if the RNA fragment is too large, raw data become more and more complex as the number of intermediate states usually increases significantly—and thus their interpretation becomes problematic. In our hands, we have been able to thoroughly analyze force vs. displacement data in terms of RNA 3D structure for RNA fragments with sizes ranging from ~20 nucleotides [12] up to ~200 nucleotides [3, 13].

Acknowledgments

This work was supported by the CNRS, ESPCI, University Paris Diderot, and by a Human Frontier Science Program grant [RGP008/2014] to UB.

The authors acknowledge that Fig. 3 and a few other sentences of this chapter were first originally published in RNA [10].

References

1. Bockelmann U, Thomen P, Essevaz-Roulet B, Viasnoff V, Heslot F (2002) Unzipping DNA with optical tweerers: high sequence sensitivity and force flips. Biophys J 82:1537–1553
2. Gross P, Laurens N, Oddershede LB, Bockelmann U, Peterman EJG, Wuite GJL (2011) Quantifying how DNA stretches, melts and changes twist under tension. Nat Phys 7:731–736
3. Geffroy L, Mangeol P, Bizebard T, Bockelmann U (2018) RNA unzipping and force measurements with a dual optical trap. Methods Mol Biol 1665:25–40
4. Seidel R, Dekker C (2007) Single-molecule studies of nucleic acid motors. Curr Opin Struct Biol 17:80 86
5. Bronson JE, Fei J, Hofman JM, Gonzalez RL, Wiggins CH (2009) Learning rates and states from biophysical time series: a Bayesian approach to model selection and single-molecule FRET data. Biophys J 97:3196–3205
6. Schuster-Böckler B, Bateman A (2007) An introduction to hidden Markov models. Curr Protoc Bioinformatics 3:3A
7. Bustamante C, Marko JF, Siggia ED, Smith S (1994) Entropic elasticity of lambda-phage DNA. Science 265:1599–1600
8. Odijk T (1995) Stiff chains and filaments under tension. Macromolecules 28:7016–7018
9. Cheng W, Arunajadai SG, Moffitt JR, Tinoco I, Bustamante C (2011) Single base-pair unwinding and asynchronous RNA release by the

hepatitis C virus NS3 helicase. Science 333:1746–1749

10. Geffroy L, Bizebard T, Aoyama R, Ueda T, Bockelmann U (2019) Force measurements show that uL4 and uL24 mechanically stabilize a fragment of 23S rRNA essential for ribosome assembly. RNA 25:472–480
11. Zuker M (2003) Mfold web server for nucleic acid folding and hybridization prediction. Nucleic Acids Res 31:3406–3415
12. Bercy M, Bockelmann U (2013) Hairpins under tension: RNA versus DNA. Nucleic Acids Res 43:9928–9936
13. Mangeol P, Bizebard T, Chiaruttini C, Dreyfus M, Springer M, Bockelmann U (2011) Probing ribosomal protein-RNA interactions with an external force. Proc Natl Acad Sci U S A 108:18272–18276
14. Blanco M, Walter G (2010) Analysis of complex single-molecule FRET time trajectories. Methods Enzymol 472:153–178

Chapter 8

Mapping of Posttranscriptional tRNA Modifications by Two-Dimensional Gel Electrophoresis Mass Spectrometry

Laura Antoine and Philippe Wolff

Abstract

RNA modification mapping by mass spectrometry (MS) is based on the use of specific ribonucleases (RNases) that generate short oligonucleotide digestion products which are further separated by nano-liquid chromatography and analyzed by MS and MS/MS. Recent developments in MS instrumentation allow the possibility to deeply explore posttranscriptional modifications. Notably, development of nano-liquid chromatography and nano-electrospray drastically increases the detection sensitivity and allows the identification and sequencing of RNA digested fragments separated and extracted from two-dimensional polyacrylamide gels, as long as the mapping and characterization of ribonucleotide modifications.

Key words 2D Gel isolation, Posttranscriptional tRNA modifications, Nano-LC-MS/MS

1 Introduction

RNA modification mapping recently experienced a renewed interest because of the development of new sensitive detection technologies and implication of RNA modifications in the deciphering of gene expression regulation or antibiotic resistance [1, 2]. In this chapter, we provide a detailed protocol for *E. coli* tRNA modification mapping. We present the sequencing by mass spectrometry of two tRNAs, tRNA Ser GCU and tRNA Leu CAG. To obtain the largest sequence coverage, three different nucleotide-specific RNases are used: RNase T1 (cleaves at 3′ end of G), RNase A (cleaves at 3′ end of U and C), and RNase U2 (cleaves at 3′ end of purine but preferably at 3′ end of A). The RNA modification mapping and data analysis workflow are shown in Fig. 1. This protocol had been applied successfully to different organisms such as bacteria, archaea, and eukaryotes.

The MS analysis procedure presents several caveats. First, as different tRNAs share similar sequences, it may be difficult to

Véronique Arluison and Frank Wien (eds.), *RNA Spectroscopy: Methods and Protocols*, Methods in Molecular Biology, vol. 2113, https://doi.org/10.1007/978-1-0716-0278-2_8, © Springer Science+Business Media, LLC, part of Springer Nature 2020

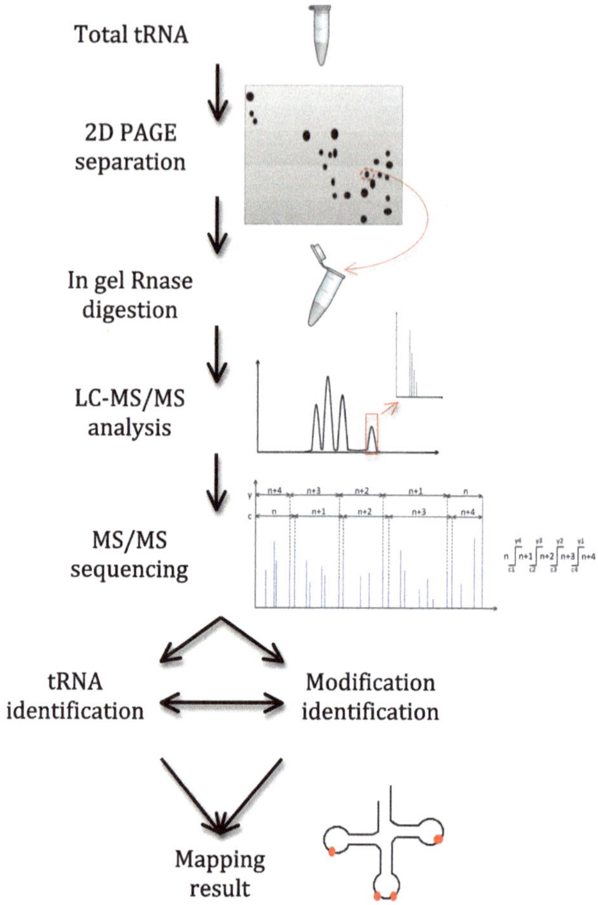

Fig. 1 Workflow for posttranscriptional tRNA modifications mapping by two-dimensional gel electrophoresis followed by mass spectrometry analysis. tRNA is separated by 2D gel electrophoresis. Individual tRNA spots are sliced out and in-gel digested by nucleoside-specific RNases. RNase digests are sequenced by LC-MS/MS in order to identify tRNA and to localize modifications

unambiguously attribute such sequences to a single tRNA. Additionally, different sequences can have the same mass (same nucleotide composition) or similar mass (there is a difference of only 1 Da between cytosine and uridine). In those cases, there is often an isotopic pattern overlay, and it can be very difficult to sequence correctly MS/MS spectra. In order to simplify MS/MS analysis and tRNA identification, two-dimensional polyacrylamide gels (2D gels) were used to separate tRNAs (Fig. 2) [3, 4].

After "in-gel RNase digestion," oligonucleotide products (passively eluted from the gel) are separated by nano-ion-pair reversed-phase high-performance liquid chromatography (nano-IP-RP-HPLC) [5]. Eluted digests are directly injected in the spectrometer

spot	tRNA
S02	Ser GCU
S03	Leu CAG
S05	Leu CAA
S08	Ile GAU
S08	Asp GUC
S09	Asn GUU
S11	Glu UUC
S10	Gln UUG
S10	Gln CUG
S09	Pro CGG
S09	Ala UGC
S07	Val UAC
S07	Gly GCC
S06	His GUG
S06	Arg ACG
S06	ProUGG
S13	Leu UAG
S03	Ser GGA
S14	Tyr GUA

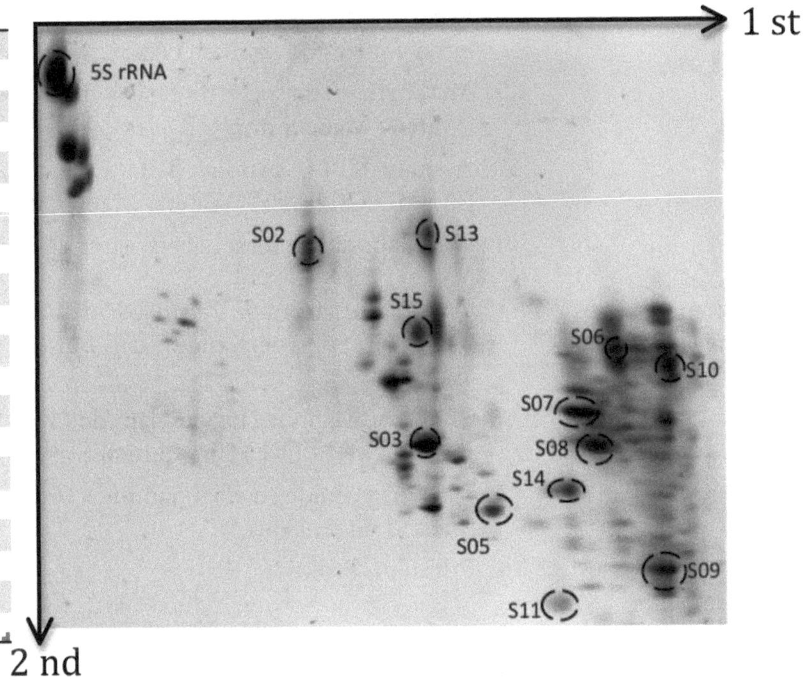

Fig. 2 2D polyacrylamide gel (20%) of *E. coli* total tRNA. 10 μg of total tRNA from *E. coli* were loaded; tRNA spots are revealed by ethidium bromide staining. The spots containing tRNA identified by MS/MS sequencing are indicated (black circles) and annotated in the table on the left

via a commercial NanoLockSpray ionization source to provide exact mass measurement. All oligonucleotides are fragmented by collision-induced dissociation (CID) and produce preferential fragments from y and c series for RNA sequencing [6]. Sequencing results are compared to genomic sequence in order to identify tRNA and to map RNA modifications in the sequence.

2 Materials

2.1 Instruments and Equipment

1. Polyacrylamide gel electrophoresis (PAGE) apparatus for 50 cm glass plates, 1 mm spacers (first dimension).
2. Polyacrylamide gel electrophoresis (PAGE) apparatus for 30 cm glass plates, 1 mm spacers (second dimension).
3. Electrophoresis power supply.
4. UV transilluminator and UV safety glasses.
5. Sterile surgical blades.
6. SpeedVac vacuum concentrator.
7. NanoAcquity UPLC system (Waters, Manchester, UK).

8. Synapt G2-S mass spectrometer equipped with NanoLock-Spray ionization source (Waters, Manchester, UK).
9. MassLynx mass spectrometry software with MaxEnt3 module (Waters, Manchester, UK).
10. Acquity UPLC peptide BEH C18 column (130 Å, 1.7 μm, 75 μm × 200 mm) (Waters, Manchester, UK).
11. Basic laboratory materials (vortex, thermoblock, etc.)

2.2 Two-Dimensional Gel Electrophoresis

1. Total tRNA from *E. coli* used in this study was prepared by Gérard Keith (personal gift), but it is also commercially available.
2. 12.5% Acrylamide/bis-acrylamide (19:1), 8 M urea, 1× Tris-Borat-EDTA (TBE) (first dimension).
3. 20% Acrylamide/bis-acrylamide (19:1), 4 M urea, 1× TBE (second dimension).
4. Ethidium bromide (EtBr).

2.3 In-Gel Digestion of Separated tRNA

1. RNase T1 and A (Thermo Fischer Scientific).
2. RNase U2 was homemade prepared as described [7].
3. Ammonium acetate 100 mM.
4. ZIP Tip C18 (Millipore).

2.4 Nano-liquid Chromatography

1. Mobile phase A.
   - 200 mM Hexafluoropropanol (HFIP).
   - 7 mM Triethylamine (TEA).
   - 7.5 mM Triethylamine acetate (TEAA) pH 7.5.
2. Mobile phase B.
   - 100% Methanol LC-MS grade.

3 Methods

3.1 Two-Dimensional Gel Electrophoresis (See Note 1)

3.1.1 First Dimension (Denaturing Conditions)

1. Cast the first-dimensional, 12.5% acrylamide/bis-acrylamide, 8 M urea.
2. Pre-run the gel in 1× TBE buffer for 30 min at 30 W (constant power).
3. Load 10 μg per well of *E. coli* total tRNA in a minimum volume (less than 10 μL).
4. Run the gel for 7 h at 30 W (constant power).
5. Stain the gel with an EtBr solution (10 μg/L) for 10 min.

6. Visualize the bands containing tRNAs under UV light (302 nm). Beware of wearing appropriate protections for handling gel containing EtBr and manipulation under UV light.
7. Excise the gel fragments containing tRNAs with a clean razor blade.

3.1.2 Second Dimension (Semi-denaturing Condition)

1. Put the excised gel lanes on the top of the second polyacrylamide gel, and cast the second-dimensional gel 20% acrylamide/bis-acrylamide, 4 M urea. The excised gel fragment from the first electrophoresis should be embedded in the gel.
2. Run the gel for 24 h at 6 W (constant power).
3. Stain the gel with an EtBr solution (10 μg/L) for 10 min.
4. Using UV transilluminator, excise the spots containing tRNAs (Fig. 2). Beware of wearing appropriate protections for handling gel containing EtBr and manipulation under UV light.
5. Dry gel spots in Eppendorf tubes under vacuum for 10 min without heating.

3.2 In-Gel RNase Digestion

Rehydrate excised gel fragments with 20 μL of RNase T1 (1 U/μL) solution or 20 μL of RNase A (0.01 U/μL) solution in 100 mM ammonium acetate (pH is not adjusted), and incubate at 50 °C for 3 h, followed by 12 h at 37 °C.

For RNase U2 digestion, rehydrate gel fragments with 50 μL of RNase U2 at 0.3 ng μL^{-1} in 100 mM ammonium acetate (pH is adjusted to 5.3), and incubate at 37 °C for 1 h.

3.3 RNA Digest Product Desalting

1. Prepare ZipTip C18 with 50% acetonitrile in water (3 × 10 μL).
2. Equilibrate the ZipTip for sample binding with 200 mM ammonium acetate (1 × 10 μL).
3. Bind the digested sample, and wash oligonucleotides with 200 mM ammonium acetate (5 × 10 μL).
4. Elute oligonucleotides with 10 μL of 50% acetonitrile in water.
5. Dry under vacuum for 20 min without heating.

3.4 LC-MS/MS

3.4.1 Nano-liquid Chromatography

Nano-LC is performed at a 0.3 μL/min flow rate for 1 h. RNase digestion products are eluted using a gradient from 15% to 35% B in 2 min followed by an increase of B to 50% in 20 min and then returning to 15% B in 25 min.

Table 1
List of natural and modified nucleotides identified in this study

Name	Short name	Symbol	Nucleotide mass (Da)	[M-H]$^-$ Np (Da)	[M-H]$^-$ N > p (Da)	Neutral loss (Da)
Guanosine	G	G	345	362	344	
Cytidine	C	C	305	322	304	
Uridine	U	U	306	323	305	
Adenosine	A	A	329	346	328	
2′-*O*-Methylguanosine	Gm	#	359	376	358	
1-Methylguanosine	m1G	K	359	376	359	
2′-*O*-Methylcytidine	Cm	B	319	336	318	
5-Methyluridine	m5U	T	320	337	319	
Dihydrouridine	D	D	308	325	307	
N6-Threonylcarbamoyladenosine[a]	t6A	6	329	346	328	145

A full list of modified nucleosides is available in the Modomics database (http://modomics.genesilico.pl/modifications/)
[a]Due to the neutral loss of threonylcarbamoyl, nucleotide mass fragmentation of t6A corresponds to the mass without modification. This modification occurs exclusively at position 37 in the anticodon loop

3.4.2 Mass Spectrometry Analysis

All experiments are performed in negative mode. The eluted RNase digests are injected on line into the mass spectrometer via Nano-LockSpray. Capillary voltage is set to 2.6 kV and sample cone to 30 V. Source is heated to 130 °C. For MS, a mass range from 500 to 1500 (m/z) is used, followed by collision-induced dissociation (CID) fragmentation of most intense signals using fast data directed acquisition (FastDDA) mode with a m/z detection range of fragments from 50 to 2000. A collision energy ramp (18–28 V at m/z 500 and 28–38 V at m/z 1500) is applied in the trap collision cell to provide the maximum fragment ions.

3.5 Data Analysis

All the fragment spectra are manually sequenced (*see* **Note 2**).

1. Deconvolute CID spectra by using MaxEnt3 software (*see* **Note 3**).
2. Using Table 1, listing the masses of the most common natural and modified nucleotides, reconstruct the sequence by following the y and/or c series (*see* **Note 4**). Three sequencing examples are shown in Fig. 3. Additionally, Fig. 4 gives a list of all the different masses obtained by nano-LC-MS/MS and the corresponding sequences.
3. Note that pseudouridine, an isomer of uridine, is a silent mass and cannot be detected by this method (*see* **Note 5**).

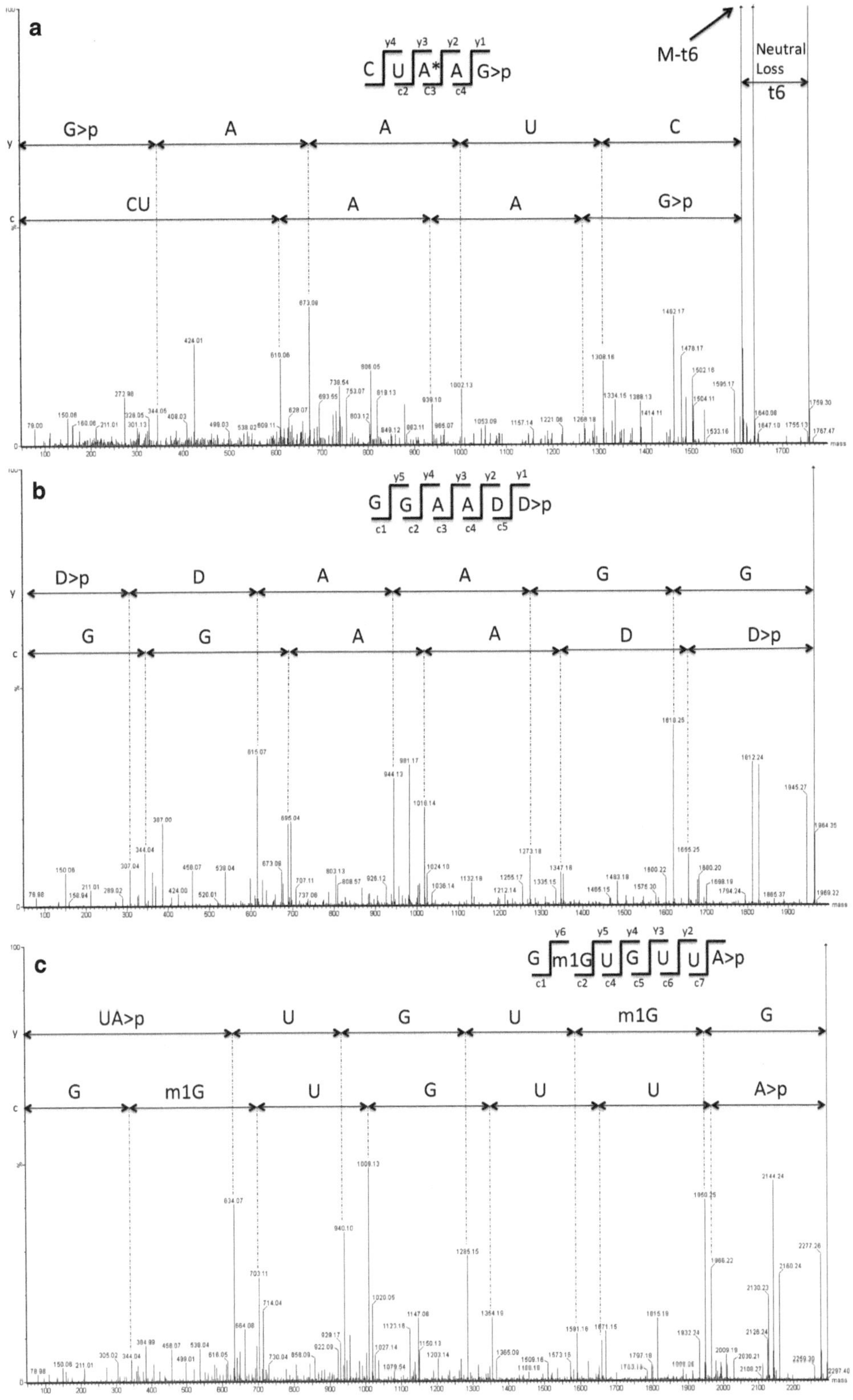

Fig. 3 Typical MS/MS spectra of RNase digested tRNA fragments. Only series y and c have been sequenced. (**a**) MS/MS sequencing spectrum of CU**t6A**AG>p after RNase T1 digestion. In the case of t6A modification,

theoretical	observed					theoretical	observed				
m/z	m/z	z	molecular mass (Da)	sequence	tRNA	m/z	m/z	z	molecular mass (Da)	sequence	tRNA
RNase T1											
878,62	878,63	-2	1758,3	CU6AG>p	Ser GCU	947,11	947,10	-2	1895,2	CUUCAG>p	Leu CAG
946,62	946,62	-2	1894,2	CAUCCG>p	Ser GCU	956,11	956,10	-2	1913,2	CUUCAGp	Leu CAG
955,62	955,62	-2	1912,2	CAUCCGp	Ser GCU	966,12	966,09	-2	1933,2	TUCAAG>p	Leu CAG
1135,15	1135,13	-2	2271,3	UCAAAAG>p	Ser GCU	975,13	975,10	-2	1951,2	TUCAAGp	Leu CAG
1144,16	1144,13	-2	2289,3	UCAAAAGp	Ser GCU	988,14	988,13	-2	1977,3	AADD#G>p	Leu CAG
1240,14	1240,1	-2	2481,2	CUCCCCUG>p	Ser GCU	997,15	997,13	-2	1995,3	AADD#Gp	Leu CAG
1249,15	1249,12	-2	2499,2	CUCCCCUGp	Ser GCU	1252,64	1252,60	-2	2506,2	UCCUUACG>p	Leu CAG
1251,66	1251,63	-2	2504,3	CCUCACCG>p	Ser GCU	1261,65	1261,62	-2	2524,2	UCCUUACGp	Leu CAG
1263,66	1263,64	-2	2528,3	AAUCCCCG>p	Ser GCU	1233,15	1233,11	-3	3701,3	UCCCCCCCCUCG>p	Leu CAG
1272,67	1272,63	-2	2546,3	AAUCCCCGp	Ser GCU	1239,15	1239,10	-3	3719,3	UCCCCCCCCUCGp	Leu CAG
RNase A											
858,11	858,11	-2	1717,2	GGGGTp	Ser GCU	662,09	662,09	-2	1325,2	AGACp	Leu CAG
991,15	991,16	-2	1983,3	AAAAGCp	Ser GCU	1326,18	1326,17	-1	1326,2	AAGUp	Leu CAG
1007,14	1007,15	-2	2015,3	GAAGGCp	Ser GCU	836,12	836,12	-2	1673,2	GGAADp	Leu CAG
1179,66	1179,69	-2	2360,3	GAGAGGCp	Ser GCU	981,13	981,14	-2	1963,3	GGAADD>p	Leu CAG
1417,20	1417,18	-2	2835,4	6AGGGAGUp	Ser GCU	1007,63	1007,64	-2	2016,3	GAAGGUp	Leu CAG
1046,47	1046,49	-3	3141,5	U6AGGGAGUp	Ser GCU	1030,64	1030,64	-2	2062,3	GGGGGTp	Leu CAG
RNase U2											
706,10	706,1	-2	1413,2	CU6A>p	Ser GCU	823,61	823,63	-2	1648,3	ADD#G>p	Leu CAG
807,11	807,11	-2	1615,2	CDGAA>p	Ser GCU	865,60	865,59	-2	1732,2	pGCGAA>p	Leu CAG
915,15	915,15	-2	1831,3	CCGCCA	Ser GCU	927,60	927,60	-2	1856,2	UCCUUA>p	Leu CAG
966,62	966,62	-2	1934,2	GCUGCA>p	Ser GCU	977,63	977,64	-2	1956,3	DD#GDA>p	Leu CAG
1240,14	1240,14	-2	2481,3	CUCCCCUG>p	Ser GCU	1142,16	1142,17	-2	2285,3	ADD#GDA>p	Leu CAG
1388,19	1388,21	-2	2777,4	CU6AGGGA>p	Ser GCU	1147,13	1147,13	-2	2295,3	GKUGUUA>p	Leu CAG
1273,15	1273,16	-3	3821,5	GGCGCUCCCCUG>p	Ser GCU	1425,66	1425,67	-2	2852,3	GUGUCCUUA>p	Leu CAG
						1479,21	1479,20	-2	2959,4	ADD#GDAGA>p	Leu CAG
						1233,15	1233,17	-3	3701,5	UCCCCCCCCUCG>p	Leu CAG

tRNA Ser GCU
pGGUGAGGUGGCCGAGAGGCDGAAGGCGCUCCCCUGCU6AGGGAGUAUGCGGUCAAAAGCUGCAUCCGGGGTUCGAAU CCCCGCCUCACCGCCA
Sequence coverage 77%
tRNA Leu CAG
pGCGAAGGUGGCGGAADD#GDAGACGCGCUAGCUUCAGKUGUUAGUGUCCUUACGGACGUGGGGGTUCAAGUCCCCC CCCUCGCACCA
Sequence coverage 82%

Fig. 4 List of all digested products experimentally observed for tRNA Ser GCU and Leu CAG, after individual digestion with RNases T1, A, and U2. The MS/MS sequenced products are indicated in red in the full tRNA sequences below the table. Although sequence coverage is close to 100%, some parts of the sequences are still missing. Thus, in the case of uncharacterized tRNA, modifications existing in vivo may be missed

4. For methyl groups, it is not possible to localize the group on the base and/or the ribose with this technique (*see* **Note 6**).
5. When a digested tRNA fragment is sequenced, the parent ion and the fragments' masses have to be checked using Mongo Oligo mass calculator (https://mods.rna.albany.edu/masspec/Mongo-Oligo). A maximum tolerance of 0.05 Da between the measured and calculated mass is routinely tolerated.
6. Comparison of the resulting sequences with *E. coli* genomic sequences (http://gtrnadb.ucsc.edu) identifies tRNA (*see* **Note 7**).

Fig. 3 (continued) collision-induced dissociation (CID) fragmentation leads to a neutral loss of 145 Da corresponding to the leaving of the threonylcarbamoyl group. ∗ corresponds to the location of the neutral loss. (**b**) MS/MS sequencing spectra of GGAA**DD**>p after RNase A digestion. (**c**) MS/MS sequencing spectra of G**m1G**UGUUA>p after RNase A digestion

7. Map modifications on the tRNA sequence. Figure 4 shows a list of all digested tRNA fragments sequenced by MS/MS.

4 Notes

1. In the first-dimension electrophoresis, tRNAs are completely heat denatured due to the dual action of gel temperature (70 °C) and urea (8 M) and are separated by their length. In the second dimension, urea concentration was decreased to 4 M, and the gel is run at room temperature. Under these conditions, tRNAs are partially refolded, and the separation is based on the partial secondary conformation.
2. Software have been developed for interpretation and annotation of MS/MS data (Ariadne [8], RoboOligo [9]), but to obtain unambiguous localization of modifications, manual interpretation of MS/MS spectra is still required.
3. Electrospray spectrum produces multiple charged ions. Typically, RNase digest products are two or three times charged. Spectrum deconvolution allows to simplify spectrum reading by transforming multiple charges spectrum (m/z) to mono charge spectrum (mass).
4. Generally, the most intense series are y and c. Mongo Oligo online calculator (https://mods.rna.albany.edu/masspec/Mongo-Oligo) could be helpful to find and/or check ion series.
5. Pseudouridine may be specifically derivatized using *N*-cyclohexyl-*N′*-(2-morpholinoethyl)carbodiimide metho-*p*-toluenesulfonate (CMCT) in order to be detectable [10].
6. LC-MS/MS of digestion products allows the localization of methylation in the correct nucleotide but does not allow the localization on the ribose and/or on the base. Using known tRNA modification sequences, it could be possible to predict the type of methylation. Modomics tRNA database (http://modomics.genesilico.pl/sequences/list/tRNA/) [11] provides a large collection of modified tRNA sequences. Furthermore, based on phylogenetic conservation of modification locations, tRNAmodviz (http://genesilico.pl/trnamodviz/) [12], an online software tool, allows the visualization of modification pattern in tRNAs.
7. As with proteomics strategy, RNase digestion generates a set of oligonucleotides, specific to a unique tRNA. tRNA identification is possible by using a genomic tRNA database.

Acknowledgments

The authors would like to thank Gérard Keith for the total *E. coli* tRNA preparation and Dominique Burnouf for critical reading of the manuscript and useful comments. We are grateful to Pascale Romby and Eric Westhof for their constant support. This work was supported by Labex NetRNA.

References

1. Helm M, Motorin Y (2017) Detecting RNA modifications in the Epitranscriptome: predict and validate. Nat Rev Genet 18:275–291
2. Gaston KW, Limbach PA (2014) The identification and characterization of non-coding and coding RNAs and their modified nucleosides by mass spectrometry. RNA Biol 11:1568–1585
3. Fradin AH, Gruhl H, Feldmann H (1975) Mapping of yeast tRNAs by two-dimensional electrophoresis on polyacrylamide gels. FEBS Lett 50:185–189
4. Dong H, Nilsson L, Kurland GC (1996) Co-variation of tRNA abundance and codon usage in *Escherichia Coli* at different growth rates. J Mol Biol 260:649–663
5. Masato T, Ikumi M, Nakayama H, Masaki S, Matsuda R, Nobe Y, Yamauchi Y, Takeda J, Takahashi N, Toshiaki I (2010) In-gel digestion for mass spectrometric characterization of RNA from fluorescently stained polyacrylamide gels. Anal Chem 82:7795–7803
6. McLuckey SA, Van Berkel GJ, Glish GL (1992) Tandem mass spectrometry of small, multiply charged oligonucleotides. J Am Soc Mass Spectrom 3:60–70
7. Houser WM, Butterer A, Addepalli B, Limbach PA (2015) Combining recombinant Ribonuclease U2 and protein phosphatase for RNA modification mapping by liquid chromatography-mass spectrometry. Anal Biochem 478:52–58
8. Hiroshi N, Akiyama M, Taoka M, Yamauchi Y, Nobe Y, Ishikawa H, Takahashi N, Isobe T (2009) Ariadne: a database search engine for identification and chemical analysis of RNA using tandem mass spectrometry data. Nucleic Acids Res 37:e47
9. Sample PJ, Gaston KW, Alfonzo JD, Limbach PA (2015) RoboOligo: software for mass spectrometry data to support manual and de novo sequencing of post-transcriptionally modified ribonucleic acids. Nucleic Acids Res 43:e64
10. Durairaj A, Limbach PA (2008) Matrix-assisted laser desorption/ionization mass spectrometry screening for Pseudouridine in mixtures of small RNAs by chemical Derivatization, RNase digestion and signature products. Rapid Commun Mass Spectrom 22:3727–3734
11. Boccaletto P, Machnicka MA, Purta E, Piatkowski P, Baginski B, Wirecki TK, de Crécy-Lagard V (2018) MODOMICS: a database of RNA modification pathways. 2017 update. Nucleic Acids Res 46:D303–D307
12. Machnicka MA, Olchowik A, Grosjean H, Bujnicki JM (2014) Distribution and frequencies of post-transcriptional modifications in tRNAs. RNA Biol 11:1619–1629

Chapter 9

Native Electrospray Ionization Mass Spectrometry of RNA-Ligand Complexes

Philippe Wolff and Eric Ennifar

Abstract

Native electrospray ionization mass spectrometry (native ESI-MS) is a powerful tool to investigate non-covalent biomolecular interactions. It has been widely used to study protein complexes, but only few examples are described for the analysis of complexes involving RNA-RNA interactions. Here, we provide a detailed protocol for native ESI-MS analysis of RNA complexes. As an example, we present the analysis of the HIV-1 genomic RNA dimerization initiation site (DIS) extended duplex dimer bound to the aminoglycoside antibiotic lividomycin.

Key words RNA, Native mass spectrometry

1 Introduction

Protein-protein [1, 2], protein-ligand [3], and protein-RNA [4–7] complexes are readily analyzed by native mass spectrometry. Electrospray ionization mass spectrometry (ESI-MS) under native conditions could preserved quaternary conformations in the gas phase of mass spectrometer and allows the mass determination of intact complexes and, by deduction, their precise stoichiometry. However, native ESI-MS of nucleic acids demands some considerations because of the polyanionic nature of the phosphate backbone [8, 9], and examples in the literature are significantly more sparse [10–16]. First, it requires strong desalting procedures to exchange sodium and potassium ions by volatile ions (ammonium acetate is a typical buffer). Second, as a consequence, nucleic acid ionization has to be performed in negative mode to obtain the better ionization efficiency.

In this chapter, we use the RNA dimerization initiation site (DIS) of HIV-1 viral genome in complex with lividomycin aminoglycoside as a model system for native ESI-MS analysis (any other 4,5 disubstituted 2-deoxystreptamine aminoglycoside could also be used alternatively).

Véronique Arluison and Frank Wien (eds.), *RNA Spectroscopy: Methods and Protocols*, Methods in Molecular Biology, vol. 2113,
https://doi.org/10.1007/978-1-0716-0278-2_9,

The HIV-1 genomic RNA DIS is a highly conserved sequence adopting a stem-loop structure in the 5′-untranslated region of the viral genome. The 8-nucleotide loop contains a 6-nucleotide self-complementary sequence which promotes genome dimerization by forming a loop-loop complex [17–19]. This kissing-loop complex is likely to be converted into a more stable extended duplex form by the viral nucleocapsid protein NCp7 [20–22]. X-ray crystal structures of the DIS kissing-loop complex [23, 24] and extended duplex [25, 26] revealed an unexpected similarity with the bacterial ribosomal decoding site (A-site), which is the target of aminoglycoside antibiotics [27, 28]. Specificity of the interaction was confirmed by X-ray structures of DIS/aminoglycoside complexes for both the loop-loop complex [29] and the extended duplex [30]. Here, we show that native ESI-MS allows to easily visualize the ratio of free RNA/RNA DIS duplex from the lividomycin aminoglycoside-bound complex.

2 Materials

2.1 Instruments and Equipment

1. AKTA Purifier (GE Healthcare) equipped with DNA Pac100 9 × 250 mm (Thermo Fisher Scientific, ref. 043011).
2. Centrifugal filter units, Amicon Ultra-4 10 K (Millipore, ref. UFC81096).
3. NanoDrop 1000 spectrophotometer (Thermo Fisher Scientific) or equivalent.
4. Micromass quadrupole time-of-flight (Q-Tof) mass spectrometer (Waters) or equivalent.
5. Basic laboratory materials (vortex, thermoblock, centrifuge, etc.)

2.2 Reagents

1. Oligoribonucleotide (ORN) is chemically synthesized at a 1 μmole scale, followed by standard desalting.
   DIS sequence used here is the following:
   5′ CUUGCUGAGGUGCACACAGCAAG 3′ ($n = 23$), MW = 7385.5 Da.
2. Lividomycin sulfate, MW = 761.8 Da (Sigma-Aldrich).
3. Ammonium acetate.

2.3 Reagent Setup

1. *ORN/DIS sequence.* Use ultrapure water to dissolve the chemically synthesized RNA to a final concentration of 0.3 mM.
2. *Lividomycin.* Use ultrapure water to dissolve lividomycin to a final concentration of 50 μM.

3 Methods

3.1 RNA Preparation for Native ESI-MS

3.1.1 Oligoribonucleotide Purification and RNA-RNA/Ligand Formation

A first ORN purification is performed by high-performance chromatography (HPLC) using a Dionex ion exchange chromatography as described in [26]. DIS RNA can be folded either as a loop-loop complex or an extended duplex depending on the renaturation protocol [31], ORN concentration, and salts being the most important factors to be considered. Here, ~80 μg of DIS ORN at ~300 μM is used and purified/desalted in the extended duplex native form by ion exchange chromatography. High-performance chromatography (HPLC) is performed at a flow rate of 1 ml/min using a gradient from 0.05 to 2 M of ammonium acetate in 30 min (*see* **Note 1**).

3.1.2 Oligoribonucleotide Concentration

HPLC fractions containing ORN are concentrated using centrifugal filter unit (Amicon Ultra-4 10 K).

- Spin at 7500 × g at 4 °C for approximately 20 min.
- Dialyze ORN against 100 mM ammonium acetate 3 × 4 ml.
- Finally, recover ORN in approximately 50 μl.
- Determine ORN concentration by OD measurements (260 nm) using a UV spectrophotometer.
- The stock solution concentration should be 1 mM.

3.1.3 Sample Preparation for Native ESI-MS

- HIV-1 DIS duplex is formed by heating purified ORN (11 μM final) 2 min at 90 °C, followed by a flash cooling on ice for 5 min, and then incubate for 30 min at 37 °C. Prepare 45 μl of renatured ORN at 11 μM in 100 mM ammonium acetate per ESI-MS injection. Theoretical MW of DIS duplex is 14,771 Da.
- Add 5 μl of lividomycin (50 μM) and incubate for 5 min at room temperature (*see* **Note 2**) before ESI-MS injection.

3.2 Mass Spectrometry Analysis

All experiments were performed in negative mode in a quadrupole time-of-flight (Q-Tof) mass spectrometer. Samples were directly introduced into the mass spectrometer via a classic electrospray probe at a flow rate of 10 μl/min. The following parameters were set:

- Capillary 2.2 kV.
- Sample cone 100 V.
- Extraction cone 0.5 V.
- Ion energy 2.0 V.
- Collision energy 2.0 V.
- Cone 20 l/h N_2.

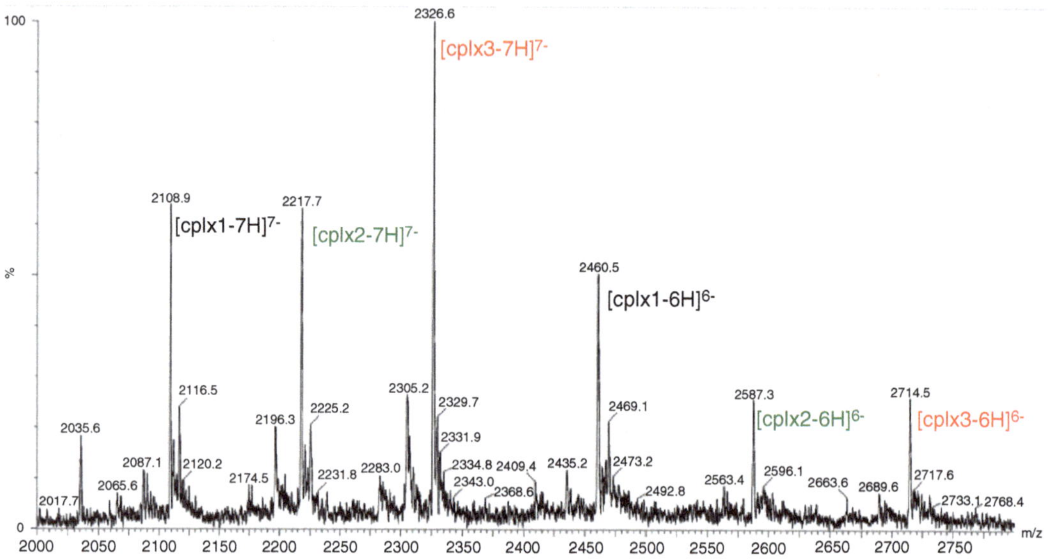

Fig. 1 Native ESI-MS spectrum of DIS duplex with lividomycin. Complex 1 (Cplx1) corresponds to the RNA/RNA duplex form, Cplx2 corresponds to the RNA/RNA duplex with 1 bound lividomycin, and Cplx3 corresponds to the duplex with two bound lividomycin. Complexes are formed by mixing 11 μM of RNA with 5 μM of lividomycin in 100 mM NH_4OAc

- Source temperature 40 °C.
- Desolvation 300 l/h N_2.
- Desolvation temperature 60 °C.

Data were acquired using a scan time of 5 s and a m/z range of 1000–3000.

Figure 1 shows a native ESI-MS spectrum of DIS duplex incubated with lividomycin.

3.3 Data Analysis

Before measuring the molecular weight of the different RNA/ligand complexes, charge states of each peak must be determined.

3.3.1 Charge State Determination of the Peaks

Determine the charge state of each peak using the two formulas (*see* **Note 3**):

$$z_1 = (p_2 + 1)/(p_2 - p_1) \quad (1)$$

$$z_2 = z_1 - 1 \quad (2)$$

with$z$ = charge of the peak$p_1 = m/z$ value for the complex peak corresponding to the first charge state$p_2 = m/z$ value for the complex peak corresponding to the second charge state.

Example for complex 1 corresponding to the RNA-RNA DIS duplex (Fig. 1):

m/z value of cplx1 for the first charge state (z_1):

$$p_1^{[\text{cplx1}(z_1)]} = 2108.9,$$

m/z value of cplx1 for the second charge state (z_2):

$$p_2^{[\text{cplx1}(z_2)]} = 2460.5.$$

According to Eq. 1:

$z_1 = (2460.5 + 1)/(2460.5\text{–}2108.9)$,

$z_1 = 7$.

According to Eq. 2:

$z_2 = z_1 - 1$,

$z_2 = 7 - 1$,

$z_2 = 6$.

3.3.2 MW Calculation from ESI Spectra

The mass of each peak was calculated by using the formula (*see* **Note 4**):

$$MW = (p \times z) + z. \tag{3}$$

Example for Cplx1(z_1):

$$\text{MW Cplx1}(z_1) = (2108.9 \times 7) + 7$$

$$\text{MW Cplx1}(z_1) = 14769.3 \text{ Da}$$

Table 1 provides the list of the theoretical and calculated molecular weights of the different complexes. For Cplx1, the observed mass is 14.769 Da, corresponding to ligand-free DIS duplex. For Cplx2, a mass of 15.530 Da is determined which corresponds to the mass of the DIS duplex plus one lividomycin. Finally, for Cplx3, the mass, 16.293 Da, corresponds to the mass of the DIS duplex plus two lividomycin.

Table 1
List of theoretical and experimental masses for the three forms of DIS-lividomycin complexes observed in solution

	Dis/Dis (Cplx1)		Dis/Dis/Liv (Cplx2)		Dis/Dis/Liv/Liv (Cplx3)	
Charge state	Theoretical m/z	Observed m/z	Theoretical m/z	Observed m/z	Theoretical m/z	Observed m/z
6	2460.8	2460.5	2587.8	2587.3	2714.8	2714.5
Calculated mass $z = 6$	14771.0 Da	14769.0 Da	15532.8 Da	15529.8 Da	16294.6 Da	16293.0 Da
7	2109.1	2108.9	2218.0	2217.7	2326.8	2326.6
Calculated mass $z = 7$	14771.0 Da	14769.3 Da	15532.8 Da	15530.9 Da	16294.6 Da	16293.2 Da

In our example, native ESI-MS analysis of DIS RNA complexes formed with lividomycin allows to determine the binding of one or two aminoglycoside per RNA duplex. This result is in agreement with isothermal titration calorimetry analyses [27] and X-ray crystallography structures [30]. Noteworthy, in our conditions, non-covalent interactions are preserved in the gas phase of the mass spectrometer, as observed for protein-ligand complexes [3], and provide stoichiometry information.

4 Notes

1. ORN purification by anion exchange eliminates the incomplete sequences resulting from aborted chemical synthesis. Furthermore, the use of ammonium acetate as buffer allows to exchange sodium and potassium as ammonium counterions of the phosphate backbone. This procedure yields a "high-quality" ORN for mass spectrometry analysis.
2. In our example, we use a ratio of one ORN duplex per lividomycin to see all the different forms of the complex (without lividomycin, with one lividomycin, and with two lividomycin molecules). Depending on the affinity, the ratio between the ORN and the ligand should be adjusted. It is also possible to determine the affinity constant by ESI-MS titration experiments [3, 32].
3. Correct charge state assignment is important to determine accurate mass of complexes [33]. Negative ESI-MS generated multiple charged ion series due to the deprotonation of the sample during the ionization process. In an ESI-mass spectrum, two peaks corresponding to the same complex have a difference charge state of one.
4. Theoretical mass of ORN can be calculated by using Mongo Oligo online calculator (https://mods.rna.albany.edu/masspec/Mongo-Oligo).

Acknowledgments

Authors would like to thank Dominique Burnouf for critical reading of the manuscript and useful comments and are grateful to Philippe Dumas for his support.

References

1. Boeri Erba E, Petosa C (2015) The emerging role of native mass spectrometry in characterizing the structure and dynamics of macromolecular complexes. Protein Sci 24:1176–1192
2. Leney AC, Heck AJ (2017) Native mass spectrometry: what is in the name? J Am Soc Mass Spectrom 28:5–13
3. Wolff P, Da Veiga C, Ennifar E, Bec G, Guichard G, Burnouf D, Dumas P (2017) Native ESI mass spectrometry can help to avoid wrong interpretations from isothermal titration Calorimetry in difficult situations. J Am Soc Mass Spectrom 28:347–357
4. Chen L, Tanimoto A, So BR, Bakhtina M, Magliery TJ, Wysocki VH, Musier-Forsyth K (2019) Stoichiometry of triple-sieve tRNA editing complex ensures fidelity of aminoacyl-tRNA formation. Nucleic Acids Res 47:929–940
5. McKay AR, Ruotolo BT, Ilag LL, Robinson CV (2006) Mass measurements of increased accuracy resolve heterogeneous populations of intact ribosomes. J Am Chem Soc 128:11433–11442
6. Schneeberger EM, Breuker K (2017) Native top-down mass spectrometry of TAR RNA in complexes with a wild-type tat peptide for binding site mapping. Angew Chem Int Ed Engl 56:1254–1258
7. van de Waterbeemd M, Fort KL, Boll D, Reinhardt-Szyba M, Routh A, Makarov A, Heck AJ (2017) High-fidelity mass analysis unveils heterogeneity in intact ribosomal particles. Nat Methods 14:283–286
8. Collie GW, Parkinson GN, Neidle S, Rosu F, De Pauw E, Gabelica V (2010) Electrospray mass spectrometry of telomeric RNA (TERRA) reveals the formation of stable multimeric G-quadruplex structures. J Am Chem Soc 132:9328–9334
9. Shah S, Friedman SH (2008) An ESI-MS method for characterization of native and modified oligonucleotides used for RNA interference and other biological applications. Nat Protoc 3:351–356
10. Hagan N, Fabris D (2003) Direct mass spectrometric determination of the stoichiometry and binding affinity of the complexes between nucleocapsid protein and RNA stem-loop hairpins of the HIV-1 psi-recognition element. Biochemistry 42:10736–10745
11. Hagan NA, Fabris D (2007) Dissecting the protein-RNA and RNA-RNA interactions in the nucleocapsid-mediated dimerization and isomerization of HIV-1 stemloop 1. J Mol Biol 365:396–410
12. Porrini M, Rosu F, Rabin C, Darre L, Gomez H, Orozco M, Gabelica V (2017) Compaction of duplex nucleic acids upon native electrospray mass spectrometry. ACS Cent Sci 3:454–461
13. Stephenson W, Asare-Okai PN, Chen AA, Keller S, Santiago R, Tenenbaum SA, Garcia AE, Fabris D, Li PT (2013) The essential role of stacking adenines in a two-base-pair RNA kissing complex. J Am Chem Soc 135:5602–5611
14. Turner KB, Brinson RG, Yi-Brunozzi HY, Rausch JW, Miller JT, Le Grice SF, Marino JP, Fabris D (2008) Structural probing of the HIV-1 polypurine tract RNA:DNA hybrid using classic nucleic acid ligands. Nucleic Acids Res 36:2799–2810
15. Turner KB, Hagan NA, Fabris D (2006) Inhibitory effects of archetypical nucleic acid ligands on the interactions of HIV-1 nucleocapsid protein with elements of psi-RNA. Nucleic Acids Res 34:1305–1316
16. Turner KB, Kohlway AS, Hagan NA, Fabris D (2009) Noncovalent probes for the investigation of structure and dynamics of protein-nucleic acid assemblies: the case of NC-mediated dimerization of genomic RNA in HIV-1. Biopolymers 91:283–296
17. Laughrea M, Jette L (1994) A 19-nucleotide sequence upstream of the 5′ major splice donor is part of the dimerization domain of human immunodeficiency virus 1 genomic RNA. Biochemistry 33:13464–13474
18. Paillart JC, Skripkin E, Ehresmann B, Ehresmann C, Marquet R (1996) A loop-loop "kissing" complex is the essential part of the dimer linkage of genomic HIV-1 RNA. Proc Natl Acad Sci U S A 93:5572–5577
19. Skripkin E, Paillart JC, Marquet R, Ehresmann B, Ehresmann C (1994) Identification of the primary site of the human immunodeficiency virus type 1 RNA dimerization in vitro. Proc Natl Acad Sci U S A 91:4945–4949
20. Laughrea M, Jette L (1996) Kissing-loop model of HIV-1 genome dimerization: HIV-1 RNAs can assume alternative dimeric forms, and all sequences upstream or downstream of hairpin 248-271 are dispensable for dimer formation. Biochemistry 35:1589–1598
21. Muriaux D, De Rocquigny H, Roques BP, Paoletti J (1996) NCp7 activates HIV-1Lai RNA dimerization by converting a transient

loop-loop complex into a stable dimer. J Biol Chem 271:33686–33692

22. Takahashi KI, Baba S, Chattopadhyay P, Koyanagi Y, Yamamoto N, Takaku H, Kawai G (2000) Structural requirement for the two-step dimerization of human immunodeficiency virus type 1 genome. RNA 6:96–102
23. Ennifar E, Dumas P (2006) Polymorphism of bulged-out residues in HIV-1 RNA DIS kissing complex and structure comparison with solution studies. J Mol Biol 356:771–782
24. Ennifar E, Walter P, Ehresmann B, Ehresmann C, Dumas P (2001) Crystal structures of coaxially stacked kissing complexes of the HIV-1 RNA dimerization initiation site. Nat Struct Biol 8:1064–1068
25. Ennifar E, Walter P, Dumas P (2010) Cation-dependent cleavage of the duplex form of the subtype-B HIV-1 RNA dimerization initiation site. Nucleic Acids Res 38:5807–5816
26. Ennifar E, Yusupov M, Walter P, Marquet R, Ehresmann B, Ehresmann C, Dumas P (1999) The crystal structure of the dimerization initiation site of genomic HIV-1 RNA reveals an extended duplex with two adenine bulges. Structure 7:1439–1449
27. Bernacchi S, Freisz S, Maechling C, Spiess B, Marquet R, Dumas P, Ennifar E (2007) Aminoglycoside binding to the HIV-1 RNA dimerization initiation site: thermodynamics and effect on the kissing-loop to duplex conversion. Nucleic Acids Res 35:7128–7139
28. Ennifar E, Paillart JC, Marquet R, Ehresmann B, Ehresmann C, Dumas P, Walter P (2003) HIV-1 RNA dimerization initiation site is structurally similar to the ribosomal a site and binds aminoglycoside antibiotics. J Biol Chem 278:2723–2730
29. Ennifar E, Paillart JC, Bodlenner A, Walter P, Weibel JM, Aubertin AM, Pale P, Dumas P, Marquet R (2006) Targeting the dimerization initiation site of HIV-1 RNA with aminoglycosides: from crystal to cell. Nucleic Acids Res 34:2328–2339
30. Freisz S, Lang K, Micura R, Dumas P, Ennifar E (2008) Binding of aminoglycoside antibiotics to the duplex form of the HIV-1 genomic RNA dimerization initiation site. Angew Chem Int Ed Engl 47:4110–4113
31. Ennifar E, Bernacchi S, Wolff P, Dumas P (2007) Influence of C-5 halogenation of uridines on hairpin versus duplex RNA folding. RNA 13:1445–1452
32. Gulbakan B, Barylyuk K, Schneider P, Pillong M, Schneider G, Zenobi R (2018) Native electrospray ionization mass spectrometry reveals multiple facets of Aptamer-ligand interactions: from mechanism to binding constants. J Am Chem Soc 140:7486–7497
33. Liepold L, Oltrogge LM, Suci PA, Young MJ, Douglas T (2009) Correct charge state assignment of native electrospray spectra of protein complexes. J Am Soc Mass Spectrom 20:435–442

Chapter 10

Application of FTIR Spectroscopy to Analyze RNA Structure

Frédéric Geinguenaud, Valeria Militello, and Véronique Arluison

Abstract

Fourier transform infrared (FTIR) spectroscopy has been widely used for the analysis of both protein and nucleic acid secondary structure. This is one of the vibration spectroscopy methods that are extremely sensitive to any change in molecular structure. While numerous reports describe how to proceed to analyze protein and deoxyribonucleic acid (DNA) structures using FTIR, reports related to the analyses of ribonucleic acids (RNAs) are few. Nevertheless, RNAs are versatile molecules involved in a multitude of roles in the cell. In this chapter, we present applications of FTIR for the structural analysis of RNA, including the analysis of helical parameters and noncanonical base pairing, often found in RNA. The effect of temperature pretreatment, which has a great impact on RNA folding, will also be discussed.

Key words RNA structure, Vibrational spectroscopy, Noncanonical base pairing

Abbreviations

ATR	Attenuated total reflectance
FTIR	Fourier transform infrared spectroscopy
nt	Nucleotide
ss-/dsRNA	Single/double-stranded RNA

1 Introduction

RNA is a single-stranded nucleic acid polymer built of nucleotides joined by a phosphate and ribose backbone. In contrast to double-stranded DNA, this single-stranded nucleic acid is flexible and can fold due to internal base pairing. This folding involves standard Watson-Crick base pairs, which are adenine-uracil (A-U) and cytosine-guanine (G-C), but also noncanonical base pairs such as the wobble base pair between G and U or the Hoogsteen-type base pairs [1, 2]. Using this combination of base pairs, RNA usually forms hairpins, folded in a typical α-helix with a C3'-endo sugar pucker (sugar puckers in nucleic acid are predominately C3'-endo

Véronique Arluison and Frank Wien (eds.), *RNA Spectroscopy: Methods and Protocols*, Methods in Molecular Biology, vol. 2113, https://doi.org/10.1007/978-1-0716-0278-2_10, © Springer Science+Business Media, LLC, part of Springer Nature 2020

in A-DNA or A-RNA or C2'-endo in B-DNA). Additionally, long-range interactions, usually involving noncanonical Watson-Crick base pairs, may also occur. This allows RNA to adopt a variety of three-dimensional structures.

In the cell, RNAs play a broad range of functions. The best characterized is the first step converting the information from DNA to proteins, namely, via a process called transcription into messenger RNA (mRNA). But RNAs also act as major actors in translation, in which mRNA is converted into proteins. Ribosomal RNAs (rRNAs) are structural components of the ribosome. They also act as a catalyst in the chemical reaction to join amino acids during translation [3]. Transfer RNAs (tRNA) act as the adaptor between mRNA and amino acid residues. RNAs can also act as structural molecules without any direct effect on gene expression, as the 4.5S RNA in the signal recognition ribonucleoprotein particle (SRP) [4]. Certain more recently discovered RNAs do not encode any protein, hence their name of noncoding RNA (ncRNA) (in addition to previously known tRNA and rRNAs, which are also noncoding RNA). Regulatory ncRNAs have emerged as an important class of RNA that regulates the expression of hundreds of genes in all living organisms. This family includes microRNAs and small interfering RNAs (miRNA and siRNA) in eukaryotes [5] or small noncoding RNAs (sRNA) in bacteria [6].

Understanding RNA structure thus represents a major challenge. Mainly due to phase problem, few crystallographic structures are available for RNA, and most structural information comes from biophysical and biochemical characterization. These include chemical probing. Different chemicals serve to modify specific bases at certain sites to identify single- and double-stranded regions. In parallel, optical spectroscopy characterizations such as fluorescence or circular dichroïsm allow the analysis of RNA folding in many experimental conditions (*see* Chapters 1–5, 11, and 18). Finally, vibrational spectroscopy, including FTIR and Raman spectroscopy, has also emerged as a useful tool to characterize the structure of nucleic acids and the conformational changes after specific ligand binding, for example, after interaction with proteins. While Raman spectroscopy will be presented in Chapter 12, the aim of the present chapter is to describe how to use FTIR spectroscopy to analyze RNA folding. This chapter will be divided in different parts including the preparation of RNA for FTIR analysis; the presentation of the different setups available for this analysis (reflection/transmission, Fig. 1); the acquisition and treatment of the spectra; and the information that can be obtained from spectra, in particular about base pairing and sugar re-puckering.

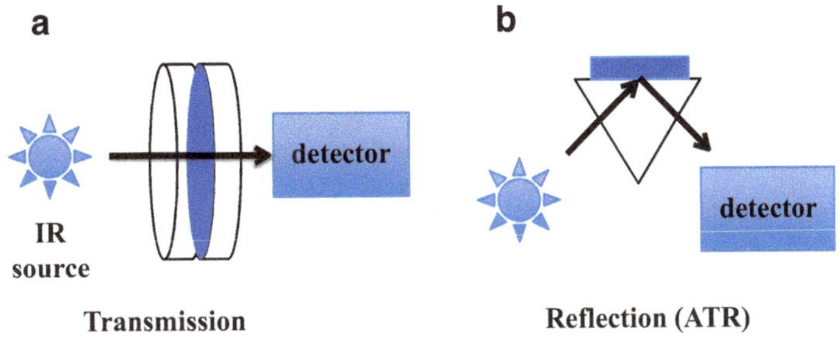

Fig. 1 FTIR sampling methods. (**a**) Transmission. (**b**) Reflection

2 Materials

2.1 Enzymes, Buffers, and Chemicals

All solutions have to be RNAse-free.

1. Sterile dd water (double distilled, endotoxin-free).
2. TAE buffer: 40 mM Tris-acetate, 1 mM ethylenediaminetetraacetic acid (EDTA) pH 8.0.
3. T7 RNA polymerase (T7 RNAP) and T7 RNAP transcription buffer (supplied with T7 polymerase).
4. Nucleoside triphosphate (NTP).
5. Denaturing gel-loading buffer: 60% formamide, 12 mM EDTA, 0.03% bromophenol blue, 0.03% xylene cyanol.
6. Phenol/chloroform/isoamyl alcohol 25:24:1 (v/v/v) mixture (SIGMA #77617), ethanol 70 and 100%.
7. SYBRgreen II (10,000×, Life Technologies ref. S7568).
8. RNeasy column for RNA purification (Qiagen).
9. Phusion® High-Fidelity DNA Polymerase (Thermo Fisher Scientific) or iProof DNA polymerase (Biorad) to produce DNA template for transcription.
10. RNase-free DNase (RQ1-DNase, Promega).
11. Sterile D_2O.

2.2 FTIR Spectroscopy

1. Interferometer-based spectrophotometer coupled with an IR source.
2. For transmission measurements, a liquid cell equipped with a ZnSe window. Different infrared window materials are available (*see* **Note 1**). The main ones used are calcium fluoride (CaF_2) and zinc selenide (ZnSe). ZnSe is resistant to most solvents (not soluble in water); its transmission range is 20,000–500 cm^{-1}. Alternatively, the hardest crystal CaF_2 is particularly resistant to high pressure. Nevertheless, its transmission range is 75,000–1000 cm^{-1}, which limits its use for RNA analysis at lower wavenumbers (*see* below). In both cases, cleaning agents such as alcohol and water can be used.

Alternatively, an attenuated total reflectance (ATR) sampling setup may be used but will not be presented herein.

3. FTIR analyses can be achieved with dedicated software. Most software packages such as OMNIC™ (Thermo Fisher Scientific) and OPUS (Bruker) provide spectral manipulation and curve-fitting (CF) algorithms.

3 Methods

3.1 In Vitro Transcription for FTIR Studies

RNA can be prepared by T7 in vitro transcription. To rule out the possibility of a read-through during transcription, polymerase chain reaction (PCR) fragments ending just at the end of RNA sequence are produced. The DNA template can be constructed by PCR as follows:

3.1.1 Design of the Primers

Forward primer: the T7 promoter (underlined) is added to the beginning of the RNA in the forward primer: 5′ TAATACGACTCACTATAGGG-RNAseq 3′.

Reverse primer, which is RNA specific, corresponds to the 3′ end of the RNA.

It is usually not necessary to add a T7 terminator if the transcripts correspond to the end of the PCR fragment.

The PCR results in a blunt end template encoding the RNA under the control of T7 promoter. It is recommended to check the quality of the PCR product on a gel, by confirming the proper size and uniformity of band. When necessary, purification must be performed to remove abortive products.

3.1.2 In Vitro Transcription

T7 transcription is performed in 20 μL of transcription buffer (supplied with T7 RNAP) with 1 pmol of the PCR-amplified DNA template, 2 mM NTP, 12 mM $MgCl_2$, and 100 U of T7 RNAP. The reaction mix is incubated for 3–6 h at 37 °C.

3.1.3 DNA Template Digestion

The DNA PCR template has then to be digested by an RNase-free DNase (RQ1-DNase, Promega). The digestion is performed with one unit of enzyme per μg of DNA for 15 min at 37 °C, followed by phenol/chloroform/isoamyl alcohol extraction and ethanol precipitation [7]. The transcript is resuspended in few μl of sterile water. It is necessary to perform the phenol-chloroform extraction to remove proteins used for in vitro transcription that can give an FTIR signal in the region of bases. RNA concentrations are measured by UV spectrophotometry at 260 nm (1 OD at 260 nm for RNA molecules = 40 ng/μL of RNA [7]).

3.1.4 RNA Purification

In order to check the efficiency of transcription, the RNA transcript is analyzed using denaturing polyacrylamide gel electrophoresis (PAGE). This can be achieved with a 15% acrylamide/bis 19:1 gel [provide dimensions?] containing 8 M urea in Tris-acetate-EDTA buffer (TAE; 10 mM Tris-acetate, 1 mM EDTA pH 8). The sample is preheated at 80 °C in denaturing gel-loading buffer containing 60% formamide, 12 mM EDTA, 0.03% bromophenol blue, and 0.03% xylene cyanol. The electrophoresis is run for 1 h with TAE buffer at ~50 °C. After migration, the RNA is stained with SYBR-green II. The visualization is achieved with a transilluminator. Two results can be observed:

If there are no abortive products, the RNA can be purified directly using the RNeasy Mini Kit (Qiagen) according to the manufacturer's protocol.

Alternatively, if there are some abortive products, the transcript needs to be purified from the PAGE. For this goal, the band corresponding to the RNA has to be cut from the gel, ground, and eluted overnight at 37 °C under agitation with two volumes of elution buffer (10 mM Tris–HCl pH 7.5 containing 0.5 M ammonium acetate, 1 mM EDTA, and 0.1% sodium dodecyl sulfate (SDS)). The supernatant is recovered, and SDS is removed by phenol-chloroform/chloroform extraction followed by ethanol precipitation (*see* Subheading 3.1.3). The transcript is resuspended in water (*see* **Note 2**).

3.1.5 RNA Folding

The transcripts are usually used after internal folding, i.e., the RNA is heated and slowly cooled down at 20 °C. Alternatively, RNA can also be rapidly cooled down by a transfer from the heating device to an ice bucket. Both conditions should be tested as they could give different results [8, 9].

3.2 RNA Chemical Synthesis

When short RNAs are analyzed, they can be chemically synthetized and purchased from different biotechnology companies, such as Eurogentec. Note that different purification processes are possible. Nevertheless, we recommend the dual HPLC purification (reverse phase and ion exchange) avoiding PAGE purification when possible (*see* **Note 2**). Similar to transcripts, the oligonucleotides are usually used after internal folding (*see* Subheading 3.1.5).

3.3 Infrared Spectra Acquisition

3.3.1 FTIR Setups

Useful spectral information is readily obtainable between 700 and 2000 cm^{-1}, through the use of an interferometer-based spectrophotometer coupled with an IR source and a triglycine sulfate (DTGS) detector. The interferograms recorded are transformed in an IR spectrum through a Fourier transform. The spectrum of the solvent has to be recorded at the same temperature and experimental conditions of the sample, to be subtracted from those of the sample after a suitable normalization.

Two main infrared sampling methods are available, transmission or reflection modes (Fig. 1). To obtain an IR spectrum in the transmission mode, the sample must be placed in a cell transparent in the IR region. The most used IR-transparent cells are ZnSe and CaF_2 windows (*see* Subheading 2.2 and **Note 1**). Usually, thin films are used (≤ 50 μm). As water has a strong IR absorbance in the base vibration region, it is impossible to have a high thickness without saturating the detector. To solve this problem, samples can be analyzed using D_2O solution. The reason why samples are prepared in D_2O is to avoid spectral overlaps between base bands and strong ν_2 absorption band of water at approximately 1650 cm^{-1}. The advantage of transmission mode is that a spectrum without the need of correction is collected (*see* below and **Note 3**).

For reflection measurement, an attenuated total reflectance (ATR) setup must be used. With an ATR setup, the infrared beam light is passed through a crystal in a way that it reflects at least once. In the latter case, the IR beam reflects from the internal surface of the crystal and creates an evanescent wave, which penetrates into the sample in contact with the ATR crystal. Some energy of evanescent wave is absorbed by the sample, and reflected radiation is returned to the detector. Note that the penetration depth of the evanescent wave (typically around few μm in the infrared) is dependent on the wavelength and that ATR spectra must be corrected in a spectrum. ATR correction removes this effect through application of mathematic correction describing the underlying physics. In addition, the ATR spectral features are usually weaker than those of transmission spectra, due to the small penetration depth of the evanescent wave (the absorbance values at ~ 1000 cm^{-1} are similar to those that would be observed in transmission with a pathlength ~3 μm). Common ATR crystal materials are ZnSe or diamond. Diamond ATR is more resistant and easy to clean up, but diamond absorbs IR between 2200 to 1900 cm^{-1}. Note that ATR is a surface technique and that the spectra may differ subtly from transmission spectra (*see* **Note 3**). For this reason, only transmission spectra will be presented herein.

3.3.2 RNA Sequence Analysis in D_2O Solution

The RNA (usually ~4 OD_{260nm} in tens of μL of RNase-free water) is dropped (10 μL by 10 μL) on a ZnSe window and dried (under a flow of dry air or nitrogen or alternatively with a hair dryer) to obtain a thin film. We recommend to clean the window with ethanol before adding RNA and to manipulate the sample and window with gloves. All the sample are in D_2O solutions to avoid the spectral overlaps with the strong ν_2 absorption band of water at approximately 1650 cm^{-1}.

Once dried, the sample is placed in a glove box under a flow of dry air, or nitrogen, and 20 μL of D_2O are added and left to evaporate (10 μL by 10 μL). When the film is dried, 0.5–2 μL of D_2O can be added to reach a concentration of strands ranging from µM to mM (the exact concentration depends on the sequence). We recommend filtering D_2O through a 0.2 µm pore size filter to generate a sterile filtrate. The first slide is covered with a second ZnSe window without spacer and placed in the FTIR spectrophotometer. Transmission spectra are recorded with 1 cm^{-1} resolution, and a total of 20 measurements are usually accumulated and averaged. The infrared spectra are then used as they are or processed using the processing software (see Subheading 3.4).

3.3.3 RNA Sequences Analyzed with an ATR

Alternatively, an attenuated total reflectance sampling setup can be used. With an ATR setup, sample in water can be analyzed. For this goal, the RNA at ~4 OD_{260nm} in ~20 µL of water is deposited on a 12-reflection ATR device. We also recommend to clean the ATR with ethanol before adding RNA and to manipulate the sample with gloves. Depending on the ATR device and the length of the RNA, 2–20 µL of solution can be added to analyze a concentration of strands ranging from µM to mM. As mentioned previously, ATR analysis can introduce artifacts and will not be presented herein.

3.4 Spectral Manipulation (Baseline Subtraction, Baseline Drift Correction)

To analyze FTIR spectra, different software can be used, such as OMNIC™ (Thermo Fisher Scientific), which is presented herein.

All absorption bands measured have to be normalized for sample concentration by dividing for the absorption intensity of the maximum peak of each sample. The manipulation of the absorption spectrum can be done following the steps below:

- Correct the baseline: option Process Baseline correct.
- When needed, smooth the spectra: option Process Smooth. Usually use 13 smooth points.
- Subtraction of the solvent: open the spectra of the sample and the solvent, option Process Subtraction, and adjust the factor of subtraction with the cursor.
- Determine the second derivative of the spectra to identify the peak positions: option Process derivative second.
- Deconvolution of the bands: the aim of this operation is to enhance the apparent resolution of a spectrum or decrease the width of all lines contributing to the investigated spectral range. Positions of the bands can be estimated with the use of the method of the second derivative for obtaining the band assignment to the specific molecular groups in RNA. For example, the estimation of the half-width, integral intensity, and real frequencies of overlapping bands of the complex spectral bands at 3800–2300 cm^{-1} (region due to stretching vibrations of OH, NH, and CH groups), at 1800–1550 cm^{-1} (region due to base vibrations), and at 1350–800 cm^{-1} (region due to phosphate and sugar vibrations) is decomposed by the fitting which is in general an option ("Curve fit") of the analysis program. In

addition, shapes of different spectral band contributions can be approximated with sums of the Lorentz and Gauss functions [10].

FTIR spectroscopy can also be used to analyze RNA complexed to proteins. In this case, both information on protein and RNA structural changes can be obtained by differential absorption spectra, which put in evidence the main differences. They are obtained by subtracting the spectrum of one of the partners from the spectrum of the complex; for instance, ΔAbs = Abs(complex) − Abs (protein) gives access to the spectrum of complexed RNA. Nevertheless, such an analysis needs a careful subtraction of the second component of the complex. When proceedings to a subtraction, it is highly recommended to acquire the spectra in the same experimental conditions and to verify that the shapes of the two spectra are similar. In order to prove the robustness of the subtraction process, it is also prescribed to test various levels of subtraction to ensure that the result is not an artifact of the processing.

3.5 RNA Spectrum Analysis

FTIR spectroscopy can then be used to analyze sugar puckering and base pairing in the RNA [11]. In particular, FTIR provides evidence for noncanonical base pairs (*see* **Note 4**) such as G·U wobble (Fig. 2) or Hoogsteen base pairs. The latter being energetically disfavored with RNA strands, they have mainly been described by FTIR with DNAs [12, 13]. However, their observations can be easily extended to the RNAs.

The three main RNA spectral domains are described below and summarized in Table 1.

3.5.1 Out-of-Plane Base Vibration Spectral Domain

The spectral domain shown in Fig. 2 (800–760 cm^{-1}) contains absorption bands due to out-of-plane vibrations of bases [14, 15]. The absorption bands can be assigned as summarized in Table 1. In the example presented here (the palindromic sequence AAGUGCUU that forms duplex RNA), all the bases are classically paired except the guanine and uracil, which display absorption bands at 778 and 772 cm^{-1}. These vibrations are assigned to the G·U wobble base pair [9].

3.5.2 In-Plane Base Vibration Spectral Domain

The spectral domain shown in Fig. 3 (1750–1450 cm^{-1}) contains absorption bands due to in-plane double bond stretching vibrations of the bases, sensitive to base pairing and base stacking [14, 16, 17]. The absorption bands of bases can be assigned as summarized in Table 1. The base absorption band at 1658 cm^{-1} can, for instance, be assigned to free uracil $C_4 = O_4$. With classical base pairing, the base intensity of the absorption bands at 1564 and 1579 cm^{-1} characteristic of the guanine ring vibrations is supposed to simultaneously decrease while conserving their relative intensity [16, 17]. This is not the case here, indicating that guanine is

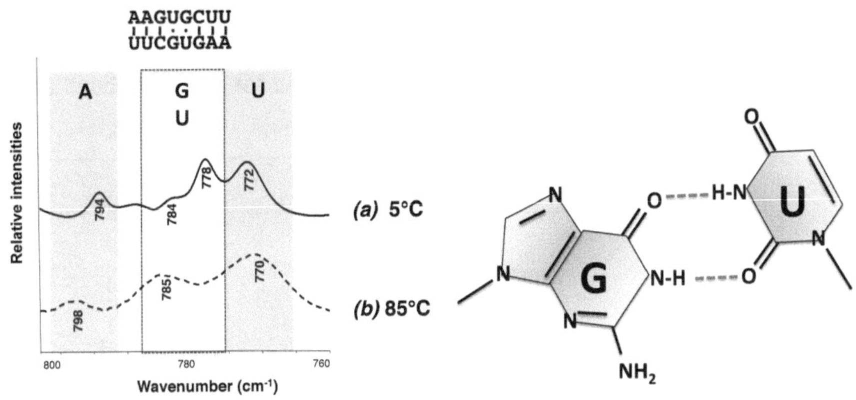

Fig. 2 FTIR spectra recorded in D_2O solution. The 800–760 cm^{-1} region contains absorption bands due to the bases. Here, the example of AAGUGCUU sequence forming dimers is shown [9] (**a**) at 5 °C (double stranded) and (**b**) at 85 °C (single stranded). The characteristic absorption bands of the base-paired adenine, cytosine, guanine, and uracil bases are, respectively, observed at 794, 784, 781, and 772 cm^{-1}, whereas a free (unpaired) guanine is detected at 785 cm^{-1} [16, 17, 22]. This evidences that all bases are classically paired, except the guanine, which displays an absorption band at 778 cm^{-1} that can be assigned to the G·U wobble base pair

engaged in base pairing only with the N_1D, leaving the N_2D_2 free, a characteristic of G·U wobble base pair [1].

3.5.3 Sugar Conformation Spectral Domain

FTIR spectra of RNA can also be analyzed in the region of sugar-backbone vibrations. When only classical N-type (C3′-endo) sugars are present in the polymer, three characteristic absorption bands located at 875, 861, and 811 cm^{-1} are observed (Fig. 4), corresponding to A form RNA. A contribution around 836 cm^{-1} is indicative of a C2′-endo conformation, less frequent in RNA (but not impossible [18]).

3.5.4 Phosphate Spectral Domain

Unlike the in-plane base vibration spectral domain, antisymmetric phosphate in A form is visible in H_2O but partially hidden by D_2O and may be seen at 1245–1235 cm^{-1}. In B form (uncommon in RNA), this antisymmetric phosphate stretching (also visible in H_2O but partially hidden by D_2O) may be expected at 1225–1220 cm^{-1}. A sugar-phosphate vibration band characteristic of the A form is observed at 1188–1175 cm^{-1} in H_2O (weak intensity).

3.6 Specific Case of Protein/RNA Complexes

FTIR spectroscopy can also be used to analyze protein secondary change (i.e., α-helices, β-sheets, β-turns, random coils) when a complex with RNA is formed. The peptide bond is an amide group (CONH) that gives rise to seven active absorption bands in the infrared. Three bands carry information on the secondary structure, but the Amide I band is the most used since it has the smallest contribution from peptide side-chain absorptions. Amide I band of proteins in H_2O solutions is centered at approximately

Table 1
The wavenumbers and corresponding spectral assignment of infrared (IR) bands characteristic of nucleic acids in D_2O or H_2O^* [14, 16, 11, 9, 12]. ss and ds stand for single stranded and double stranded, respectively

Wavenumber (cm^{-1})	Assignment
1698–1691	$C_2 = 0_2$ of U in ss or ds
1677–1672	$C_4 = 0_4$ of U in ds
1658–1653	$C_4 = 0_4$ of U in ss
1618–1615	Ring of U in ss
770	Free U
770	A-base paired U
772	G-base paired U (G·U wobble)
1689–1678	$C_6 = 0_6$ of G in ds
1673–1660	$C_6 = 0_6$ of G in ss
1590–1595 and 1568–1564	Ring of G in ss or ds
1350* and 1320*	G in *syn* conformation (GC^+ Hoogsteen base pair)
785	Free G
781	C-base paired G
778	U-base paired G (G·U wobble)
1655–1647	$C_2 = 0_2$ of C in ss or ds
1624–1616, 1585–1582, 1527–1520, 1506–1498	Ring of C in ss or ds
787	Free C
784	G-base paired C
1626–1627	ND_2 coupled to ring vibration of A (6-aminopurine)
1632–1622 and 1579–1576	Ring of A in ss or ds
798	Free A
794	T-base paired A
1245–1235*	A form/antisymmetric PO_2^-stretching (partially hidden by D_2O)
1188–1175	A form, sugar-phosphate backbone
1225–1220*	B form/antisymmetric PO_2^-stretching (partially hidden by D_2O)
845–835	Ribose S-type
880–860	Ribose N-type

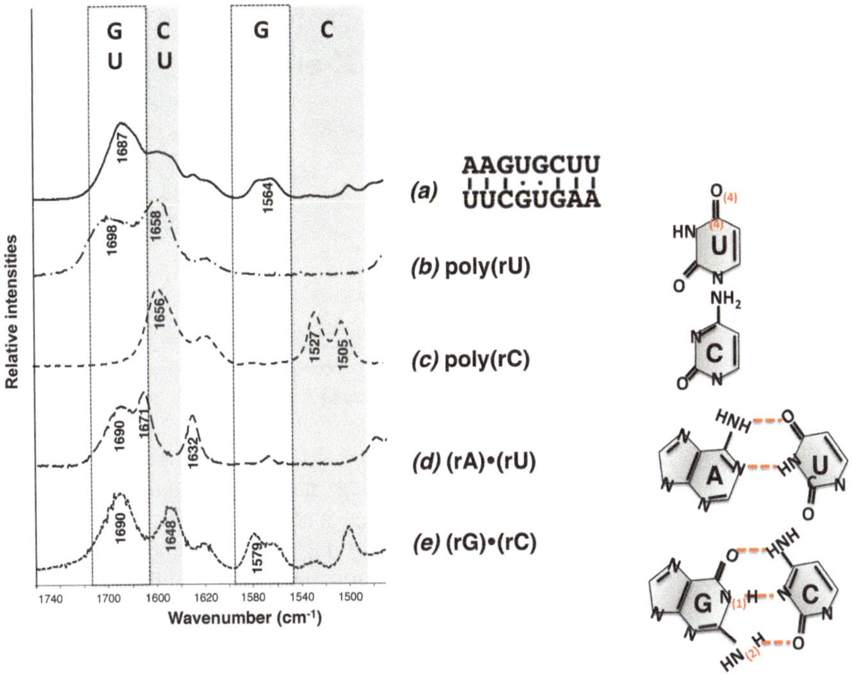

Fig. 3 FTIR spectra recorded in D_2O solution. The region between 1750 and 1450 cm^{-1} is characteristic of in-plane double bond stretching vibrations of the bases [14, 16, 17]. The example of sequence AAGUGCUU is shown [9] (**a**) at 5 °C (double stranded) and for comparison (**b**) single-stranded polyrU, (**c**) single-stranded polyrC, (**d**) double-stranded rA.rU, and (**e**) double-stranded rG.rC and. The spectra of single-stranded poly (rU) are characterized by two absorption bands assigned to the free uracil at 1698 cm^{-1} ($C_2 = O_2$) and 1658 cm^{-1} ($C_4 = O_4$). The free cytosine is identified on the spectra by the presence of three bands at 1656 ($C_2 = O_2$), 1527, and 1505 cm^{-1}; a decrease in intensity, and a small downshift for the carbonyl (1648 cm^{-1}), is observed for these bands upon duplexing. Thereby, we can conclude that the cytosine of AAGUGCUU is engaged in a base pair and that a part of the uracil's $C_4 = O_4$ is free. Two bands, at 1564 and 1579 cm^{-1}, are characteristics of the guanine ring vibrations. Upon duplexing in Watson-Crick base pairing, in which the $C_6 = O_6$, the N_1-H_1 and the N_2-H_2 are engaged in an H bond; these bands decrease together conserving their relative intensity. In the *a* spectrum, the increase of the band intensity at 1564 cm^{-1} compared to that at 1579 cm^{-1} presumes that the guanine is engaged in a nonclassical base pairing, as in a G·U wobble base pair

1660 cm^{-1} and shifts toward 1650 cm^{-1} (Amide I′) in D_2O solutions. Amide I band absorption is mainly due to the C=O stretching of the amide group vibration, but this vibration is strongly delocalized over the amide C-N and C-H bonds giving rise to a strong coupling: the trans dipole coupling. Amide I is thus attributed to an out-of-phase combination of C-O and C-N stretchings. The exact frequency of the absorption peak is thus sensitive to the angle of the amide bond, which is dependent on the local structure of the protein. Amide I band has a composite band profile, consisting of several spectral components related to the different secondary structures (Fig. 5). Besides, Amide II band is attributed to an out-of-phase combination of in-plane C-N stretching and N-H bending of amide groups (*see* **Note 5**).

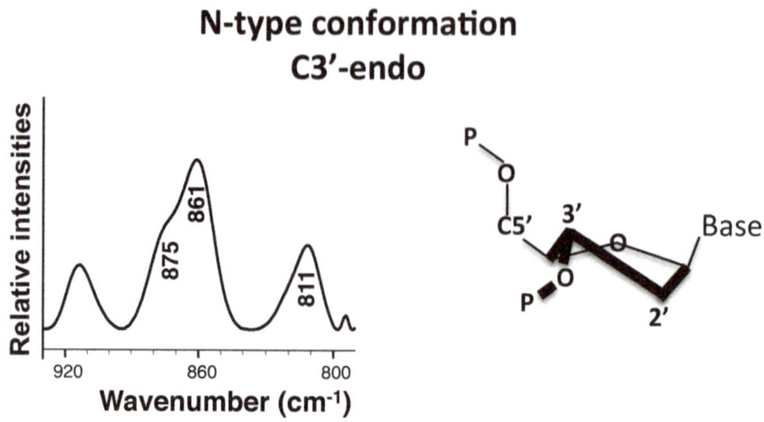

Fig. 4 FTIR spectrum recorded in D_2O solution. The region between 950 and 800 cm^{-1} is characteristic of sugar geometries. Here, the example of AAGUG-CUUCUUGCUUAAGCAAG sequence is shown [9]. The presence of three absorption bands located at 875, 861, and 811 cm^{-1}, corroborated by the lack of any contribution around 836 cm^{-1} (C2′-endo conformation), clearly demonstrates that only N type sugars (C3′-endo) are present

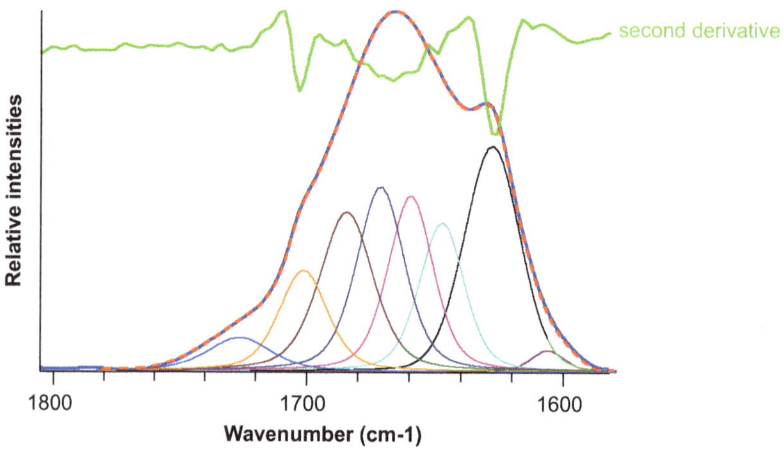

Fig. 5 Example of curve fitting of a protein spectrum (Amide I band, recorded in D_2O). Blue, original spectrum; dashed red, composite spectrum; green, secondary derivative; other colors, Gaussian/Lorentzian peaks. *See* Table 2 for protein secondary structure peak attribution

3.6.1 Analysis of the RNA-Complexed Protein Secondary Structure

The protein FTIR analysis consists of qualitative evaluation of secondary structure through the combination of second derivative, Fourier self-deconvolution (FSD), and curve-fitting (CF) analysis [19]. FSD and the second derivative can be used to find the number and exact positions of the overlapping peaks composing the Amide I band. Then CF involves fitting a series of model peaks to the experimental data. The best function for peak shape in liquid samples is usually a convolution of Lorentzian and Gaussian bands, the

Table 2
Empirical assignments for Amide I infrared (IR) band characteristic of proteins [19]

Wavenumber (cm^{-1})	Assignment
1615–1625	Intermolecular β-sheet (in particular vibrations of strongly bound intermolecular aggregated β-strands)
1630–1640	Intramolecular native β-sheet
1640–1650	Disordered random coil
1650–1660	α-Helix
1660–1695	β-Turn
1675–1695	Intermolecular β-sheet (in particular antiparallel amyloid β-sheets)

Voigt profile [10]. Overlapping peaks composing the Amide I band are fitted with Voigt profiles with a bandwidth varying between 15 and 25 cm^{-1}. The goodness of fit is generally evaluated by comparing the rest with the noise.

The assignment of a peak at a given wavenumber to a given secondary structure should be done accordingly to Table 2. Briefly, a peak near 1645 cm^{-1} is indicative of random coil, 1655 cm^{-1} of α-helix, and 1630 cm^{-1} and 1690 cm^{-1} of β-sheet.

3.6.2 Using OMNIC™: The Manipulation of the Spectrum Can be Done Following the Steps Below

- Select the option Process and then Baseline Correct to correct the baseline.
- Select the option: Analyze and then Peak Resolve.
- Check that in the left side of the screen, the options Original, Result, Peaks, and second Deriv are selected.
- Select the icon Selection tool in the bottom left side of the screen. Taking into account the (negative) bands in the second derivative spectrum, select the peaks to be considered for the deconvolution.
- In the icon Peaks (left side of the screen), you have the option to manually correct the following parameters: intensity of the band, full width at the half-height (FWHH), and wave number. In addition, you can eliminate and/or add bands for the manual deconvolution.
- In the top left side of the screen, select the option Gaussian/Lorentzian for the peaks and choose a high sensitivity (sensitivity: high).
- Finally, select the option Fit Peaks in the top right side of the screen in order to obtain the result from deconvolution.
- Select the option Peaks again (left side of the screen). You will be able to observe the area corresponding to each peak after

deconvolution. From these areas, calculate the percentage corresponding to each type of secondary structure taking into account the information in Table 2.

Figure 5 shows a typical spectrum of a protein analyzed by FTIR spectroscopy.

4 Notes

1. Quartz windows, which are mostly used in the UV-Vis and NIR spectral regions, are not used for FTIR analysis as the transmission range of the quartz is 25,000–2200 cm^{-1}.
2. We recommend avoiding PAGE purification when possible, as this can introduce contaminants that can be detected by FTIR spectroscopy.
3. ATR is a surface technique, and spectra may differ subtly from transmission spectra.
4. Minimal sequences should usually be preferred for the analysis of noncanonical base pairs because the spectrum corresponds to an averaged signal of Watson-Crick and unconventional base pairs. Using minimal sequences thus increases the specific signal due to noncanonical base pairs to evidence them.
5. Information about tertiary structure can also be obtained through the Amide II band. Amide II band of proteins in H_2O solutions is centered at approximately 1550 cm^{-1} and shifts toward 1450 cm^{-1} (Amide II') in D_2O solutions. Precisely, tertiary structure information may be obtained by monitoring the H-D exchanges between protein and solvent [20, 21]. Such an analysis will not be discussed herein but is illustrated in Malabirade et al. [21].

Acknowledgments

This work was supported by the CNRS, CEA, University Paris Diderot, University Paris 13, and University of Palermo. We are grateful to RR Sinden (South Dakota School of Mines and Technology, USA) for critical reading of the manuscript and many fruitful comments.

References

1. Varani G, McClain WH (2000) The G x U wobble base pair. A fundamental building block of RNA structure crucial to RNA function in diverse biological systems. EMBO Rep 1:18–23
2. Kimsey I, Al-Hashimi HM (2014) Increasing occurrences and functional roles for high energy purine-pyrimidine base-pairs in nucleic acids. Curr Opin Struct Biol 24:72–80
3. Rodnina MV (2013) The ribosome as a versatile catalyst: reactions at the peptidyl transferase center. Curr Opin Struct Biol 23:595–602
4. Pool MR (2005) Signal recognition particles in chloroplasts, bacteria, yeast and mammals. Mol Membr Biol 22:3–15
5. Patil VS, Zhou R, Rana TM (2013) Gene regulation by non-coding RNAs. Crit Rev Biochem Mol Biol 49:16–32
6. Gottesman S, Storz G (2011) Bacterial small RNA regulators: versatile roles and rapidly evolving variations. Cold Spring Harb Perspect Biol 3:a003798
7. Maniatis T, Fritsch EF, Sambrook J (1982) Molecular cloning: a laboratory manual. Cold Spring Harbor Laboratory Press, New York
8. Sun X, Li JM, Wartell RM (2007) Conversion of stable RNA hairpin to a metastable dimer in frozen solution. RNA 13:2277–2286
9. Cayrol B, Geinguenaud F, Lacoste J, Busi F, Le Derout J, Pietrement O, Le Cam E, Regnier P, Lavelle C, Arluison V (2009) Auto-assembly of E. coli DsrA small noncoding RNA: molecular characteristics and functional consequences. RNA Biol 6:434–445
10. Bradley M (2015) Lineshapes in IR and Raman spectroscopy. Spectroscopy 30:42–46
11. Banyay M, Sarkar M, Graslund A (2003) A library of IR bands of nucleic acids in solution. Biophys Chem 104:477–488
12. Stelling AL, Xu Y, Zhou H, Choi SH, Clay MC, Merriman DK, Al-Hashimi HM (2017) Robust IR-based detection of stable and fractionally populated G-C+ and A-T Hoogsteen base pairs in duplex DNA. FEBS Lett 591:1770–1784
13. Rangadurai A, Zhou H, Merriman DK, Meiser N, Liu B, Shi H, Szymanski ES, Al-Hashimi HM (2018) Why are Hoogsteen base pairs energetically disfavored in A-RNA compared to B-DNA? Nucleic Acids Res 46:11099–11114
14. Tsuboi M (1969) Application of infrared spectroscopy to structure studies of nucleic acids. In: Brame EGJ (ed) Applied spectroscopy reviews. Dekker, New York, pp 45–90
15. Letellier R, Ghomi M, Taillandier E (1987) Out-of-plane vibration modes of nucleic acid bases. Eur Biophys J 14:227–241
16. Liquier J, Taillandier E (1996) Infrared spectroscopy of nucleic acids. In: Mantsch HH, Chapman D (Eds) Infrared spectroscopy of biomolecules. Wiley-Liss, Inc, New York, pp. 131–158
17. Taillandier E, Liquier J (2002) Vibrational Spectroscopy of Nucleic Acids, In: Chalmers JM, Griffith PR (Eds) Handbook of Vibrational Spectroscopy. Wiley & sons, Chichester, vol. 5, Chapter 2
18. Clay MC, Ganser LR, Merriman DK, Al-Hashimi HM (2017) Resolving sugar puckers in RNA excited states exposes slow modes of repuckering dynamics. Nucleic Acids Res 45: e134
19. Byler DM, Susi H (1986) Examination of the secondary structure of proteins by deconvolved FTIR spectra. Biopolymers 25:469–487
20. Militello V, Casarino C, Emanuele A, Giostra A, Pullara F, Leone M (2004) Aggregation kinetics of bovine serum albumin studied by FTIR spectroscopy and light scattering. Biophys Chem 107:175–187
21. Malabirade A, Partouche D, El Hamoui O, Turbant F, Geinguenaud F, Recouvreux P, Bizien T, Busi F, Wien F, Arluison V (2018) Revised role for Hfq bacterial regulator on DNA topology. Sci Rep 8:16792
22. Tsuboi M, Higuchi S, Kyogoku Y, Nishimura S (1969) Infrared spectra of transfer RNA's. II. Formylmethionine transfer RNA from Escherichia coli in aqueous solution. Biochim Biophys Acta 19:23–28

Chapter 11

Application of Synchrotron Radiation Circular Dichroism for RNA Structural Analysis

Etienne Le Brun, Véronique Arluison, and Frank Wien

Abstract

Circular dichroism (CD) spectroscopy is a fast and simple technique providing important information about the conformation of nucleic acids, proteins, sugars, lipids, and their interactions between each other. This electronic absorption spectroscopy method is extremely sensitive to any change in molecular structure containing asymmetric molecules. While numerous reviews describe how to analyze deoxyribonucleic acid (DNA) structures using CD, analyses of ribonucleic acids (RNAs) are scarce. Nevertheless, RNAs are important molecules involved in a multitude of roles in the cell. In this chapter, we present applications of synchrotron radiation circular dichroism (SRCD) extending the spectral range down to 170 nm, improving structural analysis of RNA, including the analysis of helical parameters and alternative structures found in RNA. The effects of temperature to measure thermodynamic parameters and analyze ribonucleoprotein complexes will also be presented.

Key words RNA structure, Chiro-optical spectroscopy, Small noncoding RNA, DsrA

1 Introduction

Circular dichroism (CD) is a sensitive absorption spectroscopy technique, widely used for studying the conformation of biological samples such as proteins and nucleic acids (NA). Physically, CD assesses the absorption of light due to excitation of electronic transitions by the near and parts of the far-UV light (300–200 nm for near and 200–170 nm for the far UV in aqueous solutions). CD is sensitive to the differential absorption of circularly left- and right-polarized light due to electronic transitions such as the n-π or π-π* transitions of chiral molecules encountered within peptide bonds in proteins, as well as in nucleic acids including deoxyribonucleic acid (DNA) and ribonucleic acids (RNAs). For DNA and RNA consisting of phosphate sugars and bases, the latter are the chromophores. Asymmetric sugars like deoxyriboses and riboses induce a circular dichroism absorption on the symmetrical bases. Moreover, the spatial arrangement of RNA and DNA molecules in helices and

Véronique Arluison and Frank Wien (eds.), *RNA Spectroscopy: Methods and Protocols*, Methods in Molecular Biology, vol. 2113, https://doi.org/10.1007/978-1-0716-0278-2_11,

loops forming super asymmetric molecules causes a strong CD absorption involving the bases and their stacking. Hence, CD is very sensitive to the conformation of the nucleic acid macromolecules. For decades CD has been used as the method of choice to follow conformational and structural changes of nucleic acid macromolecules in solution at different pHs, ionic strengths, or temperatures. In addition, CD can detect their interactions with achiral and chiral molecules including proteins.

Quantitative CD measurements yield the absorbance difference ($\Delta A = A_{left} - A_{right}$) [1–3]. As for any absorption spectroscopy, the absorbance of the samples will obey the Beer–Lambert law, allowing for experimental adaptation of cell path length and sample concentration. Benchtop spectrophotometers consist of halogen and mercury light sources being monochromatized and circularly polarized (phase modulation) in a nitrogen-purged airtight chamber. Maintenance depending, they allow the exploration down to 190 nm at 1 nm bandwidth. Worldwide, several synchrotron light sources provide end stations for SRCD (synchrotron radiation circular dichroism) at which the spectral range of the far UV may be explored for films down to 120 nm and further under nitrogen purging conditions or ultravacuum, respectively. These big research infrastructures operate in general on peer-reviewed project demands and provide continuous spectral intensities of up to the 10^{12} photons/second from the near to far UV in a 24-h continuous mode. The interest lies in the fact that they provide high photon fluxes below 190 nm with good signal to noise ratios (>1:100), thus extending the spectral range down to 170 nm in water or low absorbing buffers. Spectral extension in the far UV below 190 nm at low bandwidth (1 nm) allows the exploitation of charge-transfer regions in proteins and nucleic acids. Indeed, conventional spectrometers scan below 200 nm and rarely below 190 nm, while NA SRCD spectra show the most informative region in the 170–220 nm range [4, 5].

In the case of nucleic acids, SRCD thus allows the qualitative identification of secondary structures, including helical conformation (B-, A-, or Z-forms) or the presence of quadruplexes [6, 7].

In contrast to medium- or high-resolution techniques such as X-ray crystallography, NMR, and cryo-electron microscopy, CD does not provide information at the atomic level, but can be used empirically to analyze and screen global NA structural changes in solution. In addition, CD spectroscopy also allows following conformational changes in real time at microseconds to hours. Thus, CD permits the facile measurement of the kinetics of folding, of the thermodynamic parameters, and to analyze the structure of ribonucleoprotein complexes. Therefore, it represents an alternative or complementary technique to solution-state NMR, which is limited by the size of RNA/protein complexes that are often very large. Nevertheless, while numerous reports describe how to proceed to analyze DNA structures using CD or SCRD, reports related to the analyses of RNA are few.

RNAs play a central role in many cellular processes including transcription and translation, act as catalysts (ribozyme), or comprise the genome for RNA viruses. Understanding RNA structures and dynamics thus represents an important field for elucidating and understanding its essential functions. Unlike most DNAs, RNA is usually a single-stranded nucleic acid polymer built of ribonucleotides, containing riboses as sugar and the pyrimidine uracil instead of thymine. Due to its single-stranded nature, RNA is very flexible. Nevertheless, the presence of internal self-complementary sequences allows it to form significant intramolecular base-pairing and to adopt three-dimensional structures. RNA internal base paring usually involves standard base pairs (adenine–uracil [A-U] and cytosine–guanine [C-G]), but also other noncanonical base pairs [8, 9]. These three-dimensional structures are important for the interactions with other NA, proteins, or small molecules (riboswitches). When internally folded, RNAs form local helices that are typically A-helices, with a C3′-endo sugar pucker (sugar puckers in nucleic acid are predominately C3′-endo in A-form). RNA may also form other 3D structures, such as quadruplexes, pseudoknots, kissing loops, or extended duplexes [10–13].

The aim of this chapter is to describe how to use SRCD spectroscopy to analyze RNA structure and folding. This chapter is divided into different parts including presentation of SRCD setup, acquisition, and treatment of CD spectra and information that can be obtained from them.

2 Materials

2.1 Chemicals

All solutions have to be RNAse-free.

1. Sterile double-distilled, endotoxin-free water (dd-water) to resuspend RNA.
2. Buffers recommended for SRCD analysis are summarized in Table 1.

   To clean cells:

3. Hellmanex 1–2% liquid alkaline solution for highly effective cleaning of cuvettes.
4. Ethanol 70%.
5. RNase decontamination solution (such as RNaseZap™, Invitrogen™) to clean the sample area.

Table 1
Recommended buffer for SRCD analysis

Buffer or salt	pH/pK_a	Transparent (nm)	Absorbance at 200 nm	Absorbance at 180 nm
Na_2HPO_4	6–7.2	210	0.05	>0.5
NaH_2PO_4		195	0	0.15
Acetate	3.6–5.6 pK_a = 4.76	220	0.17	>0.5
Glycine–HCl	2.2–3.6 pK_a = 2.35	220	0.1	>0.5
Glycine–NaOH	8.6–10.6 pK_a = 9.78			
Tris–HCl	6.8–8.5 pK_a = 8.06 *Strongly T° dependent*	220	0.13	>0.5
Hepes	6.8–8.2 pK_a = 7.55	230	0.5	>0.5
Pipes	6.1–7.5 pK_a = 6.8	230	0.5	>0.5
Mops	6.5–7.9 pK_a = 7.2	230	0.34	>0.5
Mes	5.2–7.1 pK_a = 6.15	230	0.3	>0.5
Cacodylate	5–7.4 pK_a = 6.25	210	0.2	>0.5
NaCl	NaCl	205	0.02	>0.5

Useful information for buffer preparation may also be found here: https://www.aatbio.com/resources/buffer-preparations-and-recipes

2.2 SRCD Spectroscopy

1. SRCD end station such as that of DISCO beamline at SOLEIL synchrotron (France), UV-1 at ISA (Denmark), B23 at Diamond (UK), or HISOR (Japan) [14–19].
2. Manually loaded circular demountable cells [19] (Fig. 1). CaF_2 and low OH quartz cells are transparent material down to the far UV. CaF_2 optical paths range from 1 to 50 μm. A few microliters of sample are used according to the optical path chosen: from 0.5 to 5 μL for cells of 2.5 and 50 μm, respectively. Quartz cells (low OH cells or high-performance quartz QS cells) are used for longer path lengths in view of lower sample concentrations. For quartz cells, the transmission ranges from the visible down to 190 nm, which limits its use for RNA analysis at lower wavelength. They usually need

Fig. 1 Demountable circular CaF_2 cells. The sample is deposited in a quartz or CaF_2 cell with an adapted optical path length (few μL). The cell is then deposited in the sample holder having a diaphragm to control the size of the beam. The cell is tightened thanks to joints and a screw in the sample holder and put in the experimental chamber purged with nitrogen

greater volumes, consume more sample, and are more susceptible to air bubble formation. Alternatively, CaF_2 may be used as its transmission range includes the visible down to 120 nm for films and 170 nm in low absorbing buffers within short path lengths, e.g., few microns. Even though they consume very little sample (microliter range) and they will not pressurize the samples while heating (e.g., in thermal denaturation assays), they are very fragile crystals and therefore break easily.

In either case a cleaning procedure using Hellmanex, alcohol, and dd-water must be used.

3. For spectral data treatment, the following software programs are used, CDtool or the new CDtoolX, for averaging, smoothing, subtraction of buffer baselines from sample spectra, calibration ((+)-camphor-10-sulfonic acid [CSA]), and normalization [20].
4. BestSel open-access software (http://bestsel.elte.hu/) is used for protein secondary structure content analysis [21, 22].

5. Kinetics and thermodynamic analyses can be achieved with dedicated software such as MathLab, IGOR, Kaleidagraph (Synergy software), or SigmaPlot (Systat Software).

3 Methods

3.1 In Vitro Transcription of RNA

RNA can be prepared by in vitro transcription as described in Chapter 10 or 20. When short RNAs are analyzed, they can be chemically synthetized and purchased from different biotechnology companies, such as Eurogentec. Note that in this case, different purification processes are possible. Nevertheless, we recommend avoiding PAGE purification (*see* **Note 1**). Transcripts or oligonucleotides are usually used after internal folding, i.e., the RNA is heated and slowly cooled to 20 °C. Alternatively, RNA can also be rapidly cooled down by a transfer from the heating device to an ice bucket. Both conditions should be tested as they could give different results [23, 24].

3.2 Buffers Compatible with SRCD Analysis

Many buffers and salts absorb in the vacuum ultraviolet (VUV) region and must be avoided. Table 1, adapted from [25], summarizes the properties of common salts and buffers used for SRCD analysis in biology.

For thermodynamic analyses, phosphate buffer is preferred, being more temperature stable and deep-UV transparent compared with Tris-based buffers (Table 1).

The choice of buffer content is also a function of how much chloride, which absorbs in far UV, will be considered essential, but 10–50 mM NaCl allows obtaining spectra down to 185 nm. If more salt is required, chloride is readily replaced by fluoride allowing twice the initial chloride concentration.

In addition, as RNA folding is coupled to the binding of divalent cations, such as Mg^{2+}, $MgSO_4$ (preferred to $MgCl_2$) may be added to the RNA solution. Usually few mM of Mg^{2+} are sufficient to allow proper folding.

3.3 RNA Sample Loading

1. The sample area should be cleaned prior to any manipulation with a commercially available RNase decontamination solution. Before RNA loading, we recommend to clean the windows with Hellmanex and 70% ethanol and rinse with sterile dd-water and to manipulate the sample and window with gloves. The RNA (usually ~ 0.8 OD_{260nm} in few μL) is dropped on the CaF_2 window (Fig. 1).
2. *Other sample preparation.* Proteins are usually prepared in a transparent buffer (Table 1) or water at 5–10 mg/mL. As a first test we suggest to use a ratio of 1 protein for 20 RNA base pairs, but this must be optimized for each complex. Note that

protein spectra are usually normalized to the mean residue weight ellipticity $[\Theta]_{mrw} = (\Theta \times 100)/Cln$, in deg. cm^2/dmol, where C = protein concentration, l = cell path length, and n = number of residues/amino acids.

3. Sample concentration, buffer composition, and path length choices for SRCD data acquisition should be handled with care and are strongly dependant on RNA and protein secondary structure contents (*see* **Notes 2** and **3**). For proteins, α-helix-rich protein concentrations should in general be about 1–2 g/L, for β-sheet-rich proteins about 3–4 g/L, and for unordered structures above 5 g/L in a 20 μm path length cell.

3.4 SRCD Setup and Data Acquisition

Useful spectral information is readily obtainable between 170 and 320 nm (*see* **Note 4**).

Difference in absorption $\Delta A = A_{left} - A_{right}$ is converted into ellipticity Θ:

$$\Theta = 33,000\,\Delta A \text{ (in mdeg)}.$$

Acquisitions of triplet SRCD spectra are recorded with 1 nm steps at 1.2 s integration time, between 320 and 170 nm at a 1 nm bandwidth. Averaged sample spectra are generally smoothed with the Savitzky–Golay algorithm with a factor between 5 and 15. Sample spectra are subtracted from their corresponding buffer baselines. Spectral magnitudes are calibrated using a solution of (+)-camphor-10-sulphonic acid (CSA) at a known concentration. A standard solution of CSA of 6 mg/mL in a 100 μm path length (quartz) cell should provide a CD maximum at 290.5 nm of 21 mdeg ellipticity. At 192.5 nm, a negative CD minimum should be close to −42 mdeg giving a ratio of 2 between the two peaks. Spectra shall be calibrated to this later ratio, so that spectra obtained from different sources will provide same amplitudes [26].

The CD spectrum should be normalized for sample concentration (in mol/L), cell path length (in cm), and the number of RNA nucleotides for comparison between RNAs of different sizes. Following the Beer–Lambert law, NA concentrations and cell path lengths should be chosen in accordance, optimizing for CD amplitude size. It is important to choose UV-transparent buffers and short path lengths to allow the spectral band extension down to 170 nm for SRCD. As a rule of thumb, in a 50 μm path length cell, a 1 mg/mL concentration of NA in 10–50 mM phosphate results in an absorbance of ~1 OD at 178 nm.

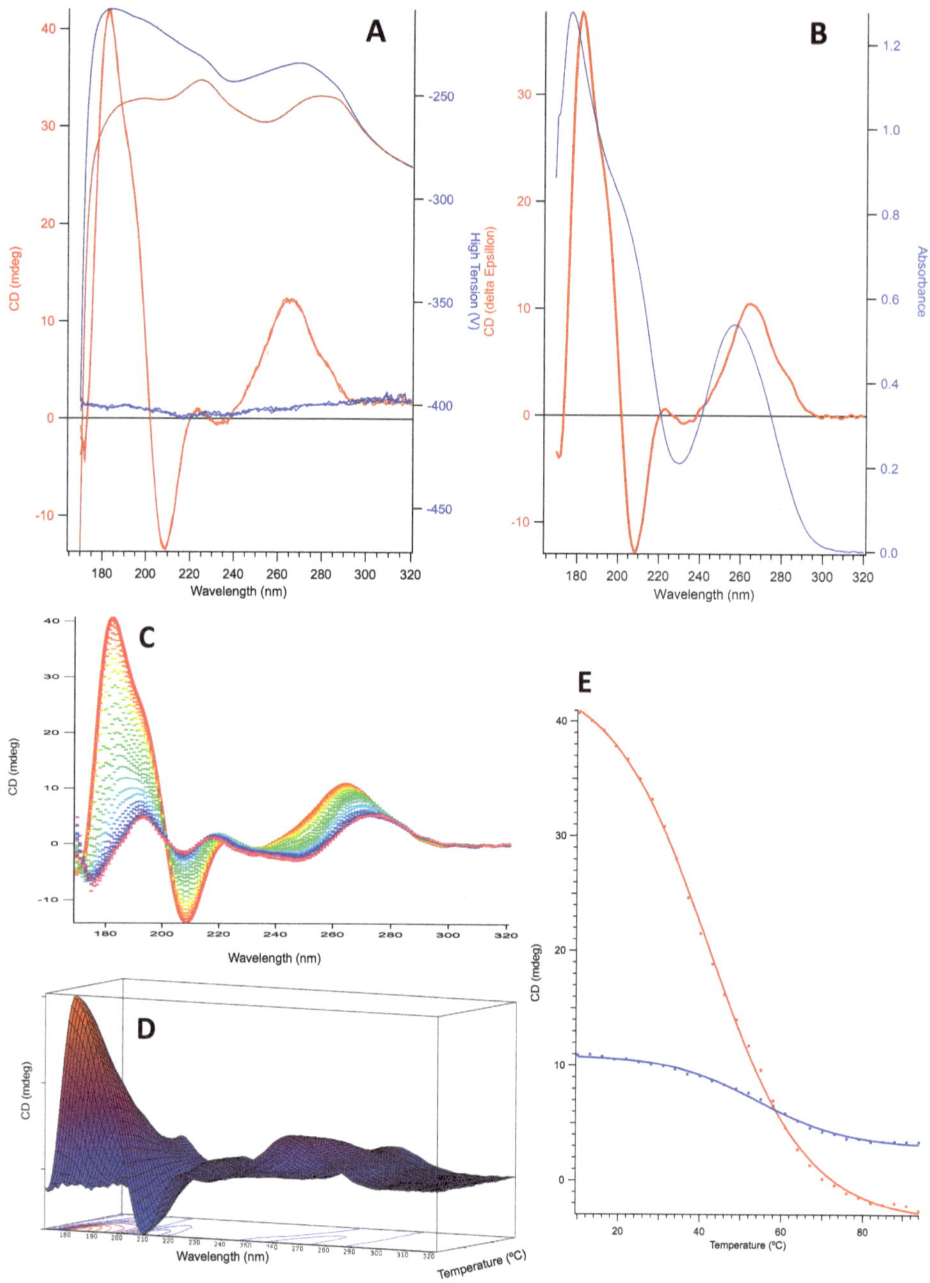

Fig. 2 Example of RNA SRCD spectra (DsrA noncoding RNA fragment [24]). (**a**) The triplet raw data shown for the RNA (red) and its buffer solution (blue) align between 290 and 320 nm indicating a good realignment of the CaF_2 cells for the two loadings (*see* **Notes 5**). The high tension (HT) of the detector (red and blue lines above the triplet spectra), which is proportional to the absorption of the sample, must be measured to detect any problem of aggregation (*see* **Notes 2** and **3**). The spectral region contains several CD absorption bands, mainly one positive band around 260–268 nm, which is influenced by base-stacking and base-pairing, and one positive band around 185 nm and one negative band between 200 and 210 nm, indicative for right-handed RNA molecules (A- and B-forms). (**b**) Smoothed (Savitzky–Golay algorithm using 5 points) SRCD spectrum (red line) and absorption band (blue line) of same RNA. (**c**, **d**) Thermal scan carried out every 3 °C from 10 °C to

3.5 Spectral Data Treatment

To analyze SRCD spectra, different software can be used such as CDtool [20], which is presented herein.

The manipulation of the absorption spectrum can be done following the steps below:

- In the near UV a spectral band of minimum 5 nm should be chosen for which the sample spectra and baseline (buffer) spectra coincide resulting in a zero baseline after subtraction (Fig. 2a, b) (*see* **Note 5**).
- Smoothing the spectra, usually seven points in the Savitzky–Golay smoothing algorithm will provide good results without shifting peak positions or reducing amplitudes of the rather broadband peaks.
- The second derivative of the spectra allows precise identification of the peak positions.

3.6 RNA Spectra Analysis

SRCD spectroscopy is an empirical technique based on comparison between spectra obtained of novel samples and standardized known spectra of determined samples with high-resolution techniques such as NMR or X-ray crystallography.

The spectral region between 320 and 170 nm contains several absorption bands, most prominently the 260–268 nm positive CD signal which is influenced by intra- and intermolecular and base-stacking and base-pairing, including Watson–Crick, Hoogsteen, and cross-strand (CS) base pairs (Fig. 2) [7, 27]. This band lies within a broader band between 210 and 300 nm, including absorptions due to interactions between the π-π* oscillators of the bases and the sugar phosphate backbone and their n-π* transitions [28, 29]. Comparatively in this band sequence differences of RNA molecules are related to spectral differences revealing how the sequence affects the base–base interactions.

At the far UV end of the spectrum, an intense positive band at 180 nm and a negative band at 190 nm are indicative for left-handed RNA molecules (similar to the DNA Z-form [30]), whereas a positive band around 185 nm and a negative one between 200 and 210 nm are indicative for right-handed RNA molecules (A- and B-forms, Fig. 2) (*see* also **Note 6**).

Fig. 2 (continued) 95 °C. Note the isosbestic points, indicating the existence of an equilibrium between two coexisting species. (**e**) Melting curves obtained from the thermal scan at two wavelengths 182 nm and 265 nm. T_m are obtained with a Boltzmann sigmoid equation. The sigmoidal for the 182 nm (red) reveals a melting point at 42.1 °C and that for 265 nm (blue) a melting point at 54.3 °C with standard deviations of 0.3 and 0.6 °C, respectively. The difference in melting point temperature is due to structural changes occurring at different temperatures. These changes may include unwinding of the helix form (right handed) and disruption of the base-pairing

3.7 Melting Curves

3.7.1 Thermal Scan Acquisition

Melting curves of RNA can be obtained from thermal scans using a Peltier-controlled sample holder (Fig. 2c, d). General thermal scans are carried out every 3 °C from 10 to 95 °C (*see* **Note 7**).

One characteristic of a successful temperature scan is the observation of isosbestic points (Fig. 2c). These points of spectral convergence indicate first of all that the solution did not contain air bubbles and second that the heating of the sample occurred in a continuous and smooth way. They also hint at the existence of equilibria between two states or species of protein or NA folds.

Melting points or the transition midpoint T_m are extracted from melting curves (Fig. 2e), as described in Subheading 3.7.2.

T_m are often determined with hypochromicity measurements, by comparing the absorbance of the non-denatured and denatured NA. Nevertheless in this case, only an averaged T_m corresponding to the whole denaturation process of the NA can be measured [24]. As shown on Fig. 2e, SRCD curves allow measuring different T_m (*see* Subheading 3.7.2).

Note that T_m may also be very sensitive to buffer conditions, to the presence of divalent ions, and very dependent on RNA concentration for intermolecular base-pairing [24].

Finally, in order to verify the reversibility, a last acquisition should be taken at the initial temperature after melting.

3.7.2 Interpretation of Thermal Scans

With the help of a Boltzmann sigmoid equation, melting curves can be fitted.

$y = \text{bottom} + (\text{top} - \text{bottom})/(1 + e((T_m - x)/\text{slope}))$ (Fig. 2e).

A Boltzmann sigmoid function is characterized by a bottom, a top plateau, and a characteristic value describing the point where the x value is exactly between the top and bottom value, here T_m.

Reduction of the amplitude of the 265 nm peak is related to the disruption of the base-stacking, while amplitude of the 182 nm peak is related to A-helix melting.

3.8 Specific Case of Protein/RNA Complexes

As protein may also greatly influence RNA structure, ribonucleoprotein complexes can also be analyzed (Fig. 3). It is important to work with identical protein and RNA concentrations in the complex and for the RNA and protein alone.

After baseline subtraction of the protein/RNA complex, RNA and protein spectra are subtracted from the complex spectra. Interactions are observable by comparing the protein or RNA spectra alone with the complex subtracted RNA or protein spectra. Figure 3 shows a typical spectrum of a nucleoprotein complex analyzed by SRCD spectroscopy (and corresponding controls, RNA and protein alone).

A concentration range of the RNA (or protein) may be chosen to confirm concentration dependencies (*see* **Notes 8** and **9**).

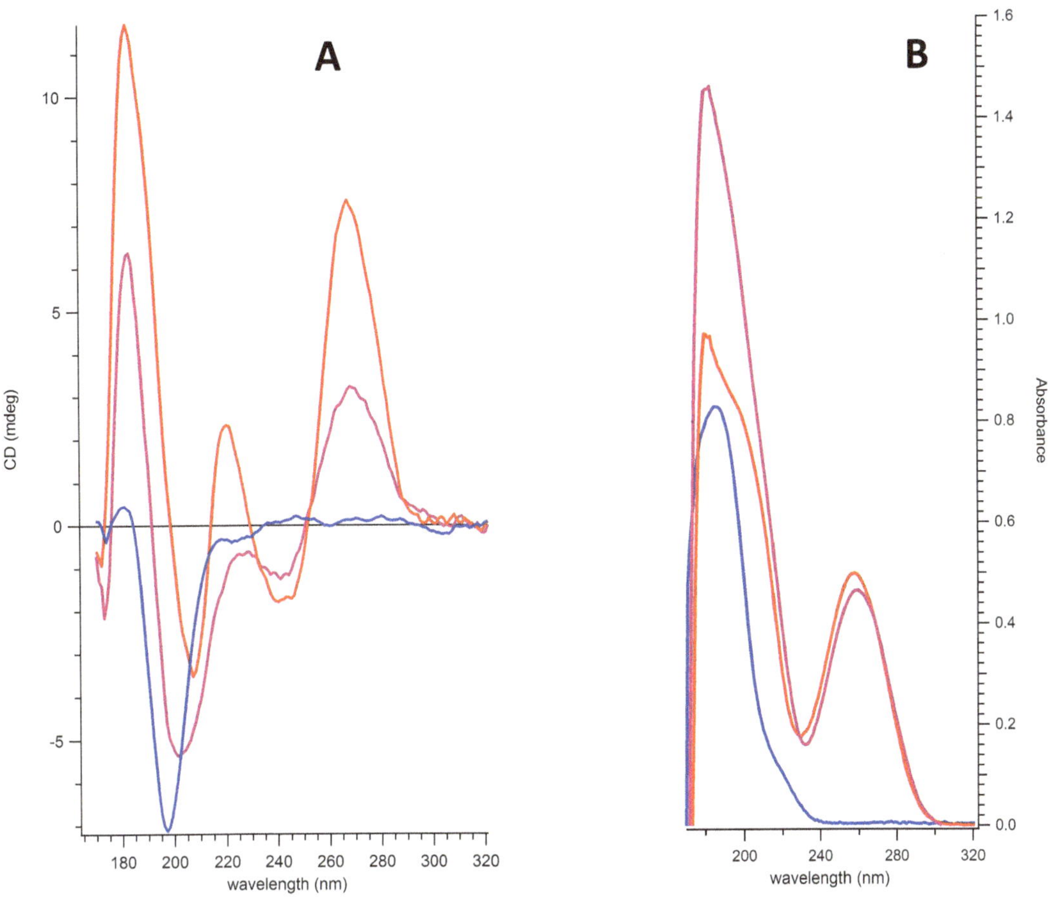

Fig. 3 (**a**) SRCD spectra of a ribonucleoprotein complex (purple line) and corresponding controls (DsrA noncoding RNA fragment in red, Hfq protein in blue [33]). (**b**) Absorption of the samples (proportional to the high-tension detector) showing absorptions adding up. Spectra have been taken from solutions with equal concentrations. The 265 nm and the 225 nm CD band are definitely reduced for the complex similar to a sharp decrease of the 182 nm band. This is the result of a structural change of the protein for which the 197 nm negative CD signal (blue) has been shifted to 202 nm (purple) and most likely to 210 nm, which is hidden in the spectrum of the complex. Also the RNA (red) has changed upon complex formation, as can be seen from the strong reduction of the positive 265 nm peak (more than half the amplitude) most likely revealing a change in the base-pairing and base-stacking. The annihilation of the 220 nm positive CD band of the RNA and the of the 182 nm band indicates that the helical twisting seems also to be affected which means that the backbone and the base-pairing seem to change upon binding with the protein

Thermal denaturation of the complex (*see* Subheading 3.7) and its constituents will provide melting points, which may be shifted to higher T_m upon complex formation if the protein stabilizes the complex, or oppositely to lower T_m if the protein melts the complex.

Once the spectrum of the RNA has been subtracted from the complex, the resulting spectra of the protein shall be compared to the protein spectra (Fig. 3). For further protein secondary structure determination, we recommend to use BestSel software. This publicly accessible software allows for determination of secondary structure content in proteins including the distinction between parallel and antiparallel β-sheets. Further, it gives an idea of the folding by comparison with similar folded proteins obtained from the pdb databank. Normalized root-mean-square deviation (NRMSD) indicates the most accurate fit for each spectrum; values of <0.15 are considered significant.

4 Notes

1. We recommend avoiding PAGE purification after RNA synthesis when possible, as this can introduce contaminants that can be detected by SRCD spectroscopy.
2. The high tension (HT) of the detector, which is proportional to the absorption of the sample, should be used to observe any problem concerning aggregation. If, e.g., in the far UV the HT would not rise exponentially up to the HT limit, but rather evolve shallowly, this is indicative of aggregation of the sample. The wavelength would then be close to the macromolecular size and therefore be scattered or diffused. Ultimately, this impacts the amplitude measured in this region.
3. Between the absorption edge, corresponding to the midpoint of the rising HT, and the closest CD maximum or minimum, a distance of at least 5 nm should be observed. Otherwise a redshift of the later is inevitable.
4. The effect of the far-UV light especially between 190 and 170 nm upon RNA molecules should be investigated. In this region the UVC is known to cause alterations in the NA structure leading to single-strand and even double-strand breakage. To avoid the impact of this kind, an appropriate sample surface size of 4×4 mm should be chosen in which the beam photon flux of less than 10^{11} photons/s should not induce any structural changes. Verification by means of an action spectrum is certainly recommended [31].
5. Due to the high sensitivity of the method, we strongly recommend to record the baseline spectra in the same cell than that of the sample.
6. RNA quadruplexes can also be analyzed using SRCD [11, 32].
7. For melting curve analysis, we recommend avoiding low-volume cells to reduce significant sample evaporation.
8. The CD signal is influenced by the alignment of any dichroic chromophore, which would produce a linear dichroism. Therefore, attention should be paid to avoid alignment of the RNA molecules; this may occur in high concentrated solutions of

RNA, which are mounted in very short path length cells, or when a protein is added. Turning the cell holder should indicate any linear dichroism contribution.

9. Note that in addition, CD is very sensitive to diffusion due to molecule aggregation, which may also occur in high concentrated solutions or when a complex forms.

Acknowledgments

This work was supported by Synchrotron SOLEIL, CNRS, and CEA. SRCD measurements on DISCO beamline at SOLEIL Synchrotron were performed under proposal # 20181037. We are grateful to R.R. Sinden (South Dakota School of Mines and Technology, USA) for critical reading of the manuscript.

References

1. Berova N, Nakanishi K, Woody RW (2000) Protein characterisation by synchrotron radiation circular dichroism spectroscopy. In: Circular dichroism principles and applications, vol 4. Wiley, New York, pp 317–370
2. Fasman GD (1996) Circular dichroism and the conformational analysis of biomolecules. Springer, New York, pp 317–370
3. Nordén B, Rodger A, Daffron T (2010) Linear dichroism and circular dichroism a textbook on polarized-light spectroscopy. RCS Publishing, Cambridge, pp 317–370
4. Johnson WC (1990) Electronic circular dichroism spectroscopy (CD) spectroscopic of nucleic acids. In: Biophysics, vol 1. Springer Verlag, Berlin/Heidelberg, pp 2275–2280
5. Wallace BA (2000) Synchrotron radiation circular-dichroism spectroscopy as a tool for investigating protein structures. J Synchrotron Radiat 7:289–295
6. Gray DM, Ratliff RL, Vaughan MR (1992) Circular dichroism spectroscopy of DNA. Methods Enzymol 211:389–406
7. Holm AIS, Nielsen LM, Hoffmann SV, Nielsen SB (2010) Vacuum-ultraviolet circular dichroism spectroscopy of DNA: a valuable tool to elucidate topology and electronic coupling in DNA. Phys Chem Chem Phys 12:9581–9596
8. Varani G, McClain WH (2000) The G x U wobble base pair. A fundamental building block of RNA structure crucial to RNA function in diverse biological systems. EMBO Rep 1:18–23
9. Kimsey I, Al-Hashimi HM (2014) Increasing occurrences and functional roles for high energy purine-pyrimidine base-pairs in nucleic acids. Curr Opin Struct Biol 24:72–80
10. Miao Z, Westhof E (2017) RNA structure: advances and assessment of 3D structure prediction. Annu Rev Biophys 46:483–503
11. Millevoi S, Moine H, Vagner S (2012) G-quadruplexes in RNA biology. Wiley Interdiscip Rev RNA 3:495–507
12. Peselis A, Serganov A (2014) Structure and function of pseudoknots involved in gene expression control. Wiley Interdiscip Rev RNA 5:803–822
13. Batey RT, Rambo RP, Doudna JA (1999) Tertiary motifs in RNA structure and folding. Angew Chem Int Ed Engl 38(16):2326–2343
14. Refregiers M, Wien F, Ta HP, Premvardhan L, Bac S, Jamme F, Rouam V, Lagarde B, Polack F, Giorgetta JL, Ricaud JP, Bordessoule M, Giuliani A (2012) DISCO synchrotron-radiation circular-dichroism endstation at SOLEIL. J Synchrotron Radiat 19:831–835
15. Giuliani A, Jamme F, Rouam V, Wien F, Giorgetta JL, Lagarde B, Chubar O, Bac S, Yao I, Rey S, Herbeaux C, Marlats JL, Zerbib D, Polack F, Refregiers M (2009) DISCO: a low-energy multipurpose beamline at synchrotron SOLEIL. J Synchrotron Radiat 16:835–841
16. Miles AJ, Hoffmann SV, Tao Y, Janes RW, Wallace BA (2007) Synchrotron radiation circular dichroism (SRCD) spectroscopy: new Beamlines and new applications in biology. Spectroscopy 21:245–255

17. Wallace BA, Gekko K, Hoffmann SV, Lin Y-H, Sutherland JC, Tao Y, Wien F, Janes RW (2011) Synchrotron radiation circular dichroism (SRCD) spectroscopy: an emerging method in structural biology for examining protein conformations and protein interactions. Nucl Instrum Methods Phys Res A649:177–178
18. Wallace BA (2009) Protein characterisation by synchrotron radiation circular dichroism spectroscopy. Q Rev Biophys 42(4):317–370
19. Wien F, Wallace BA (2005) Calcium fluoride micro cells for synchrotron radiation circular dichroism spectroscopy. Appl Spectrosc 59:1109–1113
20. Lees JG, Smith BR, Wien F, Miles AJ, Wallace BA (2004) CDtool-an integrated software package for circular dichroism spectroscopic data processing, analysis, and archiving. Anal Biochem 332:285–289
21. Micsonai A, Wien F, Kernya L, Lee YH, Goto Y, Refregiers M, Kardos J (2015) Accurate secondary structure prediction and fold recognition for circular dichroism spectroscopy. Proc Natl Acad Sci U S A 112: E3095–E3103
22. Micsonai A, Wien F, Bulyaki E, Kun J, Moussong E, Lee Y-H, Goto Y, Refregiers M, Kardos J (2018) BeStSel: a web server for accurate protein secondary structure prediction and fold recognition from the circular dichroism spectra. Nucleic Acids Res 46:W315–W322
23. Sun X, Li JM, Wartell RM (2007) Conversion of stable RNA hairpin to a metastable dimer in frozen solution. RNA 13:2277–2286
24. Cayrol B, Geinguenaud F, Lacoste J, Busi F, Le Derout J, Pietrement O, Le Cam E, Regnier P, Lavelle C, Arluison V (2009) Auto-assembly of E. coli DsrA small noncoding RNA: molecular characteristics and functional consequences. RNA Biol 6:434–445
25. Schmid F (1989) Protein structure—a practical approach. IRL Press, Oxford
26. Johnson WC (1990) CD spectra for nucleic acid monomers vol 1C. Spectroscopic and kinetic data. In: Physical data I; Landolt-Börnstein – group VII. Biophysics. Springer-Verlag, Berlin Heidelberg, pp 1–24
27. Moore DS, Williams AL Jr (1986) CD of nucleic acids: III. Calculated CD of RNAs from new A, U, G, and C transition-moment parameters. Biopolymers 25:1461–1491
28. Cech CL, Tinoco I Jr (1977) Circular dichroism calculations for double-stranded polynucleotides of repeating sequence. Biopolymers 16:43–65
29. Rizzo V, Schellman JA (1984) Matrix-method calculation of linear and circular dichroism spectra of nucleic acids and polynucleotides. Biopolymers 23:435–470
30. Herbert A (2019) Z-DNA and Z-RNA in human disease. Commun Biol 2:7
31. Wien F, Miles AJ, Lees JG, Vronning Hoffmann S, Wallace BA (2005) VUV irradiation effects on proteins in high-flux synchrotron radiation circular dichroism spectroscopy. J Synchrotron Radiat 12:517–523
32. Randazzo A, Spada GP, da Silva MW (2013) Circular dichroism of quadruplex structures. Top Curr Chem 330:67–86
33. Hwang W, Arluison V, Hohng S (2011) Dynamic competition of DsrA and rpoS fragments for the proximal binding site of Hfq as a means for efficient annealing. Nucleic Acids Res 39:5131–5139

Chapter 12

Application of NIR Raman Spectroscopy to Probe the Flexibility of RNA Structure

Hui Bon Hoa Gaston

Abstract

NIR Raman spectroscopy has great potential for the detection of very weak Stokes-shifted Raman scattering emitted by biomolecules. Reports relating Raman spectroscopy analyses of ribonucleic acids (RNAs) are minimal. Nevertheless, correlation between Raman vibrational spectra and specific structural features of RNA from *Avocado sunblotch* viroids (ASBVds) has been clearly established. In this chapter, we discuss how to obtain NIR Raman RNA spectra and applications of Raman spectroscopy for the structural analysis of RNA, based on the analysis of ASBVd RNA. This chapter includes the analysis of spectral changes upon thermal denaturation, deuteration, and self-cleavage perturbations.

Key words Raman spectroscopy, Plant viroids, Sugar-phosphodiester Raman markers, RNA's self-cleavage, RNA physico-chemical perturbations

1 Introduction

RNAs are single-stranded ribonucleic acids that are integral components of the cellular machinery and play essential roles in biological processes, including gene expression regulation [1, 2] and for catalysis purpose [3]. mRNA, for instance, acts as a messenger between DNAs and proteins, while DNA, a double-stranded molecule, is responsible for storing and transferring genetic information. The capability of RNAs to act as a functional molecule in addition to its coding capacity led to the RNA world hypothesis that RNA was a key functional molecule in early evolution [4–7].

Although single-stranded, RNAs have the ability to fold into stable structures and undergo conformational changes when physicochemical conditions change. Indeed, numerous studies of RNA folding have contributed to our understanding of how RNA functions in the cell. For decades, chemical and enzymatic probing have been powerful assays available to obtain structural information on RNA at the nucleotide resolution [8–10]. Nevertheless, recent developments and improvements of novel physical

Véronique Arluison and Frank Wien (eds.), *RNA Spectroscopy: Methods and Protocols*, Methods in Molecular Biology, vol. 2113, https://doi.org/10.1007/978-1-0716-0278-2_12,

instrumentations [10–15] permit deeper insights into RNA folding [16] and the effects of specific metal ion binding that regulate its structure and function [17, 18].

Raman spectroscopy [19–24] provides information about energies of normal modes of molecular vibration, which are associated with changes in polarizability in the vicinity of the vibrating nuclei. Thus, vibrations of multiply bonded or electron-rich groups generally produce more intense Raman bands than do vibrations of singly bonded groups. Raman spectra of nucleic acids are thus dominated primarily by bands caused by vibrations localized either within the heterocyclic bases or in backbone phosphate groups.

Infrared spectroscopy, which is based on absorption of light, can also be used to give information about the vibrational and rotational modes of molecules and other low-frequency modes of molecules. The difference is that in IR spectroscopy the molecule must undergo a dipole moment change during the vibration (i.e., a nonsymmetric molecule), whereas Raman spectroscopy relies on molecules which can undergo a polarizability change during the vibration (i.e., the electron cloud must undergo a positional change). Raman spectroscopy offers some advantage over IR spectroscopy. It is more versatile, and water can be used as solvent allowing aqueous macromolecular solutions to be analyzed. Additionally, samples require minimal preparation.

However, Raman scattering is typically very weak and the difficulty is to separate the weak inelastic scattering light from the intense Rayleigh scattering laser light, resulting in higher cost of the Raman spectrometer. A modern near-infrared (NIR) Raman instrumentation is required which employs a double-photon excitation light. Such a state-of-the-art Raman spectroscopy system has been applied to the study and characterization of the *Avocado sunblotch* viroid (ASBVd) RNA in response to external perturbations [25].

2 Materials

2.1 Enzymes, Buffers, and Chemicals

All solutions have to be RNAse-free.

1. Sterile dd water (double distilled, endotoxin-free).
2. TBE buffer: 89 mM Tris–borate, 2 mM ethylenediaminetetraacetic acid (EDTA), pH 8.3.
3. T7 RNA polymerase (T7 RNAP) and T7 RNAP transcription buffer (supplied with T7 polymerase).
4. Nucleoside triphosphates (NTPs).
5. Gel-loading buffer: 50% formamide, 0.05% xylene cyanol in water.
6. Ethanol 70 and 100%.

7. RNase-free DNase (RQ1-DNAse, Promega).
8. Gel elution: 300 mM sodium acetate, pH 5.2, filtered through a 0.22-μM filter.
9. Standard Raman buffer in H_2O: 20 mM sodium cacodylate, pH 7.2, 150 mM KCl, without Mg^{2+}.
10. Standard Raman buffer in D_2O: 20 mM sodium cacodylate, pH 6.8, 150 mM KCl, no Mg^{2+}.

2.2 Raman Spectrometer

As Raman scattering is typically very weak and the difficulty is to separate the weak inelastic scattering light from the intense Rayleigh scattering laser light, it is necessary to use a modern near-infrared (NIR) Raman instrumentation which employs a double-photon excitation light at 780 nm and two Semrock RazorEdge long-pass filters to separate the laser intense Rayleigh light. The sensitive detection system is composed of an Acton SpectraPro 2500i spectrograph coupled to a N_2 liquid-cooled NIR CCD. The advantage of such setup is to increase the signal to noise ratio and to avoid potentially interfering fluorescence of target molecules or trace contaminants. NIR Raman spectroscopy allows the study not only of highly concentrated fibrous forms but also of dilute nucleic acid solutions under near-physiological conditions. The instrument used in this chapter was home-built.

Raman spectra analyses can be achieved with dedicated software such as WinSpec (Princeton Instruments) and Igor Pro (WaveMetrics).

3 Methods

3.1 Preparation of Viroids for Raman Experiments

3.1.1 ASBVd RNA Transcription

1. Plant RNA viroids such as *Avocado sunblotch* viroid (ASBVd) belong to the *Avsunviroidae* family. They are small single-stranded covalently closed circular RNA molecules with a 246–401 nucleotide chain length. The ASBVds adopt a branched or quasi rodlike secondary structure in their native state, with three helical junctions (helix I, II, III), a core of invariant nucleotides, and a well-conserved single strain [26]. They exist in two polarities, (−) and (+) viroids.
2. ASBVd RNA-encoding plasmid has been constructed by cloning the monomeric coding sequence of ASBVd RNA under the control of a T7 promoter [27].
3. The transcription reaction is performed using the T7 RNA polymerase as described in Chapters 10, 14, or 20.
4. The transcription is stopped by treatment with RNase-free DNase for 30 min at 37 °C to remove the DNA template.
5. RNAs are precipitated with ethanol and resuspended in water containing 0.05% xylene cyanol and 50% deionized formamide.

The resulting mixture is denatured for 2 min at 65 °C prior to separation on a denaturing 10% polyacrylamide gel electrophoresis (PAGE, 19:1 acrylamide/bisacrylamide) containing 7 M urea and using TBE buffer for migration.

6. Transcripts are detected by UV and the bands corresponding to the 247 nt full-length fragments of both polarities are excised.
7. The RNA is eluted overnight in 300 mM sodium acetate, pH 5.2, ethanol precipitated, and dried.
8. After dissolving in ultrapure water, the RNA concentrations are determined by UV absorbance (NanoVue, GE Healthcare) and the samples are stored at −20 °C.

3.1.2 ASBVd RNA Preparation for Raman Experiments

1. For Raman experiments, 800 μg of dry RNA is suspended in 10 μL of water and denatured at 95 °C for 45 s and then slowly (3 °C/min) renatured by cooling to 20 °C.
2. Salts and buffer are adjusted to the experimental conditions in a total volume of 10–20 μL.
3. The final viroid concentration in the sample cell varied between 0.63 and 1 mM, in standard Raman buffers at 20 °C (hydrogenated or deuterated).
4. The sample cell is thermostated within ±1 °C.
5. For measurements at elevated temperatures, a thin layer of oil is poured on top of the sample solution to prevent evaporation.

3.2 Conditions for the Analysis of Self-Cleaved Viroids

1. ASBVd polarity (−) and (+) viroids in the presence of 20 mM Mg^{2+} are incubated at 45 °C, directly in the optical cell during 4 h for completion of the reaction and then brought back to 20 °C for Raman spectroscopic measurements.
2. At the same time, kinetic profile of the viroid cleavage is followed by measuring the reaction products as a function of time by loading each reaction product onto a denaturing gel for the determination of the fraction of cleaved products using gel electrophoresis analysis.

3.3 Basic Theory of Raman Inelastic Scattering

The Raman effect occurs when light encounters an obstacle or nonhomogeneity scattering material (molecules, solid, liquid, or gas) and interacts with the electron cloud and the bonds of that molecule. The polarizability (α) of a molecule is the ability of the electron cloud (constituted by positive and negative charges around the atom) to be deformed through the interaction with an electric field creating a dipole moment (P).

When starting with a classical interpretation by considering a simple diatomic molecule as mass on a spring, where m represents the atomic mass, x the displacement, and K the bond strength, the displacement of the molecule can be expressed by using Hooke's law, and by introducing the reduced mass $\mu = (m_1 \, m_2/m_1 + m_2)$ and the total displacement $q = (x_1 + x_2)$, the equation can be simplified to

$$\mu d^2 q/\mathrm{d}t^2 = -Kq \tag{1}$$

by solving the solution for q

$$q = q_0 \cos(2\pi \nu_\mathrm{m} t) \tag{2}$$

where ν_m is the molecular vibration and is defined as

$$\nu_\mathrm{m} = (1/2\pi)(\mathrm{K}/\mu)^{1/2} \tag{3}$$

From Eqs. 2 and 3, it is shown that the molecule vibrates in a cosinus pattern with a frequency proportional to the bond strength and inversely proportional to the reduced mass.

Each molecule has its own vibration signatures which are determined not only by the atoms in the molecules, but also by the characteristics of the individual bonds and the ability of each atom in a molecule to move in three different directions (x, y, and z). Consequently, the $3N$ degrees of freedom are partitioned into molecular translational, rotational, and vibrational motion. These molecular vibrational modes include stretching and bending motions of chemical bonds of the molecule.

Through the Raman effects, these vibrational frequencies of a simple diatomic molecule can be measured due to the fact that the polarizability (α) of the molecule is a function of the displacement q and can be described (by using small amplitude approximation) as

$$\alpha = \alpha_0 + q(\delta\alpha/\delta t)_{q=0} + \dots \tag{4}$$

The induced dipole moment (P) is then equal to the polarizability (α) of the molecule and the electric field with intensity E_0 and frequency ν_0 of the incident light source:

$$P = \alpha E_0 \cos(2\pi \nu_0 t) \tag{5}$$

Combining Eqs. 4 and 5 results in two terms for the dipole moment:

$$P = \alpha_0 E_0 \cos(2\pi \nu_0 t) + q_0 E_0 (\delta\alpha/\delta t)_{q=0} [\cos(2\pi\{\nu_0 - \nu_\mathrm{m} t) + \cos(2\pi\{\nu_0 + \nu_\mathrm{m}\} t)] \tag{6}$$

The first term is the so-called Rayleigh scattering effect with no change in the frequency of the incident light (the dominate effect), and the second term is the Raman inelastic component which shifts the frequency of the incident light by plus or minus the frequency of the molecular vibration. The increased in frequency is known as an anti-Stokes shift and the decrease in frequency is known as a

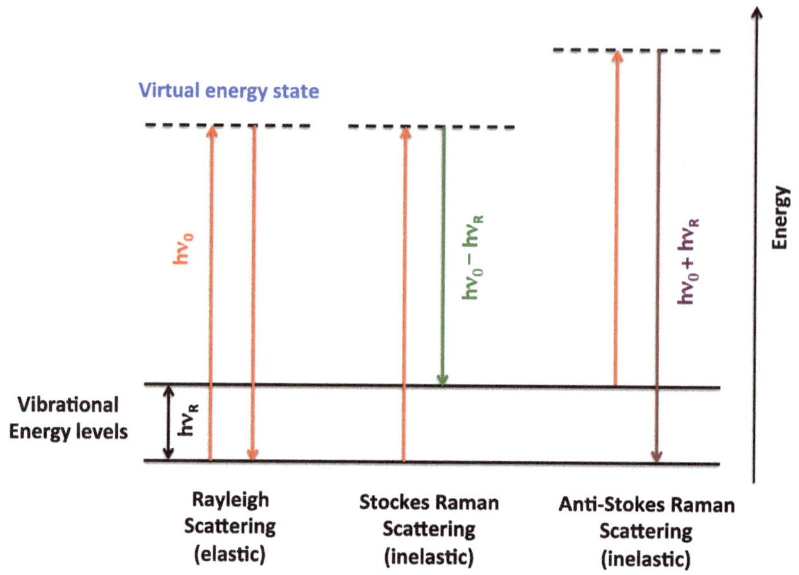

Fig. 1 Jablonski diagram representation of quantum energy transitions for Rayleigh and Raman scattering

Stokes shift. This later term is used as a direct measurement of the vibrational frequency of a molecule.

When using the quantum particle interpretation as shown in Fig. 1 (the Jablonski diagram), the Raman effect is described as inelastic scattering of a photon off of a molecular bond which results from the incident photon exciting the molecule into a virtual energy state.

When this occurs, there are three different potential outcomes.

1. First, the molecule can relax back down to the ground state and emits a photon of equal energy to that of the incident photon. This is referred to as an elastic process: the "Rayleigh scattering."
2. Second, the molecule can relax to a real phonon state and emits a photon with less energy than the incident photon: this is called "Stokes-shifted Raman scattering."
3. The third potential outcome is that the molecule, which is already in an excited phonon state, is excited to a higher virtual state and then relaxes down to the ground state, emitting a photon with more energy than the incident photons. This is called "anti-Stokes Raman scattering" (*see* **Note 1**).
4. It can be shown that the power of the scattered light (P_s) is equal to the product of the intensity I_o of the incident light (or incident photons) and a value known as the Raman cross-section (σ_R):

$$P_s \propto I_0 \sigma_R = I_0/\lambda^4 \quad (7)$$

where the Raman cross-section (σ_R) expresses the Raman scattering efficiency of a molecule in a manner analogous to Beer's law in the absorption spectroscopy. (σ_R) is then proportional to the probability that an interaction will occur ($\propto 1/\lambda^4$).

3.4 NIR Raman Instrument Design

Our home-built near-infrared (NIR) Raman instrumentation setup is shown in Fig. 2. The excitation light at 780 nm is provided by a continuous-wave Ti/sapphire laser (Spectra-Physics, model 3900S) pumped by an Argon-ion laser (Spectra-Physics Stabilite 2017).

Figure 3 shows that the NIR laser light is focused into the specifically designed quartz sample cell (internal dimensions: 2 × 2 × 22 mm) filled with 10–20-μL viroid solution, by an infinity-corrected backscattering objective (Olympus MA10, $M = \times 10$, NA = 0.25). The laser excitation power is attenuated to ≈300 mW in the sample. The collected backscattering laser Raman signal is delivered onto the spectrograph (Acton SpectraPro 2500i) coupled with deep-depletion back-illuminated NIR CCD (Princeton Instruments SPEC-10400BR/LN) and focused on the

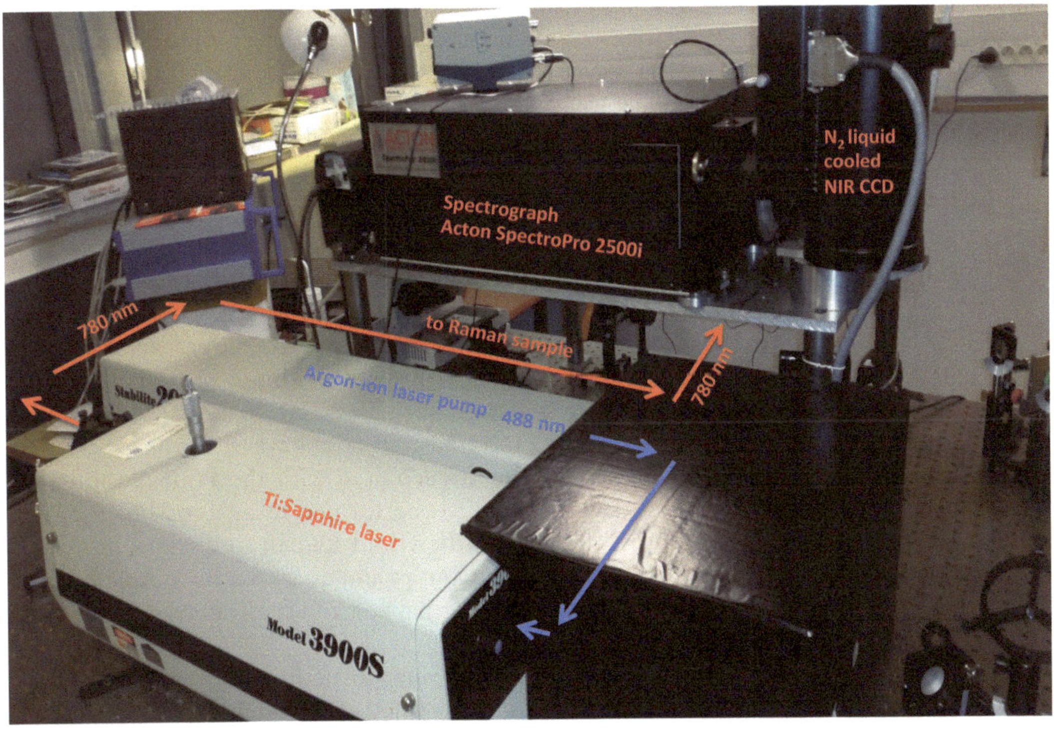

Fig. 2 NIR Raman instrumentation, argon-ion pump at 488 nm (in blue). Ti/sapphire laser (in red) with a nonlinear crystal frequency doubler which gives a wavelength output of 780 nm. A spectrograph Action SpectraPro 2500i coupled with a N_2 liquid-cooled NIR CCD detector

Fig. 3 NIR Raman excitation light at 780 nm of a sample contained in a small-volume quartz optical cell, the internal dimensions of which are 2 × 2 × 22 mm. The temperature of the optical cell is controlled and regulated by a HT thermostated block (5–80 °C). The Raman inelastic Stokes scattering lights are collected through a backscattering objective and two Semrock RazorEdge long-pass filters to separate the laser intense Rayleigh light. The resulting backscattering lights are directed through high-quality laser mirrors to the condenser of an Acton SpectraPro 2500i spectrograph coupled to a N_2 liquid-cooled NIR CCD detector

spectrograph entrance (slit width 30 μm) by an achromatic lens with $f = 75$ mm.

The Raman signal is separated from the laser light by two Semrock RazorEdge long-pass filters (grade U). One filter, with $I_{laser} = 780$ nm, is placed perpendicular to the optical beam just before the focusing lens. Another one, with $I_{laser} = 830$ nm, is used as a dichroic beam splitter at an angle of incidence of 45°. It reflects laser light at 780 nm and transmits all the wavelengths longer than 785 nm. Raman spectra are acquired with the WinSpec software, and further data treatment is performed using Igor Pro for Windows software. To obtain the viroid Raman spectrum, the contribution from the cacodylate buffer and quartz cell is subtracted from the recorded spectrum, normalized on the water bending band around 1640 cm^{-1}.

Spectra resolution of Raman experiments is ≈5 cm^{-1}, and frequency calibration is performed using Raman lines of toluene with absolute accuracy better than ±2 cm^{-1} and the relative frequency position accuracy better than ±1 cm^{-1}. The total accumulation time for one Raman spectrum is 10–30 min depending on the sample studied.

3.5 Assignments of Raman Markers

RNA is a polymer of ribonucleotides that contains a long 5′ to 3′ chain of charged phosphate groups that are linked to the ribose sugars (*see* Fig. 4).

The Raman frequency of the phosphodioxy stretching vibration modes occurs at 1100 cm^{-1}. The covalent sugar-phosphate bonds (phosphodiester bonds) link the 5-carbon of sugar molecule and the 3-carbon of the next. The phosphodiester vibration modes $\upsilon_{S\ (\text{-C-O-P-O-C-})}$ occur at 813 cm^{-1}. The nitrogenous bases of RNA, such adenine (A), cytosine (C), guanine (G), and uracil (U), are attached to the 1′ carbon of the ribose. The A and G base ring stretching modes are respectively at 727 cm^{-1}, 1485 cm^{-1}, and 669 cm^{-1}. The bases (A and G) H-bonding modes are at 1570 cm^{-1}.

Figure 4 shows that the nucleotide bases can rotate with respect to the sugar through the *N*-glycosidic bond, leading to either *syn* or the preferred *anti*-conformations for RNA.

Sugars have Raman stretching vibration modes at 820–950 cm^{-1}. Ribose sugars can adopt various puckered conformation in which C-3′ and C-2′ bonds are in either *endo* or *exo* positions relative to the base and C-5′ bond. The corresponding vibration modes exhibit characteristic Raman line at 669 cm^{-1} and 1250 cm^{-1} for the sugar C-3′ endo mode (or N-type sugar

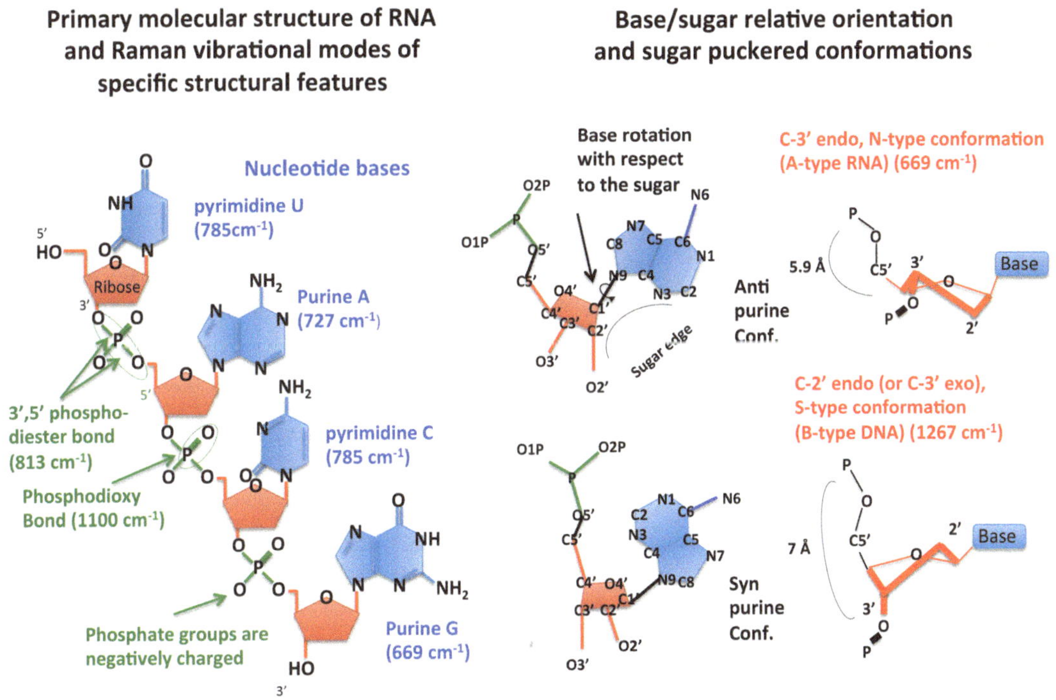

Fig. 4 Assignments of Raman vibration modes of specific structural features of the RNA primary molecular structure and of the base/sugar relative orientation and the sugar pucker conformations

puckering A-RNA conformation) and at 1267 cm^{-1} for the C-3′ exo (or C-2′ endo) mode corresponding to S-type sugar puckering B-DNA conformation.

3.6 Raman Spectra of Viroids

Figure 5 shows the Raman spectra of ASBVd RNA at 20 °C (two polarities), in the absence of Mg^{2+} [28, 29]. Such Raman spectra exhibit about 30 characteristic lines that arise from the vibration of the sugar-phosphate backbone of the RNA and bases. Different types of RNA possess rather similar lines, except that their intensities and frequency positions are different, depending on each cooperative docked structure [28, 29].

The characteristic Raman markers of the primary structure of both ASBVd polarities exhibit small differences in secondary structure, as evidenced by the symmetric stretching vibration of the phosphodiester linkage intensity $I_{(\text{-C-O-P-O-C-})}$ at 813 cm^{-1}. Polarity (+) viroid exhibits 78% double helical content, while polarity (−) viroid exhibits about 76%. Their conformational structures, based on the intensities of the 785 cm^{-1} lines of the ring breathing

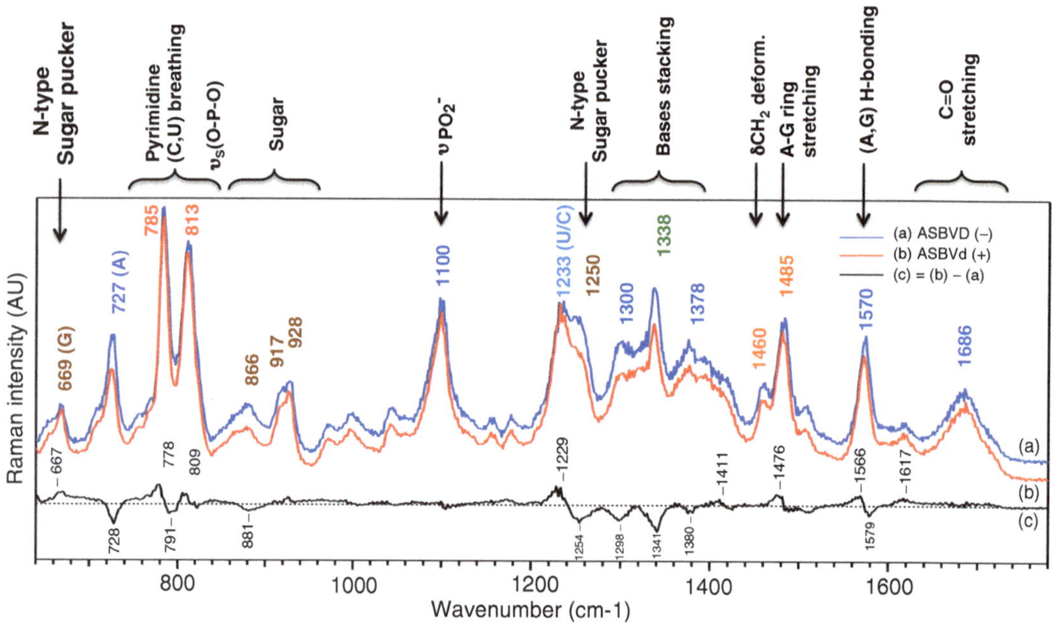

Fig. 5 Assignment of major Raman lines of ASBVd polarity (−) and (+) species. Spectra of ASBVd (−) (blue curve (**a**)) and ASBVd (+) (red curve (**b**)) are obtained in aqueous cacodylate buffer in the absence of Mg^{2+} and at 20 °C. Both spectra are normalized using the Raman line of the phosphodioxy stretch vibration mode at 1100 cm^{-1} taken as an internal intensity marker standard. The contribution from the cacodylate buffer and the quartz cell are subtracted. The black curve (**c**) is the difference Raman spectrum obtained by (1:1) subtraction of the spectrum (**a**) from the spectrum (**b**). The difference spectrum is multiplied by a factor of 3 to visually enhance the resulting spectra changes [25]. Note that the doublet (785–813 cm^{-1} lines) is very sensitive to the backbone bond angle deformation, and the sugar lines at 669 cm^{-1} and 1250 cm^{-1} are representative of the N-type sugar pucker conformations of RNA

pyrimidine (C, U) bases, were the same ($r_{conf} = 1.14$). At the same time, a slight frequency downshift in the 813 cm^{-1} as well as in the 785 cm^{-1} lines was observed, indicating a slight difference in the rigidity of the phosphodiester backbone of ASBVd (+). Phosphodiester conformations of both polarities are of the A-type structures. The "N-type sugar pucker" conformations are revealed by the existence of two characteristic Raman lines at 669 and 1250 cm^{-1}. The phosphodioxy stretching mode at 1100 cm^{-1} remains the same, both in intensity and frequency.

3.7 Quantification of Raman Markers

RNA is a single-stranded polymer which allows extensive intrastrand complementary sequence that generates a secondary structure. It is possible to quantify the intensities as well as the frequency positions of the Raman vibrational markers in order to allow comparison between the strength of several specific vibration modes in the same RNA structure and in various nucleic acids.

1. The symmetric stretching vibration of the phosphodioxy group ($\upsilon_{S\,(PO_2^-)}$) is sensitive to changes in the electrostatic environment, but insensitive to the base composition of each RNA. Thus in physiological conditions, the strong Raman line at 1100 cm^{-1} can be used as an internal Raman marker. The normalization of the intensities of the other Raman lines on this phosphodioxy Raman line allows determination of the double helical content and the nucleic acid conformation.
2. Double helical content is estimated from Eqs. 8 and 9:

$$r_2 = I_{813}/I_{1100} = 1.23 \text{ for ASBVd } (-)\ (76\%) \tag{8}$$

$$r_2 = I_{813}/I_{1100} = 1.28 \text{ for ASBVd}(+)(78\%) \tag{9}$$

(*see* **Note 2**).

Thomas and Hartman [30] have shown that for completely ordered polyribonucleotides, the value of $r_2 = 1.64$, while for completely disordered structures [31, 32], the r_2 value approaches zero. Thus the ratio r_2 observed for ASBVd (−) and (+) can be converted into percentage of ordered polyribonucleotide RNA by using the factor 1.64 as a coefficient of division.

3. Nucleic acid conformation is estimated from Eq. 10:

$$r_{conf.} = I_{785}/I_{813} = 1.14 \tag{10}$$

(*see* **Note 3**)

4. The degree of A/G base-stacking is estimated from Eq. 11:

$$r_{stack} = (I_{1300} + I_{1378})/I_{1338} = 1.35 \tag{11}$$

(*see* **Note 4)** [33].

3.8 Effects of High Temperatures on RNA Spectra

Increasing temperature from 20 to 65 °C induces several changes of the Raman vibrational markers in the spectrum of ASBVd RNA (Fig. 6a).

1. The phosphodiester mode at 815 cm^{-1} decreases in intensity and downshifts by $\Delta\upsilon_S = -15\ cm^{-1}$. At the same time, a broad shoulder at 779 cm^{-1} appears.
2. The pyrimidine stretching mode at 785 cm^{-1} downshifts by $\Delta\upsilon_S = -5\ cm^{-1}$ and shifts to 780 cm^{-1} and the intensity increases. The calculated r_2 ratio gives a loss of 27% ordered

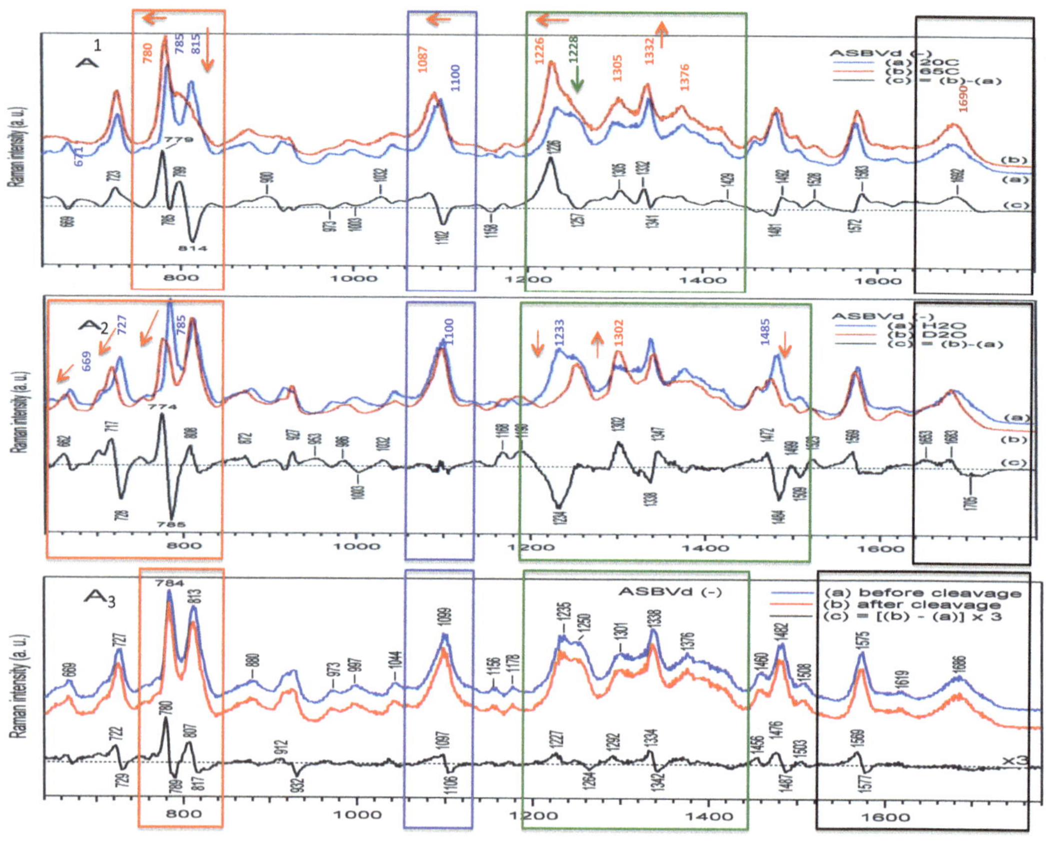

Fig. 6 Raman spectra of ASBVd (−) in response to extreme external environmental changes [25]. (**a**) Thermal unfolding effects: Raman spectra are recorded in aqueous cacodylate buffer, pH 7.2, in the absence of Mg^{2+}. Blue curve (*a*) at 20 °C. Red curve (*b*) at 65 °C. The difference spectrum (*c* = *b* − *a*, black curve) is multiplied by a factor of 3 in order to visually enhance the resulting spectral changes. (**b**) D_2O solvent perturbation: Raman spectra are recorded in cacodylate buffer, pH 7.2, in the absence of Mg^{2+}. Blue curve (*a*) at 20 °C, in aqueous solution. Red curve (*b*) at 20 °C, in the presence of D_2O solvent. The difference spectrum (*c* = *b* − *a*, black curve) is multiplied by a factor of 3 in order to visually enhance the resulting spectral changes. (**c**) Self-cleavage activity effect: Raman spectra of ASBVd (−) are recorded in aqueous cacodylate buffer. The cleavage of the viroid is initiated by the injection of 20 mM of Mg^{2+}. Blue curve (*a*) is recorded at 20 °C, immediately after Mg^{2+} injection, while in the red curve (*b*), the product is accumulated during 4 h at 45 °C prior to recording the Raman spectrum at 20 °C. The difference spectrum (*c* = *b* − *a*, black curve) is multiplied by a factor of 3 in order to visually enhance the resulting spectral changes (*see* **Note 8**)

helical content and the calculated $r_{conf.}$ ratio indicates a loss of A-type structure (*see* **Note 5**).

3. The phosphodioxy symmetric stretching line at 1100 cm^{-1} frequency downshifts to 1087 cm^{-1} ($\Delta \upsilon_S = -13\ cm^{-1}$), presumably due to temperature-induced changes in hydration of charged phosphates.
4. In base-stacking frequency intervals, the "N-type sugar pucker" line at 1228 cm^{-1} decreases and an intense line at 1226 cm^{-1} appears. At the same time an increase of the calculated r_{stack} ratio from 1.36 to 1.45 is observed. Such results are indicative of loss of base-sugar conformations and base-stacking.
5. The carbonyl region around 1690 cm^{-1} shows moderate intensity increase reflecting some rupture of hydrogen bonds between bases.

3.9 D_2O Effects

Deuteration experiments are useful tools to perturb chemical reactions which implicate acid/base catalysis. The consequence of which is to decrease proton transfer. Application of such technique on the self-cleavage activity of ASBV shows a decrease of the rates of proton transfer by a factor of 3.5 in D_2O solvent for polarity (−) viroid and a higher decrease (factor of 66.7) for polarity (+) viroid (not shown). The results show that both active protons are accessible to D_2O perturbation. H-bonding of ASBVd (+) nucleotide bases is more accessible.

Figure 6b shows that D_2O can have a second structural effect on the Raman vibration modes of ASBVd (−) as well as (+). In the 640–820 cm^{-1} frequency intervals, the purine A base at 727 cm^{-1} and the purine G base at 669 cm^{-1} exhibit huge changes in their Raman lines showing a decrease of their intensities and frequency downshifts ($\Delta \upsilon_S = -11\ cm^{-1}$). The causes of such changes are due to removal of coupling between ribose and purine bases upon deuteration. The pyrimidine line at 785 cm^{-1} downshifts to 717 cm^{-1} with an intensity decrease. The results indicate an appearance of a new A-type structure of the viroid. The calculated $r_{conf.} = 0.9$ decreases by 25% and the r_2 ratio increased from 1.28 to 1.40. Interestingly, the symmetric phosphodioxy vibration line at 1100 cm^{-1} is not perturbed. In the base-stacking frequency intervals, the U/C base-stretching vibrations at 1233 cm^{-1} disappears upon deuteration and an intense purine A line at 1302 cm^{-1} appears with a concomitant increase of r_{stack} from 1.32 to 1.66 (+34%). The (A/G) purine ring stretching at 1485 cm^{-1} greatly decreases. These results indicate that the external "in-plane" C–N stretching vibrations of purine are very sensitive to D_2O perturbations (*see* **Note 6**).

3.10 Effects Due to Self-Cleavage Activity

Figure 6c shows the Raman spectrum of ASBVd at 20 °C after completion of the cleavage reaction obtained at 45 °C in the presence of 20 mM Mg^{2+}. Interestingly several stretching Raman lines are downshifted. The pyrimidine line at 784 cm^{-1} downshifts by 6% ($\Delta \upsilon_S = -10\ cm^{-1}$). The symmetric phosphodiester line at 813 cm^{-1} also downshifts by 6% and the intensity $I_{(C\text{-}O\text{-}P\text{-}O\text{-}C)}$ increases by 4%. The phosphodioxy (PO_2^-) line also downshifts by $\Delta \upsilon_S = -9\ cm^{-1}$ (*see* **Note 7**).

4 Notes

1. At room temperature, most of the molecules are found in the ground state, and as a consequence, there is a much lower probability that a photon will be anti-Stokes scattered. Most Raman measurements are thus performed considering only the Stokes-shifted light. ($h\nu_R$) is the excess energy originated from the vibrational relaxation (R) energy of the ground state.
2. I_{813} is the intensity of the symmetric stretching vibration of the phosphodiester linkage ($\upsilon_{S\ -C-O-P-O-C-}$), while I_{1100} is the intensity of the symmetric stretching of the phosphodioxy group ($\upsilon_{S\ PO_2^-}$).
3. I_{785} is the intensity of the ring breathing of pyrimidine (C and U) bases, while I_{813} is the intensity of the symmetric stretching vibration of the phosphodiester linkage ($\upsilon_{S\ -C-O-P-O-C-}$). This 813 cm^{-1} line is characteristic of A-type RNA of the viroid, while in B-DNA, a weak broad line with shoulder appears at 835 cm^{-1}.
4. An increase of this r_{stack} value is indicative of a loss of base-stacking. The purine base ring systems at 1300 cm^{-1} (base A) and 1378 cm^{-1} (base G) and the imidazole ring vibration alone at 1338 cm^{-1} of the pyrimidine [34] constitute a base-coupled nucleotide vibration system that is sensitive to ring electronic structure and base-stacking rearrangements. One accurate way to measure the degree of base-stacking is to consider the implications of the intensity changes of these three bases as an ensemble, following Eq. 5 [33].
5. It is clear that temperature-induced unfolding considerably perturbs the conformational structure of the sugar-phosphate backbone, destabilizing the helical structure.
6. The D_2O-induced changes in the Raman spectra of ASBV are quite different from those of temperature-induced perturbations. It is suggested that D_2O perturbed internal loop interactions, base-base interactions, and increased helical rearrangement to new nucleic acid conformations.

7. All these frequency downshifts are indicative of vibrational energy decreases. Such conformational changes in the phosphate backbone are a consequence of a local cyclization reaction of the PO_2^- groups.
8. For ASBVd (+) species, the behavior of the Raman curves vs. physicochemical perturbations is similar, except that the intensity changes were smaller.

Acknowledgments

Marie-Christine Maurel from UPMC Université, Paris 6, France, is thanked for initiating and planning this work. Jacques Verne and Hussein Kaddour are thanked for plasmid constructions and purification of the ASBVds. Hussein Kaddour who was the recipient of a fellowship from the Center National de la Recherche Scientifique (CNRS) did most of the Raman experiments. Sergei Kruglik from UPMC, Paris 6 and Sorbonne Universités, Laboratoire Jean-Perrin, 75005 Paris, is thanked for initiating the NIR Raman spectroscopy and all the data processing. This study was supported by Université Paris 6, the CNRS (grant PID EPOV), and the Centre National d'Etudes Spatiale. Richard Sinden is thanked for careful reading and helpful comments on this manuscript.

References

1. Eddy SR (2001) Non-coding RNA genes and the modern RNA world. Nat Rev Genet 2:919–929
2. Hannon GJ (2002) RNA interference. Nature 418:244–251
3. Doudna JA, Cech TR (2002) The chemical repertoire of natural ribozymes. Nature 418:222–228
4. Gilbert W (1986) The RNA world. Nature 319:618
5. Maurel MC (1992) RNA in evolution. J Evol Biol 2:173–188
6. Maurel MC, Decout JL (1999) Origins of life: molecular foundations and new approaches. Tetrahedron, 55, 3141–3182
7. Pérez-Villa A, Saitta AM, Georgelin T, Lambert JF, Guyot F, Maurel MC & Pietrucci F, (2018) Synthesis of RNA nucleotides in plausible prebiotic conditions from an initio computer simulations. J Phys Chem Lett, Sep 6;9 (17):4981–4987
8. Weeks KM (2010) Advances in RNA structure analysis by chemical probing. Curr Opin Struct Biol 20:295–304
9. Ding F, Lavender CA, Weeks KM, Dokholyan NV (2012) Three-dimensional RNA structure refinement by hydroxyl radical probing. Nat Methods 9:603–608
10. Sattin BD, Zhao W, Travers K, Chu S, Herschlag D (2008) Direct measurement of tertiary contact cooperativity in RNA folding. J Am Chem Soc 130:6085–6087
11. Zhuang X, Bartley LE, Babcock HP, Russell R, Ha T, Herschlag D, Chu S (2000) A single molecule study of RNA catalysis and folding. Science 288:2048–2051
12. Cantara WA, Olson ED, Musier-Forsyth K (2017) Analysis of RNA structure using small-angle X-ray scattering. Methods 113:46–55
13. Greenfeld M, Herschlag D (2009) Probing nucleic acid–ion interactions with buffer exchange-atomic emission spectroscopy. Methods Enzymol 469:375–389
14. Qin Peter Z, Dieckmann TH (2004) Application of NMR and EPR methods to the study of RNA. Curr Opin Struct Biol 14(3):350–359
15. Solomatin SV, Greenfeld M, Chu S, Herschlag D (2010) Multiple native states of an RNA enzyme reveal persistent ruggedness of an RNA folding landscape. Nature 463:681–684
16. Das R, Travers K, Bai Y, Herschlag D (2005) Determining the Mg^{2+} stoichiometry for

folding an RNA metal ion core. J Am Chem Soc 127:8272–8273

17. Bai Y, Das R, Millet IS, Herschlag D, Doniach S (2005) Probing counterion modulated repulsion and attraction between nucleic acid duplexes in solution. Proc Natl Acad Sci U S A 102:1035–1040
18. Georges J, Jr T (1999) Raman spectroscopy of protein and nucleic acid assemblies. Annu Rev Biophys Biomol Struct 28:1–27
19. Peticolas WL, Evertsz E (1992) Conformation of DNA *in vitro* and *in vivo* from laser Raman scattering. Methods Enzymol 211:335–352
20. Lafleur L, Rice J, Thomas GJ Jr (1972) Raman studies of nucleic acids. VII. Poly a. poly U and poly G-poly C. Biopolymers 11:2423–2437
21. Prescott B, Gamache R, Livramento J, Thomas GJJR (1974) Raman studies of nucleic acids. XII. Conformations of oligonucleotides and deuterated polynucleotides. Biopolymers 13:1821–1845
22. Guan Y, Wurrey CJ, Thomas GJ Jr (1994) Vibrational analysis of nucleic acids. I. The phosphodiester group in dimethyl phosphate model compounds: $(CH_3O)_2PO_2^-$, $(CD_3O)_2PO_2^-$, and $(^{13}CH_3O)_2PO_2^-$. Biophys J 66:225–235
23. Luoma Greg A, Marshall Alan G (1978) Laser Raman evidence for new cloverleaf secondary structures for eukaryotic 5.8S RNA and prokaryotic 5S RNA. Proc Natl Acad Sci U S A 75 (10):4904–1905
24. Wartell RM, Harrell JT (1986) Characteristics and variations of B-type DNA conformations in solution—a quantitative analysis of Raman band intensities of 8 DNAs. Biochemistry 25:2664–2671
25. Hui-Bon-Hoa G, Kaddour H, Vergne J, Kruglik Sergei G, Maurel MC (2014) Raman characterization of *Avocado Sunblotch Viroid* and its response to external perturbations and self-cleavage. BMC Biophys 7:2–15
26. Gora-Sochacka A (2004) Viroids: unusual small pathogenic RNAs. Proc Natl Acad Sci U S A 101:6792–6797
27. Daros JA, Flores R (2004) Arabidopsis thaliana has the enzymatic machinery for replicating representative viroid species of the family Pospiviroidae. Proc Natl Acad Sci U S A 101:6792–6797
28. Tsuboi M, Takahashi MS, Kajiura T, Nishimura S (1971) Raman spectrum of a transfer RNA. Science 174:1142–1144
29. Luoma Grec A, Marshall Allan G (1978) Laser Raman evidence for new cloverleaf secondary structures for eukaryotic 5.8S RNA and prokaryotic 5S RNA. Proc Natl Acad Sci U S A 75 (10):4901–4905
30. Medeiros G, Thomas GJ (1971) Raman studies of nucleic acids. 4. Vibrational spectra and associative interactions of aqueous inosine derivatives. Biochim Biophys Acta 247:449–462
31. Thomas GJ, Hartman KA (1973) Raman studies of nucleic acids. 8. Estimation of RNA secondary structure from Raman scattering by phosphate group vibrations. Biochim Biophys Acta 312:311–322
32. Hartman GJ, McDonald-Ordzie PE, Kaper JM, Precott B, Thomas GJJ (1978) Studies of virus structure by laser Raman spectroscopy. 4. Turnip yellow mosaic-virus and capsids. Biochemistry 17:2118–2123
33. Weaver JL, Williams RW (1988) Raman spectroscopy measurement of base stacking in solutions of adenosine, AMP, ATP and oligoadenylate. Biochemistry 27:8899–8903
34. Valdemaras R (2001) Interactions of cyclic AMP and its dibutyryl analogue with model membrane: X-ray diffraction and Raman spectroscopy study using cubic liquid-crystalline phase of monoolein. Biophys Chem 90:75–87

Chapter 13

Small-Angle Neutron Scattering of RNA–Protein Complexes

Audrone Lapinaite, Teresa Carlomagno, and Frank Gabel

Abstract

Small-angle neutron scattering (SANS) provides structural information on biomacromolecules and their complexes in dilute solutions at the nanometer length scale. The overall dimensions, shapes, and interactions can be probed and compared to information obtained by complementary structural biology techniques such as crystallography, NMR, and EM. SANS, in combination with solvent H_2O/D_2O exchange and/or deuteration, is particularly well suited to probe the internal structure of RNA–protein (RNP) complexes since neutrons are more sensitive than X-rays to the difference in scattering length densities of proteins and RNA, with respect to an aqueous solvent. In this book chapter we provide a practical guide on how to carry out SANS experiments on RNP complexes, as well as possibilities of data analysis and interpretation.

Key words Small-angle neutron scattering (SANS), Contrast variation, Deuterium labeling, Structural biology, Macromolecular complex, RNA, Protein, RNP

1 Introduction

Small-angle neutron scattering (SANS) has been used since the 1970s (with the advent of high-flux neutron sources) to study a variety of isolated RNA and RNA–protein (RNP) complexes in solution. Contrary to X-rays, neutrons are scattered by the nuclei of atoms and are sensitive to nuclear scattering length density (SLD) fluctuations in the nanometer range. Therefore, SANS probes the conformational state of solubilized biomacromolecules and their complexes. Pioneering SANS studies on RNP complexes include the internal topology of RNA viruses [1], the solvent-dependent interaction of tRNA with tRNA-synthetase [2, 3], and a mapping of proteins within the bacterial ribosome subunits [4–6] and are based on methods of contrast variation and triangulation, previously developed for X-rays and neutrons [7–9]. Remarkably, these early SANS experiments allowed to determine the relative positions and arrangements of protein and RNA subunits within complexes whose atomic-resolution structures were resolved only decades later. In the mid-1970s, SANS was also applied to DNA–

Véronique Arluison and Frank Wien (eds.), *RNA Spectroscopy: Methods and Protocols*, Methods in Molecular Biology, vol. 2113, https://doi.org/10.1007/978-1-0716-0278-2_13,

protein complexes, in particular in a series of experiments on nucleosome particles, again decades before atomic-resolution models became available [10–12]. A more complete review of early SANS studies can be found elsewhere [13, 14].

The structural information obtained from SANS studies on biomacromolecules in the 1970s and 1980s was mainly analyzed and interpreted in terms of the molecular mass, the radius of gyration, and the interaction and stoichiometry of partners, as well as by comparing experimental SANS curves with those of simple geometrical bodies [15, 16]. The increase of computing power and software development in the 1990s allowed a direct and rapid calculation of SANS (and SAXS, small-angle X-ray scattering) curves from atomic models and their fits against experimental data [17, 18]. Concomitantly, so-called ab initio programs allowed to propose low-resolution envelopes from (multiple) SANS datasets [19, 20]. In recent years, SANS data from RNA–protein complexes have been interpreted by generating docking models of the complex from atomic-resolution structures of building blocks, often in combination with restraints provided by other structural biology techniques, such as NMR [21–24]. Here we provide a short overview of basic SANS theory and of experimental aspects of RNA–protein samples, illustrate the process of structural analysis on the basis of pedagogical examples, and discuss the state-of-the-art and limitations of current modeling approaches as well as the combination of SANS with complementary techniques.

2 Materials

2.1 Neutron Sources and Instruments

Biological SANS experiments are usually carried out with cold neutrons (wavelengths $\lambda \sim 5$–10 Å), which are well suited to probe the dimensions of biomacromolecules (~several nm) and have a favorable interaction potential with the light atomic nuclei of which biological molecules consist. In contrast to X-ray scattering, neutron "home sources" are not available and experiments need to be carried out at large-scale facilities. Two major types of neutron sources are available worldwide: reactors and spallation sources. These produce high-energy neutrons, which need to be subsequently moderated to yield a maximum flux near the desired wavelength in the Ångstrom range [25]. An overview of existing neutron facilities can be found at the following website (https://www.ncnr.nist.gov/nsources.html). Apart from these major large-scale facilities, a new generation of medium-size, high-brilliance neutron sources are currently in development; these sources would provide access to neutrons at a regional level and support local user training [26].

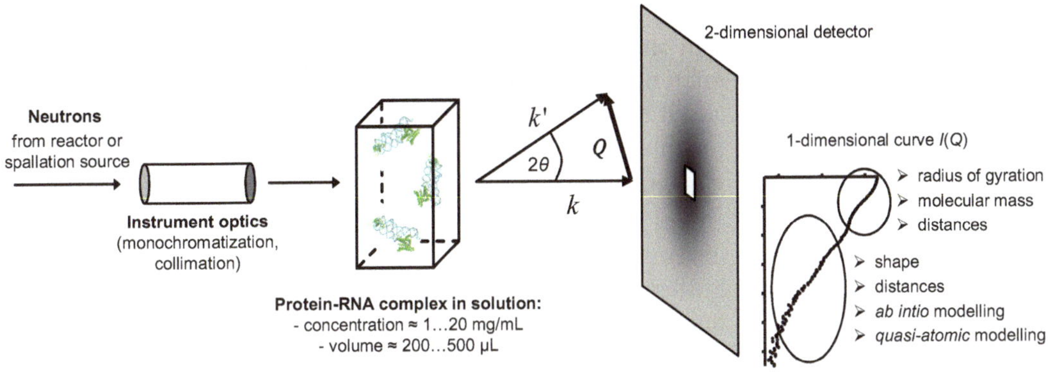

Fig. 1 Schematic representation of a SANS experiment on an RNA–protein complex. The small white rectangle in the center of the detector represents the beam stop, composed of a strongly neutron-absorbing material, and protects the detector from the direct neutron beam

A schematic presentation of a typical SANS instrumental beamline is provided in Fig. 1. An overview of available SANS instruments and their technical characteristics is available under the following link: https://www.ill.eu/users/scientific-groups/large-scale-structures/more/. Access to instrumental beamlines usually requires submitting a proposal (typically twice per year) describing the methods and aims of the scientific project. Following a review process carried out by scientific committees, the beamtime is attributed on the basis of the scientific merit of the project, as well as the expected impact of neutrons in it.

2.2 Sample Checklist for SANS Experiments

1. **RNP in solution**: The reconstituted RNA–protein (RNP) complex should be prepared in the buffer of choice in a concentration range of ~1–20 mg/mL. There are no general restrictions on solvent, apart from the fact that it should not contain exotic elements that absorb neutrons strongly like ^{5}B, ^{64}Gd, and others [27] (*see also* **Notes 1**, **2**, and **3**). A minimum sample volume of 200 μL should be available for the measurement. Sample concentrations should be measured using optical absorbance at 280 and 260 nm with the appropriate calibrations.
2. **RNP buffers**: The reference buffer samples should match the buffer containing the RNP complex as faithfully as possible, in particular regarding the H_2O/D_2O ratio when contrast variation is used. If the sample preparation is finished by a dialysis step, the buffer used during the last dialysis step should be kept as reference for the SANS measurements. If the sample is prepared using size-exclusion chromatography (SEC), an aliquot of the buffer of the chromatography run should be used as reference (*see* **Note 4**). If the sample is concentrated (or if the buffer in the sample is exchanged) using molecular weight

(MW) cutoff centrifugal filters, the flow-through of the last spin should serve as the reference buffer. The volume of the reference buffer(s) should be the same as the sample volume, and it should be measured in an identical cell as the one used for the RNP sample(s).

3. **Sample holder reference**: The sample holder reference (e.g., empty quartz cuvette) needs to be measured in the same instrumental conditions and geometry as the sample(s) and buffer(s) in order to account for its contribution to the scattered signal.
4. **Detector background measurement**: A strongly neutron-absorbing material (boron composite such as BF_4 or cadmium) is usually measured in order to account for the electronic background signal on the detector once the direct neutron beam is blocked. It is usually provided by the neutron facility.
5. **Water (H_2O) reference**: A pure water sample (same volume and geometry as samples and buffers) can be measured in order to calibrate the detector efficiency and put the data on an absolute scale [28]. It has to be measured in the same instrumental setup as the other samples and buffers.

   The measured neutron scattering intensity is the sum of the scattering intensities of all molecules and atoms that comprise the sample (biomolecules, buffer components, water, air, the sample cuvette, etc.) (Fig. 1). Thus, the quality of the structural information obtained for the biomacromolecule in solution depends on the quality of the sample as well as on the accurate correction for background scattering. When preparing samples for SANS experiments, two key points must be considered: the quality (purity and monodispersity) and the quantity of the sample. Good general practices of SAXS/SANS experiments on proteins in solution have been reviewed in detail recently [29–31].

2.3 Detailed Recommendations on Sample Quality

The quality of the structural information depends highly on the quality of the measured sample. The contribution of contaminants to the scattering intensities $I(Q)$ is proportional to the square of the contaminants' volume and to the concentration (*see Basic SANS Theory* below). Thus, one must avoid high-molecular-weight contaminants and aggregates. The most widely used approaches to assess the purity and monodispersity of the RNA–protein complex in the sample are gel electrophoresis (SDS-PAGE and native PAGE), size-exclusion chromatography (SEC), dynamic light scattering/static light scattering (DLS/SLS), and analytical ultracentrifugation (AUC) [32].

SDS-PAGE is the most common approach to monitor the composition and purity of the RNA–protein complex during preparation. The presence of components with different MWs from the molecule(s) of interest indicates the need for further purification. If the purity of the sample is >95%, the complex is considered to be suitable for SANS experiments. However, it should be kept in mind that the presence of a single band (or a set of bands corresponding to the sizes of each component of the complex) in an SDS-PAGE is not a guarantee of monodispersity. Native PAGE (without denaturing agent), SEC, AUC, or SLS/DLS should be used to assess the polydispersity of the sample and estimate the size of the RNP of interest. To detect the RNA in reconstituted RNP complexes, the native PAGE can be stained with ethidium bromide (or other types of stains suitable for the detection of nucleic acids), followed by Coomassie blue staining to visualize the protein components. On native PAGE, the band visualized with the nucleic acid stain and the band visualized with Coomassie blue must overlap, as they should both stem from the intact RNP complex.

SEC (also known as gel filtration) is used both for further purification and to assess the monodispersity of the RNP sample. During SEC, RNP complexes are separated based on their MW and shape. The column resin is chosen according to the MW of the species to be separated; column manufacturers specify which resin is suitable for a given MW range. During SEC, absorbance at both 280 nm (highest for protein) and 260 nm (highest for RNA) should be detected to monitor the elution profile of the reconstituted RNP complex. Generally, a single, symmetric peak (ratio of 260/280 > 0.5) at the expected elution volume indicates the presence of a monodisperse RNP complex. When the RNP complex multimerizes or decays to multiple subcomponents, two scenarios can occur: if the rate of exchange between the intact RNP complex and its multimers (or subcomponents) is slower than the rate of separation through the column, the SEC profile shows several peaks, with large aggregates coming off the column in the void volume. In this case the intact RNP can be isolated by fractionation. However, if the rate of exchange is faster than the rate of separation through the column, the SEC profile shows only one peak at higher or lower elution volume with respect to the intact RNP. The presence of fast exchange between the intact RNP and its multimers can be confirmed by varying the concentration of the injected sample. Since the degree of multimerization depends on concentration, the peak resulting from fast exchange between the desired RNP and its multimers moves towards lower elution volumes (higher percentage of multimers) upon increasing the concentration. In the case of fast exchange, the intact RNP complex cannot be separated from its multimers or subcomponents by fractionation.

The light scattering methods (SLS/DLS) are used to investigate the properties (e.g., MW, size, diffusion) of particles in solution. The scattering intensity is especially sensitive to variations in size and can be exploited to detect aggregates [33]. SLS/DLS requires small amounts of sample (~10 μL) and can be employed to optimize sample conditions by screening multiple conditions (e.g., various buffer compositions, pH, temperature, sample concentrations). Moreover, DLS can be used to guide the storage and handling of RNP samples prior to SANS measurements by assessing the integrity of the samples over time or during freeze-thaw cycles in various buffers and in the presence or absence of stabilizing additives (e.g., glycerol, TCEP, DTT).

The main shortcoming of SLS/DLS methods is their low resolution, which may not allow distinguishing between particles with small differences in shape, such as monomers and dimers. In this case, a high-resolution method such as multiangle laser light scattering (MALLS) may be helpful. MALLS is commonly used in line with SEC (SEC-MALLS) and allows the determination of the absolute mass of particles. This method detects the presence of high-molecular-weight aggregates or subcomponents of the RNP (as long as they can be separated by SEC) and returns the stoichiometry of the complexes.

As an alternative to light scattering, the homogeneity, MW, and stability of biomacromolecules can be characterized at nearly physiological conditions by analytical ultracentrifugation (AUC). In either equilibrium or velocity mode, AUC relies on observing the distribution of biomacromolecules in real time during the application of a centrifugal force. The method is very sensitive and it enables the identification of trace quantities of aggregates, oligomers, and even multiple coexisting complexes of various stoichiometries. Moreover, through analysis of the hydrodynamic properties of the sample, AUC is able to reveal the presence of different conformations of the same particle.

In general, a good SANS sample should be monodisperse and remain such over the length of the measurement (several hours to 1–2 days). This requisite can be verified by the most appropriate method among those described above.

2.4 Detailed Recommendations on Sample Quantity

In order to obtain a good signal-to-noise ratio, it is important to consider the quantity of the sample. The required volume depends on the sample holder and typically ranges from 200 to 500 μL. The concentration depends on the MW of the RNP of interest and is usually in the range of 1–20 mg/mL. For SANS data analysis (see dedicated section below), it is important to estimate the RNP sample concentration as accurately as possible. The estimation of the concentration of large RNP complexes is challenging due to the overlap of the absorption of RNA and proteins. In addition, in the case of very large complexes, light scattering also influences the absorption spectra. We advise to measure the absorptions of RNP samples at 260 and 280 nm (at least three times) and use the following equations [34]:

$$C_{\mathrm{RNA}} = \frac{A_{260} - \left(\frac{\varepsilon_{\mathrm{prot},260}}{\varepsilon_{\mathrm{prot},280}}\right) A_{280}}{\left(\varepsilon_{\mathrm{RNA},260} - \left(\frac{\varepsilon_{\mathrm{prot},260}}{\varepsilon_{\mathrm{prot},280}}\right) \varepsilon_{\mathrm{RNA},280}\right) L} \quad (1a)$$

$$C_{\mathrm{proteins}} = \frac{A_{260} - \left(\frac{\varepsilon_{\mathrm{RNA},260}}{\varepsilon_{\mathrm{RNA},280}}\right) A_{280}}{\left(\varepsilon_{\mathrm{prot},260} - \left(\frac{\varepsilon_{\mathrm{RNA},260}}{\varepsilon_{\mathrm{RNA},280}}\right) \varepsilon_{\mathrm{prot},280}\right) L} \quad (1b)$$

where L is the cell optical path length in cm; A_{260} and A_{280} are the absorptions measured at 260 and 280 nm, respectively; $\varepsilon_{\mathrm{RNA},260}$ is equal to the average extinction coefficient of a single nucleotide at 260 nm (8000 M^{-1} cm^{-1}) multiplied by the number of nucleotides in a single RNP particle; and $\varepsilon_{\mathrm{RNA},280}$ is equal to $\varepsilon_{\mathrm{RNA},260}/2$ (based on the fact that the A_{260}/A_{280} ratio of pure RNA is 2). The extinction coefficient for all the proteins within the complex at 280 nm ($\varepsilon_{\mathrm{prot},280}$) should be calculated according to the sequences of all the proteins, while the $\varepsilon_{\mathrm{prot},260}$ is estimated using the $\varepsilon_{\mathrm{prot},260}/\varepsilon_{\mathrm{prot},280}$ ratio of 0.6 for pure proteins.

2.5 Sample Preparation and Deuteration

Using an appropriate H_2O/D_2O ratio within the solvent, the scattering signal of proteins or RNA can be masked (*see Basic SANS Theory*), so that the resulting SANS curve is dominated by either the protein or the RNA, within the assembled RNP complex. For unlabeled (^1H) proteins, the contrast match point (CMP) is between 40 and 45% (vol/vol) of D_2O in solution, and for unlabeled (^1H) RNA at around 70% (Fig. 2). The precise values depend on the exact amino acid/nucleotide composition, the number of exchangeable protons, and the specific volumes.

The desired H_2O/D_2O ratio can be achieved via RNP sample dialysis, or buffer exchange by SEC. When using dialysis, the RNP sample is split into two parts: one-half is dialyzed overnight (or at least for 8 h) against a buffer containing 0% (vol/vol) of D_2O, while the other half is dialyzed against a buffer containing 100% (vol/vol) of D_2O. After the dialysis, the two samples are mixed in the correct proportion to obtain the required H_2O/D_2O ratio. When using SEC, the gel filtration column has to be equilibrated with at least two column volumes of the buffer that contains the appropriate H_2O/D_2O ratio. It is beneficial to use SEC because it also allows monitoring the integrity of the RNP sample during buffer exchange.

SANS data recorded from RNP samples where the scattering of proteins is matched, and only the RNA component contributes to the scattering curve, often has a low signal-to-noise ratio. In this case, the quality of RNA scattering data can be improved either by increasing the sample concentration or by increasing the contrast through deuteration of the RNA (Fig. 2). The RNA can be uniformly deuterated by in vitro transcription using T7 RNA polymerase, ^2H-NTPs, and synthetic DNA template [35].

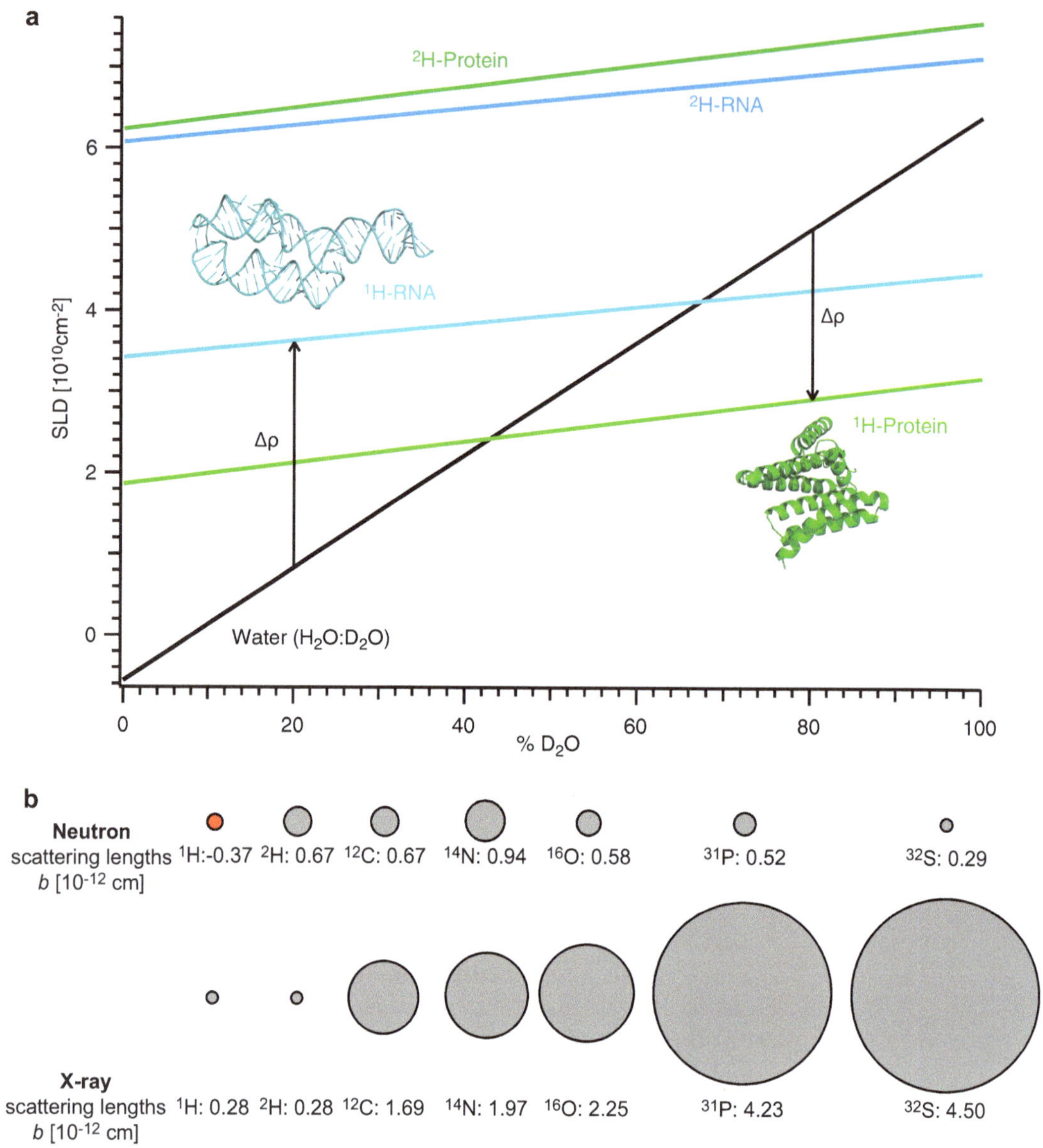

Fig. 2 Neutron scattering length densities and X-ray and neutron scattering lengths of selected atoms. (**a**) Scattering length densities (SLDs) for hydrogenated (^{1}H) and perdeuterated (^{2}H), i.e., fully deuterated, proteins and RNA. Contrasts ($\Delta\rho$) are shown for a ^{1}H-RNA in 20% D_2O and for a ^{1}H-protein in 80% D_2O. Please note that the sign of contrast (with respect to solvent) can be positive or negative. (**b**) The circles illustrate the relative size of the X-ray and neutron cross sections (i.e., a measure of the size of atoms as they appear to the respective radiation), calculated from the respective scattering lengths of light atoms composing proteins and RNAs. The red color of ^{1}H indicates a change of sign of the scattering length with respect to other atoms or isotopes and is the basis of solvent contrast variation and deuterium labeling in biological SANS

In the case of multiprotein RNP complexes, where the CMPs of all proteins are very close, an artificial contrast between proteins can be achieved *via* selective deuteration (^{2}H-labeling) of individual proteins. Uniform deuteration of recombinant proteins is achieved using the *Escherichia coli* bacterial expression host cultured in M9 medium containing D_2O and a deuterated carbon source (usually glucose). *E. coli* can tolerate deuterated medium, despite at the expense of reduced growth rates and lower protein yields. To optimize the amount of protein produced by a definite amount of cell culture, the bacteria are adapted to grow in deuterated medium by repeated subculturing in progressively higher amounts of D_2O. The resulting D_2O-tolerant cells are then used to express the protein of interest in M9 medium containing 99% D_2O and ^{2}H-glucose (all M9 components must be dissolved in 100% D_2O) [36]. The uniformly deuterated protein expressed in these conditions is purified following the protocol optimized for the unlabeled protein, using H_2O containing buffers. The deuteration levels can be confirmed by ^{1}H-NMR or mass spectrometry (e.g., LC/electrospray ionization-TOF-MS).

It has been reported that deuteration may affect the protein biochemical characteristics. It may decrease the hydrophobicity of the molecule and thus change the optimal storage and handling conditions [37]. Moreover, the presence of D_2O in the solvent promotes protein aggregation and may change the protein-ligand interactions [38]. Thus, when preparing RNP samples for contrast matching and variation experiments, one must investigate the integrity of the deuterated RNP complex. First of all, we advise to perform a small-scale ^{2}H-protein expression and purification to assess its solubility and stability and to optimize purification, storage, and handling conditions using the previously described SDS-PAGE, SEC, DLS/SLS, and AUC approaches. Next, it is important to assess whether the deuteration of subunits or the presence of D_2O in the buffer affects the dissociation constant (K_d) of the RNP complex. Electrophoretic mobility shift assay (EMSA), double-filter binding assay, and isothermal titration calorimetry (ITC) are the most popular methods to determine the K_d of RNA-protein interactions [39–41]. We advise to use either EMSA or filter binding assay with radioactively (^{32}P) labeled RNA, because these approaches are very sensitive and do not require large amounts of sample. It is important to keep in mind that, even if there is no significant effect on the K_d of the RNP complex, the k_{on} and k_{off} rates may be affected by the deuteration of subunits and/or presence of D_2O in the buffer. Thus, it may be necessary to estimate the k_{on} and k_{off} of the RNP complex in these specific conditions. The off rate can be easily determined via EMSA-based competition binding assay, where the RNP complex, reconstituted using the radioactively labeled RNA, is incubated over time in the presence of an excess of unlabeled RNA. Finally, using SDS-PAGE, SEC, DLS/SLS, or AUC, one must ensure that the deuterated RNP sample in the D_2O containing buffer is monodisperse and stable throughout the data acquisition.

2.6 Sample for Ligand Titration SANS

The concentration for SANS samples (1–20 mg/mL) is above the K_d of most ligand binding events ($K_d < 10$ μM). Thus, upon ligand binding the complexes exist in a single, fully bound state (holo state). Under these conditions, SANS can be useful to investigate the conformational changes of complexes upon interaction with ligands [42]. The measurements can be performed on two independently prepared apo and holo RNP samples or on a single sample by titrating the ligand to the apo RNP complex after SANS data acquisition. If the second setup is chosen, there are several points to consider: first, the ligand must be dissolved in the same buffer as the sample and reference buffer. Second, the ligand must be dissolved at as high concentration as possible, to be able to add only small volumes (~1–10 μL) to the initial solution and avoid excessive dilution. If the affinity of the ligand for the RNP is moderate ($K_d > 1$ μM), the ligand must be added in excess, to ensure that the majority of the RNP complex is in holo state. In this case one must consider the contribution of the free ligand to the scattering curve. In alternative, one can use an appropriate H_2O/D_2O ratio within the sample solvent to mask the ligand scattering.

3 Methods

3.1 Schematic Representation of a SANS Experiment

A SANS experiment records the scattered neutrons from a biological sample in solution (Fig. 1). To this end, a well-defined incoming beam is prepared by a series of monochromators and optical elements that select neutrons of a specific speed (i.e., wavelength λ) and direction (wave vector $\vec{k}$). After interaction with the sample, the neutrons are scattered in a direction $\vec{k'}$. The diffraction pattern is recorded on a neutron-sensitive detector as a function of the momentum transfer $Q = \left|\vec{k'} - \vec{k}\right| = (4\pi/\lambda)\sin\theta$ (where 2θ is the scattering angle). For isotropic solutions (i.e., the orientation of one biomacromolecule does not influence the orientation of others), the two-dimensional scattering pattern on the detector is symmetric around the direct beam. Given the typical fluxes (10^6–10^8 n cm^{-2} s^{-1}) and wavelengths (some Å) of cold neutrons, sample-detector distances (≈1–40 m), detector dimensions ($\approx 1 \times 1$ m^2), and exposure times (minutes to hours), the sensitivity of SANS experiments from dilute (<1%v/v) biomacromolecular solutions ranges from a few kDa to several MDa and the structural length scales from a few Å to several 1000 Å.

3.2 Instrumental Setup for RNA–Protein Complexes

The instrumental setup (in particular the sample-detector distance) depends on the linear dimensions of the RNP complex to be measured. The resulting Q-ranges need to cover both low angles in order to seize the accurate maximum dimension of the complex, as well as higher angles for accurate solvent (buffer) subtraction. Often, both requirements cannot be fulfilled with a single sample-detector setup but require at least two. The lowest Q-value, Q_{min}, which is limited by the size of the beam stop (Fig. 1), has to allow a proper extraction of the radius of gyration R_g via the Guinier approximation (*see Basic Data Analysis*). This requires fulfilling the condition $Q_{min} \approx 0.3/R_g$ [43]. R_g can be estimated from PDB models of RNP complexes of comparable size by using programs such as CRYSON [18]. Alternatively, if the maximum dimension D_{max} of the complex is known, e.g., from (Cryo-)EM, a first rough estimate of the radius of gyration can be obtained from $R_g \approx 0.3D_{max}$. For example, a small RNP complex (PAZ domain in complex with a nine nucleotide long RNA, PDB ID 1si3, 20 kDa) has an R_g of 15.2 Å (in H_2O) and requires $Q_{min} < 0.02$ Å^{-1} for the proper application of the Guinier approximation. On the other hand, a very large complex (large ribosomal subunit of *Haloarcula marismortui*, PDB ID 1ffk, 1.6 MDa) with an $R_g \approx 66$ Å (in H_2O) requires a $Q_{min} < 0.005$ Å^{-1}. The upper limit of the Q-range, Q_{max}, should ensure a proper analysis of short-distance information and accurate solvent subtraction. In practice, the decrease of signal/noise at higher angles limits Q_{max} to 0.5–0.7 Å^{-1} on most SANS instruments.

3.3 Basic SANS Theory, Solvent Contrast Variation, and Deuteration

In the case of an ideal (i.e., isotropic and noninteracting) solution of N macromolecules, the two-dimensional scattering pattern is centrosymmetric around the direct beam and the scattered intensity I can be described as a function of the modulus of the momentum transfer $\vec{Q}$ (Fig. 1) [15, 43, 44]:

$$I(Q) \propto N \left| \int_V \Delta\rho\left(\vec{r}\right) e^{i\vec{Q}\cdot\vec{r}} \mathrm{d}V \right|^2 \tag{2}$$

The integral runs over the entire (solvent-excluded) volume V of a particle with $\vec{r}$ being a vector pointing to a volume element dV of scattering contrast $\Delta\rho\left(\vec{r}\right)$, within V:

$$\Delta\rho\left(\vec{r}\right) = \frac{\sum_i b}{\mathrm{d}V} - \rho_S \tag{3}$$

In the case of neutron scattering, b_i are the (coherent) scattering lengths (Fig. 2b) of the atomic nuclei within a volume element

dV of the particle, and ρ_S is the average scattering length density (SLD) of the solvent, e.g., of a certain H_2O/D_2O ratio (Fig. 2a). Values b_i of individual atoms can be retrieved from literature [13]. Figure 2b lists the ones of the most common atoms found in proteins and RNA and compares them to the respective values for X-rays. Average contrasts $\Delta\rho$ of proteins or RNA at a given H_2O/D_2O ratio (Fig. 2a) can be calculated by summing up scattering lengths of all protein residues [13] (or RNA nucleotides [45]) and by dividing by the solvent-excluded volume V of the molecules. Importantly, exchangeable hydrogens need to be taken into account in SANS. By using appropriate software tools, it is possible to calculate the contrast of RNP complexes directly from their primary sequences [46, 47].

Figure 2a illustrates that deuteration (exchange of hydrogen, ^{1}H, against deuterium, ^{2}H) in proteins or RNA modifies the contrast. The artificial contrast conditions created by deuteration allow distinguishing two chemically similar macromolecules by SANS, e.g., a ^{1}H- from a ^{2}H-protein (or two RNAs) within multi-RNA–protein complexes [42, 48]. Several methods have been developed to biosynthetically deuterate proteins or RNA (see *Sample Preparation and Deuteration for Contrast Variation Experiments*). The interested reader can find more detailed information in recent reviews [36, 49] and references therein.

3.4 Raw Data Reduction

The raw data in a SANS experiment of solubilized, isotropic biomacromolecular complexes are two-dimensional, pixelated centrosymmetric intensity images, recorded by a neutron detector system (Fig. 1). Both sample and buffer images need to be corrected for detector background signal. The background signal (originating from electronic, terrestrial, and cosmic as well as parasitic radiation from the direct neutron beam) is determined using a strongly neutron-absorbing sample (e.g., boron or cadmium) that blocks the direct neutron beam. The empty sample holder (often a quartz cuvette such as those used for optical density measurements) and a water (H_2O) sample are measured to correct for either sample holder signal and detector efficiency or absolute scaling, respectively.

All samples and references are also measured in transmission mode to determine the respective weight with which they contribute to the overall signal. The beam center is usually determined from the transmission experiment of the empty beam. Several software programs have been developed at the respective neutron centers to reduce the raw data. They allow defining appropriate masks to cut away parasitic scattering close to the beam center and defective pixels or artifacts at the edges of the detector. The two-dimensional scattering patterns are reduced to one-dimensional scattering curves $I(Q)$, without a loss of information for isotropically oriented protein–RNA complexes in solution, by integrating over concentric rings around the beam center.

In a final step, the one-dimensional buffer intensities need to be subtracted from the respective sample intensities using available software [50]. Due to a highly variable incoherent background signal [51], it may be necessary to align the buffer and sample levels at the highest available Q-values, prior to the subtraction. A scaling factor >2% needed for alignment can be indicative of differences in sample and buffer preparations, possibly due to slightly different H_2O/D_2O ratios; in this case, the respective samples should be tested again, e.g., by comparing their transmission values, which are very sensitive to the H_2O/D_2O ratio [28].

3.5 Basic Data Analysis and Model-Free Parameters

For ideal (i.e., isotropic orientation and no interparticle effects; *see also* comments in the *Notes* section) and monodisperse (no mixture of oligomeric states or excess of free partners) solutions, the one-dimensional SANS data can be directly related to the (conformationally averaged) structure of the RNP complex.

The two most important (and model-free) structural parameters are the radius of gyration, R_g, and the intensity scattered in the forward direction, $I(0)$. Both can be extracted by a so-called Guinier approximation, i.e., by a linear fit in a ln $I(Q)$ vs. Q^2 plot [52]:

$$\ln\left[I(Q)\right] = \ln\left[I(0)\right] - 1/3R_g^2Q^2 \quad (4)$$

The Guinier approximation is valid up to $Q_{max}R_g < 1.0...1.3$ [43, 44]. As discussed above, the lower limit of the fit range ($Q_{min}R_g$) should be <0.3.

R_g is a measure for the normalized distribution of scattering contrast volume elements around the center of scattering mass of the solubilized complex:

$$R_g^2 = \int \Delta\rho\left(\vec{r}\right) r^2 \mathrm{d}V \Big/ \int \Delta\rho\left(\vec{r}\right) \mathrm{d}V \quad (5)$$

r is the distance between a scattering element (e.g., protein residue or RNA nucleotide) and the center of contrast of the complex. It is important to note that R_g integrates over all volume elements that display a different SLD than the bulk solvent and can therefore account for solvent/solute elements in the vicinity of the surface of the complex, if they display a different scattering length density than the bulk [53, 54].

The forward scattered $I(0)$ intensity can be related to the volume $V = V_{prot} + V_{RNA}$ of a given RNA–protein complex at a concentration C (in Mol cm^{-3}), if the SANS data are recorded in absolute units (cm^{-1}):

$$I(0) = C \cdot N_A \cdot \left[\Delta\rho_{prot} \cdot V_{prot} + \Delta\rho_{RNA} \cdot V_{RNA}\right]^2 \quad (6)$$

N_A is the Avogadro constant, V_{prot} and V_{RNA} are the solvent-excluded volumes of the protein and RNA molecules (in cm^3) that compose the complex, and $\Delta\rho_{prot}$ and $\Delta\rho_{RNA}$ are their respective scattering length densities (SLDs), in cm^{-2}. Both solvent-excluded volumes and SLDs of proteins and RNA can be calculated from literature, as a function of their primary sequence [45, 55]. Equation (6) represents a quality check, with respect to whether the RNP complex is in the expected oligomeric state or there are excess partners in solution. Given the accuracy of the solvent-excluded volumes, SLDs, and measured concentrations, a relative deviation of 15–20%, with respect to the theoretical $I(0)$, can be expected for a well-assembled and monodisperse complex. Deviations >20% may indicate either systematic errors in the measured concentration or problems with sample monodispersity and/or stoichiometry.

3.6 Sophisticated Data Analysis: Low-Resolution Ab Initio Envelopes

Several approaches have been developed to interpret SANS data beyond model-free parameters (molecular weight and radius of gyration). If atomic models are available from other techniques, such as crystallography, NMR, cryo-EM, or even homology modeling, for the same RNP complex, the SANS data calculated from this model can be compared and scored against the experimental SANS data [18].

If no atomic-resolution model is available, shapes of proteins and RNAs within the assembled complex can be reconstructed using ab initio shape analysis [20]. We illustrate the capacity and the limits of this approach by employing the program MONSA [19] on a pedagogical example (Fig. 3): using back-calculated SANS curves at multiple contrasts, endowed with errors, we retrieve the shapes of the proteins SRP19 and SRP68, as well as the SRP RNA, within the assembled signal recognition particle (SRP) complex [56] (PDB ID 4p3e), and compare them to the original atomic model from which the SANS curves were generated.

If the SANS data originate from a monodisperse, properly assembled complex (Fig. 3a), the ab initio approach retrieves accurately the topology of the RNP: the densities and distances of the two protein partners, as well as the elongated shape of the RNA molecule, generated from the simulated SANS data, are in good agreement with the original atomic model. However, the orientation of the proteins, as well as the RNA secondary structure elements, remains undetermined, even though the loose RNA density suggests an open helical structure. In the case of contamination, simulated by addition of the SANS signal of 50% free SRP68 to the original SANS data, the reconstructed ab initio shapes contain artifacts (Fig. 3b).

This simulated example underlines the paramount importance of monodispersity in SANS samples. In our example, the excess of protein SRP68 leads to an erroneous model, i.e., an additional protein density (Fig. 3b, left). Monodispersity should be verified

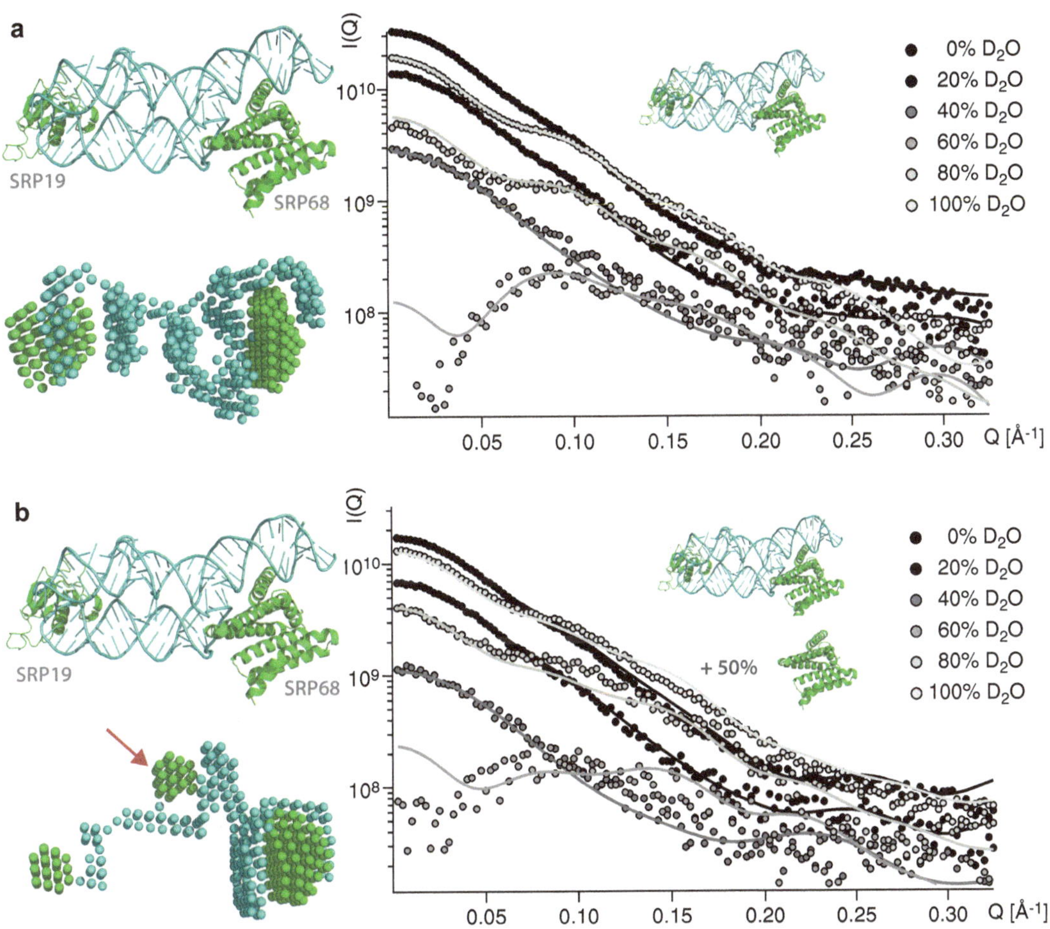

Fig. 3 MONSA ab initio models of the SRP complex as a function of data simulated for various experimental conditions (cyan, RNA densities; green, protein densities). (**a**) SANS data generated from a monodisperse solution of the complex and ab initio shapes derived from a typical MONSA run. (**b**) SANS data generated from a solution of the SRP complex and the free protein SRP68 in a 2:1 stoichiometric ratio and ab initio shapes derived from a typical MONSA run. Note the appearance of artifacts in the ab initio model in the presence of excess protein SRP68, indicated by a red arrow. Filled circles on the right represent the calculated SANS curves at multiple contrasts, including artificially added noise. Continuous lines denote the backfits of the ab initio models to the calculated data at each contrast

before the SANS experiment and importantly for all contrast conditions (i.e., H_2O/D_2O ratios), as explained in the *Sample Quality* section above.

In the past two decades, the application of small-angle scattering (SAS) to biological samples has seen a revival. The combination of NMR and SAXS is especially popular, probably due to the similarity of the sample conditions and to the complementarity of the

3.7 Sophisticated Data Analysis II: Combined Approaches Using SANS and NMR Data

structural information provided by the two techniques. NMR distance restraints cover a range from a few Ångstroms (nuclear Overhauser effect data [NOE]) to 20–30 Å (paramagnetic relaxation enhancement data [PRE]). Depending on the size of the complex, this structural information reports on local interactions (NOEs) or at most on the relative orientation of domains (PREs). SAS data inform on the global shape of the particle in solution up to a range of 1000 Å, while it rapidly loses sensitivity for detailed structural features (<15 Å). Thus, NMR and SAS can be considered fully complementary. SAXS and SANS data are commonly used to validate structures calculated by NMR, particularly in terms of domain orientation in multidomain proteins, nucleic acids, and complexes thereof [24, 57, 58]. Other more sophisticated approaches use SAS data readily during structure calculation, either by refinement against the scattering curve [59] or by restraining the conformational space available to the particle to the shape envelope calculated ab initio from the SAS data [42, 60].

A novel application of SANS in combination with NMR has been developed in our laboratories to study the structure of a large RNP complex, the box C/D RNP from Archaea [42], consisting of 12 proteins and either 6 or 2 RNAs, in its holo and apo forms, respectively (Fig. 4a). This RNP is responsible for methylation of ribosomal RNA (rRNA) at the 2′-O-ribose. The substrate sequence in the rRNA is recognized by base-pairing with guide sequences comprised in the guide RNA molecule (gRNA). The gRNA also serves as binding platform for proteins L7Ae, Nop5, and fibrillarin, among which fibrillarin catalyzes the methylation reaction. Each particle contains two copies of gRNA, four substrate RNAs (only in the holo form), and four copies of each of the three proteins, making up a MW of 390 kDa. At this molecular size, NMR focuses on the observation of methyl groups of proteins, while all other NMR signals are too broad to be detected. Methyl chemical shift perturbation (CSP) data allow detection of the interaction interfaces between proteins, while PRE effects measured on the methyl groups are used to determine interprotein distances.

NMR does not provide any structural information on the relative position of RNA secondary structure elements, as all RNA signals are broad beyond detection in this MW range. In addition, for this complex, the NMR resonances of the four copies of each protein mostly overlap, making it impossible to distinguish one copy from another. Both shortcomings of NMR could be compensated by SANS applied with contrast variation. Firstly, a SANS curve measured for the box C/D RNP containing ^{2}H-gRNA in a buffer of 42%:58% D_2O/H_2O, where the scattering length density of the proteins is matched by the solvent (Fig. 2a), was used to calculate a low-resolution RNA shape (Fig. 4b). This, together with information on RNA secondary structure elements derived from biochemical data and from the structure of subcomplexes, allowed us to

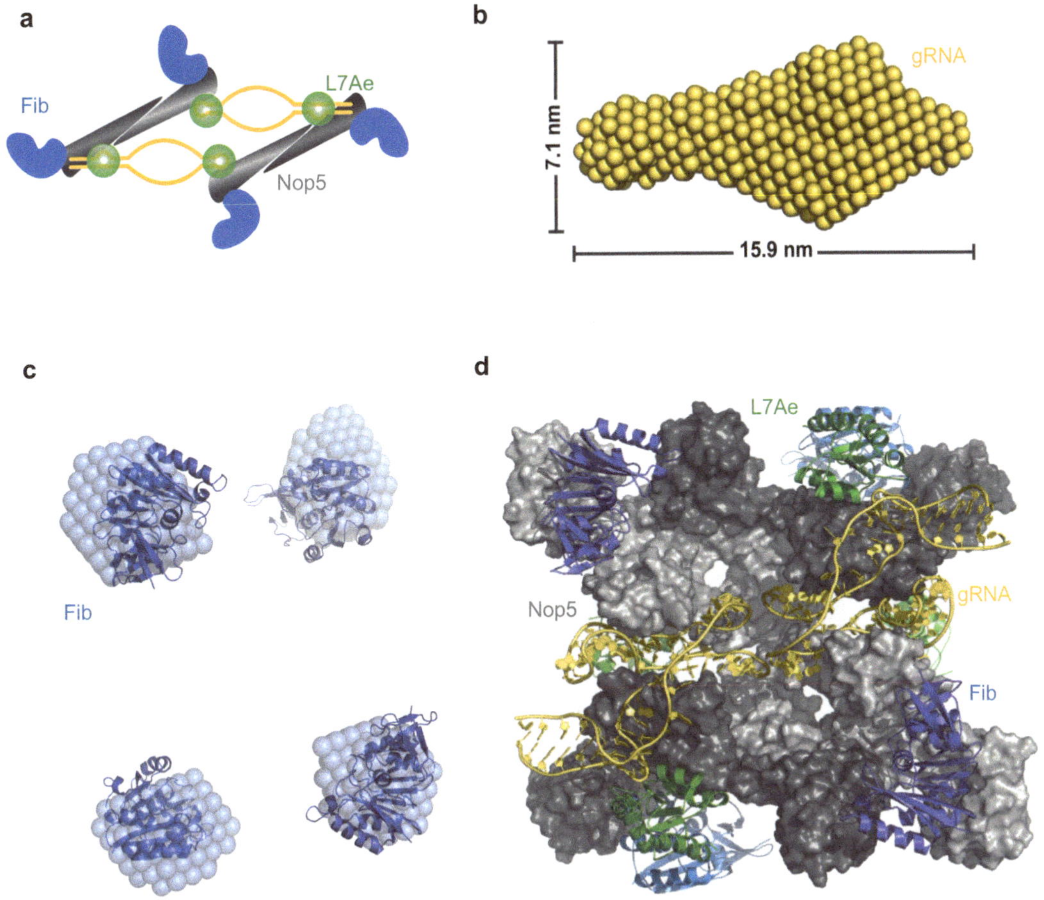

Fig. 4 SANS in contrast variation applied to the 390 kDa box C/D RNP complex. (**a**) Schematic representation of the apo form of the box C/D RNP consisting of two copies of the gRNA and four copies of each protein L7Ae (green), Nop5 (grey), and fibrillarin (Fib, blue). (**b**) Ab initio envelope of the gRNA from a single contrast, illustrating a connected and extended topology. (**c**) Ab initio envelope calculated with MONSA from multiple SANS curves from differentially labeled box C/D RNPs. The fibrillarin molecules from the final model are represented in ribbon and overlap with the envelope. (**d**) Structure of the apo box C/D RNP calculated by a combination of NMR and SANS data as described in the text [42]

calculate the structure of the gRNA in the complex. The SANS-derived RNA envelope was instrumental in this task. Secondly, the four copies of each proteins could be nicely distinguished in ab initio MONSA [20] envelopes calculated from SANS curves measured for the Box C/D RNP containing only one deuterated protein species and dissolved in a 42%:58% D_2O/H_2O buffer (Fig. 4b). This is illustrated particularly well for fibrillarin, whose four copies in the apo form of the RNP (without the four substrate RNA molecules) are well separated in space. The ab initio MONSA envelope, integrating several SANS curves from differentially ^2H-labeled Box C/D RNP complexes at multiple solvent (D_2O/H_2O)

contrasts, shows the presence of four distinct protein densities (Fig. 4c), confirming the stoichiometry of the complex and allowing an approximate estimation of the distances between the fibrillarin copies.

We then developed a structure calculation protocol that actively uses both NMR and SANS restraints to calculate the structure of the complex in two steps. The high-resolution structures of subcomponents of the complex, as well as RNA and protein secondary structure elements, were docked together under the guidance of the experimental data. In the first calculation step, the RNA envelope and an approximate estimate of the distances between copies of L7Ae and fibrillarin, derived from ab initio modeling, were used together with the NMR restraints to restrict the conformational space of the search. Subsequently, the fit of the models to one SAXS and seven SANS curves, measured in different contrast conditions, was used to select a subset of models that was subjected to the structure refinement step. At the end of the second step, the final structure ensemble was selected again based on the fit of the individual models to the eight SAS curves.

The combination of the NMR and SAS data, together with the structure calculation protocol developed to incorporate both data types early on in the calculation, allowed to solve the structure of a 390 kDa RNP in solution (Fig. 4d). This is a remarkable achievement that demonstrates the enormous potential of SANS, in the implementation with contrast variation, to determine the structure of multicomponent RNP complexes in solution. The intrinsic low resolution of the structural information provided by SANS is alleviated by: (1) use of the high-resolution structure of stable protein domains and RNA secondary structure elements in the docking protocol, (2) use of distance restraints derived by NMR, and (3) measurement of several SANS curves with contrast variation. We believe that this combination of NMR and SANS, despite being laborious, opens the way to study various RNP complexes that do not crystallize or are too fragile to be imaged by electron microscopy [61].

3.8 Concluding Remarks and Perspectives

In this chapter we have presented the application of SANS to RNP complexes, and we have given recommendations on how to prepare high-quality RNP samples for SANS measurements, how to decide on the instrumental setup, and how to process the data. For RNP complexes, we have put a special emphasis on the contrast variation approach, which allows visualizing selected combinations of RNA and protein components in the context of the intact RNP. A set of several SANS curve, measured in contrast variation with different combinations of ^{1}H- and ^{2}H-components, provides enough data to determine the relative positions of these components in the RNP. We have illustrated this concept on the example of the 390 kDa box

C/D RNP, whose structure was obtained through the joint use of NMR and SANS data.

In principle, the concept of contrast variation can be applied also to structured elements of the same molecule. ^{2}H-labeled structural elements, as part of a molecule, can be studied by the SANS contrast variation approach using a solvent that matches the scattering length density of the ^{1}H-elements. This can be useful to determine the distance between domains in multi (>2)-domain proteins or between stems in RNA molecules. Two conditions are necessary for the approach to be successful: first, each ^{2}H-labeled element must have a significant scattering mass (>~10 nucleotides for an RNA); second, the ^{2}H-labeled elements must not be randomly distributed in the RNA, i.e., their overall distribution must be distinguishable from the global RNA envelope. A number of methods are available to achieve segmental deuterium labeling of both proteins [62–65] and RNAs [66–68].

X-ray crystallography, NMR, EM, mass spectrometry (MS), SAS, fluorescence (FRET), and electron paramagnetic resonance spectroscopy (EPR) provide structural data on biomolecules at different levels of resolution and completeness. Despite the availability of so many experimental approaches, the determination of the structure of large nucleic acids and their complexes with proteins remains a challenging task. X-ray crystallography and EM may fail with highly flexible molecules such as RNA and RNPs. NMR can cope very well with dynamic processes; however, its applicability as a stand-alone structure determination method is limited to particles of ca. 50 kDa, due to the rapid decrease of the quality of the NMR signal upon increasing molecular weight. SAS, FRET, MS, and EPR deliver either low-resolution or incomplete structural information, which is insufficient for de novo structure determination. The rapid discovery of large and dynamic RNP complexes with paramount biological relevance demands a rapid answer to the challenges posed by the structure determination of RNPs. We believe that integrative structure biology methods may provide such an answer, especially if applied in solution, where the RNP complex preserves its functional dynamics. The combination of several data types allows overcoming the limitations of each technique while providing a reciprocal validation of the datasets. In this context, SANS in combination with techniques that are able to measure interatomic distances (NMR, EPR, FRET, MS) is destined to play a prominent role.

4 Notes

This section contains a list of typical problems that may be encountered during a SANS experiment and indications on how to avoid them or minimize their impact.

1. *Water condensation on quartz cuvettes*: A potential source of error can be the condensation of water vapor on quartz cuvettes. This phenomenon can happen when samples (or cuvettes) are transferred from cold temperature (ice or cold room) to room temperature. It can falsify SANS results, in particular for D_2O samples, by changing transmission values and the incoherent background. Therefore, the quartz cuvettes should be sealed (e.g., with parafilm) in order to minimize sample evaporation and condensation inside the sample holder. Condensed water on the outside surface of the sample holder should be carefully wiped off, and a sufficient time for temperature equilibration should be allowed prior to the SANS experiments.
2. *Oligomerization/aggregation in solvents containing D_2O*: The physicochemical properties of D_2O differ from those of H_2O, reinforcing the hydrophobic interaction between macromolecules [69]. While the use of heavy water is popular to increase contrast (for hydrogenated proteins) and to reduce the incoherent background, it has been experimentally observed that it increases the tendency of biomacromolecules to oligomerize, aggregate, or precipitate in solution. While very big precipitates (all linear dimensions large with respect to the RNP complex of interest) may be tolerated, lower oligomeric states impede a proper data analysis and should be avoided. Therefore, the oligomeric state of the RNP complex should be probed prior to the SANS experiments with biophysical techniques [29] in H_2O and in D_2O.
3. *Repulsive interparticle interactions*: Highly charged biomolecules such as RNA repel each other in solution, if not properly screened by an appropriate buffer. The effects of repulsive interactions can be observed in SANS experiments at low angles and influence the values of radii of gyration and molecular weight [31]. Usually, the effects tend to increase with increasing concentration and should be avoided. The presence of interparticle repulsion can be probed by a concentration series, typically in the range of 1 to 20 mg/mL. If all curves superpose (after normalization by concentration), interaction effects can in general be excluded. If interaction effects are present, two strategies can be explored to reduce them: (a) lowering the sample concentration as far as possible in order to minimize the effects and merging a low-concentration SANS curve at low angles with a high-concentration curve at higher angles [31, 53] and (b) changing the buffer properties (e.g., ionic strength, pH) in order to eliminate the long-range electrostatic effects.
4. *Solvent subtraction*: Even in a carefully prepared SANS experiment, the scattering intensity of the sample might not align

with the corresponding buffer at large angles. Since the background level at high angles originates mainly from incoherent scattering of hydrogen atoms, this discrepancy is in most cases due to slightly different levels of H_2O/D_2O in the sample and buffer. A minor scaling of the buffer level (< 2%) can remedy this situation and can be tolerated. If significantly higher discrepancies are observed, the transmission values (which depend strongly on the H_2O/D_2O ratio) of the samples and buffers should be compared.

Concluding remarks: At all stages of a SANS experiment, it is advisable to discuss with a local contact at the neutron facility (instrument responsible or experienced user) that can help out with (1) successful proposal writing and submission, (2) strategies of sample preparation, (3) measuring the SANS samples, and (4) data analysis and interpretation, in particular assessing the uniqueness of a structural model.

Acknowledgments

FG acknowledges funding from the French Agence Nationale de la Recherche (grant ANR-15-CE11-0026-01 "PROTstretch"). TC acknowledges funding from the DFG (grants CA294/3-2 and CA294/11-1).

References

1. Jacrot B, Chauvin C, Witz J (1977) Comparative neutron small-angle scattering study of small spherical RNA viruses. Nature 266:417–421
2. Giegé R, Jacrot B, Moras D, Thierry J-C, Zaccai G (1977) A neutron investigation of yeast valyl-tRNA synthetase interaction with tRNAs. Nucleic Acids Res 7(7):2421–2427
3. Dessen P, Blanquet S, Zaccai G, Jacrot B (1978) Antico-operative binding of initiator transfer RNAMet to methionyl-transfer RNA synthetase from Escherichia coli: neutron scattering studies. J Mol Biol 126:293–313
4. Hoppe W, May R, Stoeckel P, Lorenz S, Erdmann VA, Wittmann HG, Crespi HL, Katz JJ, Ibel K (1976) Neutron scattering measurements with the label triangulation method on the 50S subunit of *E. coli* ribosomes. In: Schoenborn BP (ed) Neutron scattering for the analysis of biological structures, Brookhaven, 1975. Brookhaven symposia in biology. Brookhaven National Laboratory, Upton, pp IV38–IV48
5. Stuhrmann HB, Haas J, Ibel K, Wolf B, Koch MH, Parfait R, Crichton RR (1976) New low resolution model for 50S subunit of Escherichia coli ribosomes. Proc Natl Acad Sci U S A 73(7):2379–2383
6. Engelman DM, Moore BP, Schoenborn BP (1975) Neutron scattering measurements of separation and shape of proteins in 30S ribosomal subunit of Escherichia coli: S2-S5, S5-S8, S3-S7. Proc Natl Acad Sci U S A 72 (10):3888–3892
7. Engelman DM, Moore PB (1972) A new method for the determination of biological quaternary structure by neutron scattering. Proc Natl Acad Sci USA 69:1997–1999
8. Hoppe W (1972) A new X-ray method for the determination of the quaternary structure of protein complexes. Isr J Chem 10:321–333
9. Stuhrmann HB, Kirste RG (1965) Elimination der intrapartikulaeren Untergrundstreuung bei der Roentgenkleinwinkelstreuung an kompakten Teilen (Proteinen). Z Phys Chem 46 (3–4):247–250
10. Pardon JF, Worcester DL, Wooley JC, Tatchell K, Van Holde KE, Richards BM (1975) Low-angle neutron scattering from

chromatin subunit particles. Nucleic Acids Res 2(11):2163–2176

11. Baldwin JP, Boseley PG, Bradbury EM, Ibel K (1975) The subunit structure of the eukaryotic chromosome. Nature 253:245–249
12. Bram S, Butler-Browne G, Baudy P, Ibel K (1975) Quaternary structure of chromatin. Proc Natl Acad Sci U S A 72(3):1043–1045
13. Jacrot B (1976) The study of biological structures by neurtron scattering from solution. Rep Prog Phys 39(10):911–953
14. Zaccai G, Jacrot B (1983) Small angle neutron scattering. Annu Rev Biophys Biomol Struct 12:139–157
15. Glatter O, Kratky O (1982) Small angle X-ray scattering. Academic Press, London
16. Kratky O (1972) Recent advances and applications of diffuse X-ray small-angle scattering on biopolymers in dilute solutions. Q Rev Biophys 5(4):481–537
17. Svergun DI, Barberato C, Koch MHJ (1995) CRYSOL—a program to evaluate X-ray solution scattering of biological macromolecules from atomic coordinates. J Appl Crystallogr 28:768–773
18. Svergun DI, Richard S, Koch MHJ, Sayers Z, Kuprin S, Zaccai G (1998) Protein hydration in solution: experimental observation by X-ray and neutron scattering. Proc Natl Acad Sci USA 95(5):2267–2272
19. Svergun DI, Nierhaus KH (2000) A map of protein-rRNA distribution in the 70S Escherichia coli ribosome. J Biol Chem 275 (19):14432–14439
20. Svergun DI (1999) Restoring low resolution structure of biological macromolecules from solution scattering using simulated annealing. Biophys J 76(6):2879–2886
21. Mertens HDT, Svergun DI (2017) Combining NMR and small angle X-ray scattering for the study of biomolecular structure and dynamics. Arch Biochem Biophys 628:33–41. https://doi.org/10.1016/j.abb.2017.05.005
22. Hennig J, Sattler M (2014) The dynamic duo: combining NMR and small angle scattering in structural biology. Protein Sci 23:669. https://doi.org/10.1002/pro.2467
23. Carlomagno T (2014) Present and future of NMR for RNA-protein complexes: a perspective of integrated structural biology. J Magn Reson 241:126–136
24. Madl T, Gabel F, Sattler M (2011) NMR and small-angle scattering-based structural analysis of protein complexes in solution. J Struct Biol 173(3):472–482. https://doi.org/10.1016/j.jsb.2010.11.004
25. Svergun DI, Koch MHJ, Timmins PA, May RP (2013) Small angle X-ray and neutron scattering from solutions of biological macromolecules. IUCr monographs on crystallography, vol 19. Oxford University Press, Oxford
26. Gutberlet T, Rücker U, Brückel T (2017) Towards compact accelerator driven neutron-sources for Europe. Neutron News 28 (3):20–25. https://doi.org/10.1080/10448632.2017.1342486
27. Rauch H, Waschkowski W (2001) Neutron scattering lengths. In: Dianoux A-J, Lander G (eds) Neutron data booklet. Institut Laue-Langevin, Grenoble
28. Jacrot B, Zaccai G (1981) Determination of molecular weight by neutron scattering. Biopolymers 20:2413–2426
29. Jeffries CM, Graewert MA, Blanchet CE, Langley DB, Whitten AE, Svergun DI (2016) Preparing monodisperse macromolecular samples for successful biological small-angle X-ray and neutron-scattering experiments. Nat Protoc 11(11):2122–2153. https://doi.org/10.1038/nprot.2016.113
30. Jacques DA, Trewhella J (2010) Small-angle scattering for structural biology—expanding the frontier while avoiding the pitfalls. Protein Sci 19:642–657
31. Putnam CD, Hammel M, Hura GL, Tainer JA (2007) X-ray solution scattering (SAXS) combined with crystallography and computation: defining accurate macromolecular structures, conformations and assemblies in solution. Q Rev Biophys 40(3):191–285
32. Zaccai NR, Serdyuk IN, Zaccai J (2017) Methods in molecular biophysics: structure, dynamics, function for biology and medicine, 2nd edn. Cambridge University Press, Cambridge. https://doi.org/10.1017/9781107297227
33. Philo JS (2006) Is any measurement method optimal for all aggregate sizes and types? AAPS J 8(3):E564–E571. https://doi.org/10.1208/aapsj080365
34. Porterfield JZ, Zlotnick A (2010) A simple and general method for determining the protein and nucleic acid content of viruses by UV absorbance. Virology 407(2):281–288. https://doi.org/10.1016/j.virol.2010.08.015
35. Milligan JF, Groebe DR, Witherell GW, Uhlenbeck OC (1987) Oligoribonucleotide synthesis using T7 RNA polymerase and synthetic DNA templates. Nucleic Acids Res 15 (21):8783–8798
36. Haertlein M, Moulin M, Devos JM, Laux V, Dunne O, Trevor Forsyth V (2016) Biomolecular deuteration for neutron structural biology

and dynamics. In: Kelman Z (ed) Methods in enzymology, vol 566. Academic Press, pp 113–157. https://doi.org/10.1016/bs.mie.2015.11.001

37. Hattori A, Crespi HL, Katz JJ (1965) Effect of side-chain deuteration on protein stability. Biochemistry 4(7):1213–1225
38. Makhatadze GI, Clore GM, Gronenborn AM (1995) Solvent isotope effect and protein stability. Nat Struct Biol 2(10):852–855
39. Rio DC (2012) Filter-binding assay for analysis of RNA–protein interactions. Cold Spring Harb Protoc 2012(10):pdb.prot071449
40. Ryder SP, Recht MI, Williamson JR (2008) Quantitative analysis of protein-RNA interactions by gel mobility shift. Methods Mol Biol (Clifton, NJ) 488:99–115. https://doi.org/10.1007/978-1-60327-475-3_7
41. Recht MI, Ryder SP, Williamson JR (2008) Monitoring assembly of ribonucleoprotein complexes by isothermal titration calorimetry. Methods Mol Biol (Clifton, NJ) 488:117–127. https://doi.org/10.1007/978-1-60327-475-3_8
42. Lapinaite A, Simon B, Skjaerven L, Rakwalska-Bange M, Gabel F, Carlomagno T (2013) The structure of the box C/D enzyme reveals regulation of RNA methylation. Nature 502 (7472):519–523. https://doi.org/10.1038/nature12581. http://www.nature.com/nature/journal/v502/n7472/abs/nature12581.html#supplementary-information
43. Feigin LA, Svergun DI (1987) Structure analysis by small-angle X-ray and neutron scattering. Plenum Press, New York
44. Guinier A, Fournet G (1955) Small angle scattering of X-rays. Wiley, New York
45. Voss NR, Gerstein M (2005) Calculation of standard atomic volumes for RNA and comparison with proteins: RNA is packed more tightly. J Mol Biol 346:477–492
46. Sarachan KL, Curtis JE, Krueger S (2014) Small-angle scattering contrast calculator for protein and nucleic acid complexes in solution. J Appl Crystallogr 46:1889–1893
47. Whitten AE, Cai S, Trewhella J (2008) MULCh: modules for the analysis of small-angle neutron contrast variation data from biomolecular assemblies. J Appl Crystallog 41 (1):222–226. https://doi.org/10.1107/S0021889807055136
48. Hennig J, Militti C, Popowicz GM, Wang I, Sonntag M, Geerlof A, Gabel F, Gebauer F, Sattler M (2014) Structural basis for the assembly of the Sxl-Unr translation regulatory complex. Nature 515(7526):287–290. https://doi.org/10.1038/nature13693
49. Mahieu E, Gabel F (2018) Biological small-angle neutron scattering: recent results and development. Acta Crystallogr Sect D 74 (8):715–726. https://doi.org/10.1107/S2059798318005016
50. Konarev PV, Volkov VV, Sokolova AV, Koch MHJ, Svergun DI (2003) PRIMUS: a Windows PC-based system for small-angle scattering data analysi. J Appl Crystallogr 36:1277–1282
51. Gabel F, Bicout D, Lehnert U, Tehei M, Weik M, Zaccai G (2002) Protein dynamics studied by neutron scattering. Q Rev Biophys 35(4):327–367
52. Guinier A (1939) La diffraction des Rayons X aux Tres Faibles Angles: Applications a l'Etude des Phenomenes Ultra-microscopiques. Annal Phys (Paris) 12:161–236
53. Kim Henry S, Martel A, Girard E, Moulin M, Härtlein M, Madern D, Blackledge M, Franzetti B, Gabel F (2016) SAXS/SANS on supercharged proteins reveals residue-specific modifications of the hydration shell. Biophys J 110(10):2185–2194. https://doi.org/10.1016/j.bpj.2016.04.013
54. Kim HS, Gabel F (2015) Uniqueness of quasi-atomic bio-macromolecularmodels derived from small angle scattering data. Acta Crystallogr D Biol Crystallogr 71(Pt 1):57–66
55. Jacrot B (1976) The study of biological structures by neutron scattering from solution. Rep Prog Phys 39(10):911–953
56. Grotwinkel JT, Wild K, Segnitz B, Sinning I (2014) SRP RNA remodeling by SRP68 explains its role in protein translocation. Science 344(6179):101. https://doi.org/10.1126/science.1249094
57. Wang Y-X, Zuo X, Wang J, Yu P, Butcher SE (2010) Rapid global structure determination of large RNA and RNA complexes using NMR and small-angle X-ray scattering. Methods (San Diego, CA) 52(2):180–191. https://doi.org/10.1016/j.ymeth.2010.06.009
58. Falb M, Amata I, Gabel F, Simon B, Carlomagno T (2010) Structure of the K-turn U4 RNA: a combined NMR and SANS study. Nucleic Acids Res 38(18):6274–6285
59. Grishaev A, Wu J, Trewhella J, Bax A (2005) Refinement of multidomain protein structures by combination of solution small-angle X-ray scattering and NMR data. J Am Chem Soc 127:16621–16628
60. Trabuco LG, Villa E, Mitra K, Frank J, Schulten K (2008) Flexible fitting of atomic structures into electron microscopy maps using

molecular dynamics. Structure (London, England : 1993) 16(5):673–683. https://doi.org/10.1016/j.str.2008.03.005

61. Karaca E, Rodrigues JPGLM, Graziadei A, Bonvin AMJJ, Carlomagno T (2017) M3: an integrative framework for structure determination of molecular machines. Nat Methods 14:897. https://doi.org/10.1038/nmeth.4392. https://www.nature.com/articles/nmeth.4392#supplementary-information
62. Sonntag M, Jagtap PKA, Simon B, Appavou M-S, Geerlof A, Stehle R, Gabel F, Hennig J, Sattler M (2017) Segmental, domain-selective perdeuteration and small-angle neutron scattering for structural analysis of multi-domain proteins. Angew Chem Int Ed 56 (32):9322–9325. https://doi.org/10.1002/anie.201702904
63. Michel E, Allain FHT (2015) Selective amino acid segmental labeling of multi-domain proteins. In: Kelman Z (ed) Methods in enzymology, vol 565. Academic Press, New York, pp 389–422. https://doi.org/10.1016/bs.mie.2015.05.028
64. Freiburger L, Sonntag M, Hennig J, Li J, Zou P, Sattler M (2015) Efficient segmental isotope labeling of multi-domain proteins using Sortase a. J Biomol NMR 63(1):1–8. https://doi.org/10.1007/s10858-015-9981-0
65. Michel E, Skrisovska L, Wüthrich K, Allain FH (2013) Amino acid-selective segmental isotope labeling of multidomain proteins for structural biology. Chembiochem 14(4):457–466
66. Duss O, Lukavsky PJ, Allain FH (2012) Isotope labeling and segmental labeling of larger RNAs for NMR structural studies. Adv Exp Med Biol 992:121–144
67. Nelissen FHT, van Gammeren AJ, Tessari M, Girard FC, Heus HA, Wijmenga SS (2008) Multiple segmental and selective isotope labeling of large RNA for NMR structural studies. Nucleic Acids Res 36(14):e89–e89. https://doi.org/10.1093/nar/gkn397
68. Tzakos AG, Easton LE, Lukavsky PJ (2007) Preparation of large RNA oligonucleotides with complementary isotope-labeled segments for NMR structural studies. Nat Protoc 2:2139. https://doi.org/10.1038/nprot.2007.306
69. Nemethy G, Scheraga HA (1962) Structure of water and hyrdophobic bonding in proteins. I. A model for the thermodynamic properties in liquid water. J Chem Phys 36 (12):3382–3417

Chapter 14

Structural Analysis of RNA by Small-Angle X-ray Scattering

Anne Théobald-Dietrich, Raphaël de Wijn, Kévin Rollet, Alexandra Bluhm, Joëlle Rudinger-Thirion, Caroline Paulus, Bernard Lorber, Aurélien Thureau, Magali Frugier, and Claude Sauter

Abstract

Over the past two decades small-angle X-ray scattering (SAXS) has become a popular method to characterize solutions of biomolecules including ribonucleic acid (RNA). In an integrative structural approach, SAXS is complementary to crystallography, NMR, and electron microscopy and provides information about RNA architecture and dynamics. This chapter highlights the practical advantages of combining size-exclusion chromatography and SAXS at synchrotron facilities. It is illustrated by practical case studies of samples ranging from single hairpins and tRNA to a large IRES. The emphasis is also put on sample preparation which is a critical step of SAXS analysis and on optimized protocols for in vitro RNA synthesis ensuring the production of mg amount of pure and homogeneous molecules.

Key words SEC-SAXS, RNA, tRNA, IRES, Structure, Integrative structural biology

1 Introduction

Ribonucleic acid (RNA) molecules constitute a very diverse biopolymer family in terms of sequence, size, and shape, and getting insights into their 3D structure is essential to understand their cellular functions and partnerships. Small-angle X-ray scattering (SAXS) is a powerful means to study structural and dynamic properties of RNA in solution, complementary to methods such as crystallography, nuclear magnetic resonance (NMR), or cryoelectron microscopy (cryoEM) [1]. Although SAXS does not provide atomic details, it requires no crystal or isotopic labeling and is performed in solution. The method is compatible with a wide range of solvents and gives a rapid access to structural parameters such as size, oligomeric state, shape of isolated molecules, or complexes [2, 3]. SAXS was already employed in the 60s in early structural investigations on tRNAs [4–6] and has since been applied to samples ranging from compact hairpins to particles as large as the ribosome [7] to study RNA architecture, folding, dynamics, and

Véronique Arluison and Frank Wien (eds.), *RNA Spectroscopy: Methods and Protocols*, Methods in Molecular Biology, vol. 2113, https://doi.org/10.1007/978-1-0716-0278-2_14,

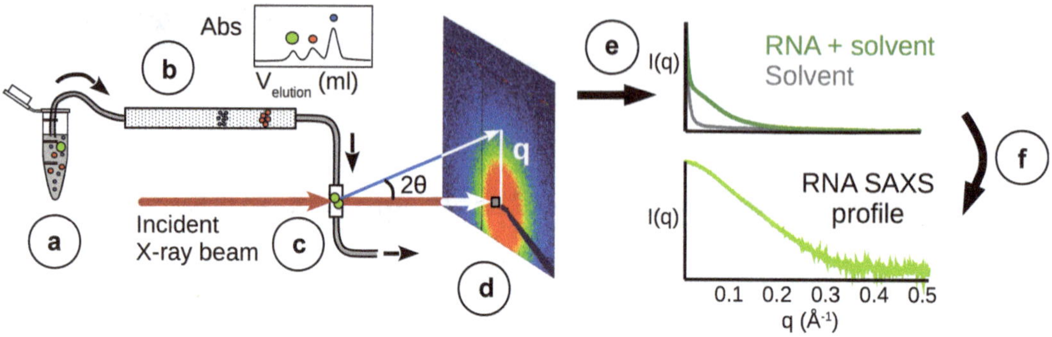

Fig. 1 Principle of a SEC-SAXS setup. The sample solution (**a**) is loaded onto a HPLC-SEC column (**b**) to separate the RNA from higher molecular mass populations (aggregates, multimers). The eluate (monitored by UV absorbance) is directly injected in the SAXS capillary cell (**c**) and scattering patterns are collected on a 2D detector (**d**). 2D patterns are radially averaged (**e**) to produce unidimensional scattering profiles for the solvent (grey curve) and for the solution containing the RNA of interest in the same solvent (green curve). The solvent signal is subtracted from that of the sample (**f**) to obtain a profile corresponding to the RNA. This profile or SAXS curve describes the variation of the scattered intensity as a function of the amplitude of the scattering vector or momentum transfer $q = 4\pi \sin\theta/\lambda$ expressed in $Å^{-1}$ or nm^{-1}, where θ is half the angle between the scattered and incident radiations and λ the wavelength. Structural information on the size and the shape of the scattering particle in solution can then be extracted from the curve $I(q)$ vs. q (*see* Fig. 7)

complexes with proteins [8–15]. Over the last two decades it has grown in popularity among the biologists thanks to the development of software packages dedicated to the characterization of biomolecules and an easy access to performant synchrotron facilities [16].

In practice, a SAXS analysis consists of shining X-rays on a solution of particles and collecting the resulting scattered patterns (Fig. 1). The signal of the solvent is recorded separately and subtracted from the signal of the RNA sample in the solvent to obtain a SAXS profile corresponding to the shape of the RNA. Solvent subtraction is a critical step in data processing. To ensure that the solvent taken as a reference is the same as the solvent in which the RNA is dissolved, it is generally recommended to carry out dialysis or size-exclusion chromatography (SEC). This chapter will indeed focus on the SEC-SAXS approach [17–20], which is available at most synchrotron facilities and implements a FPLC/HPLC-SEC directly upstream the SAXS chamber or capillary (Fig. 1) to separate molecules according to their hydrodynamic properties. The advantage of such a setup is twofold: it guarantees (1) good solvent matching and (2) monodispersity of the RNA entering the measurement cell. Because all molecular entities present in solution contribute to the scattered signal as the square of their volume, it is crucial to separate the population of interest (e.g., monomeric RNA) from aggregates or multimers (likely to form at high concentrations) prior to collecting the SAXS signal [20, 21].

Despite the progress of recent pixel detectors in terms of sensitivity and low background, SAXS, alike other structural methods, is still sample consuming. A single SEC-SAXS run typically requires 20–80 μL of pure RNA solution at 2–5 mg/mL (i.e., 40–400 μg), taking into account that SEC will dilute the sample 5–10 times. Sample preparation is another critical step and we recommend to start with at least a milligram of pure, homogeneous, and monodisperse RNA sample to work comfortably. Among RNA preparation methods detailed in [22], in vitro transcription allows the synthesis of almost any unmodified RNA sequence and has found applications in all areas of RNA research. These transcriptional systems are based on the cloning of synthetic DNA encoding the RNA sequence of interest downstream of the promoter of a phage DNA-dependent RNA polymerase. Among the different phage RNA polymerases available (SP6, T7, T5, T3), the T7 RNA polymerase has found the widest applications. Detailed protocols to produce RNA using the T7 RNA polymerase can be found in [23].

The first part of this chapter presents experimental conditions for efficient T7 transcription and practical tricks to obtain high amount of homogenous transcripts for SAXS experiments (Fig. 2). When designing an RNA molecule, several constraints have to be considered such as the nature of the 5′ sequence of the construct, its total length, and the homogeneity of the 3′ end. Three protocols adapted to the production of RNAs of different sizes by in vitro transcription are described, as well as RNA purification techniques, and the usefulness of a strict quality control of the sample to guarantee the reproducibility and the success of SAXS measurements.

The second part of this chapter is an introduction to standard SEC-SAXS analysis of RNA using synchrotron radiation. The description is based on the experimental setup of the SAXS beamline SWING [19] at SOLEIL synchrotron (Saint-Aubin, France), but the principle of data collection and analysis is applicable to any experimental setup. Besides the determination of structural parameters such as the radius of gyration (*Rg*) and maximal dimension (*Dmax*) of the particle, its Porod volume (*Vp*), and characteristic $I(0)$ directly related to its molecular mass (MM), SAXS offers a wide range of possibilities, including the study of conformational changes, flexibility, disorder, or aggregation. The experimental section illustrates data processing and analysis performed using the widespread *ATSAS* suite [24] on different RNA samples including single RNA hairpins, tRNAs and variants, and internal ribosomal entry site (IRES) present in the 5′ UTR of the mRNA of human glycyl-tRNA synthetase (GlyRS; M. Frugier, to be published). The usefulness of SAXS in an integrative structural approach is discussed in the notes and exemplified with the characterization of atypical tRNAs and tRNA/protein complexes.

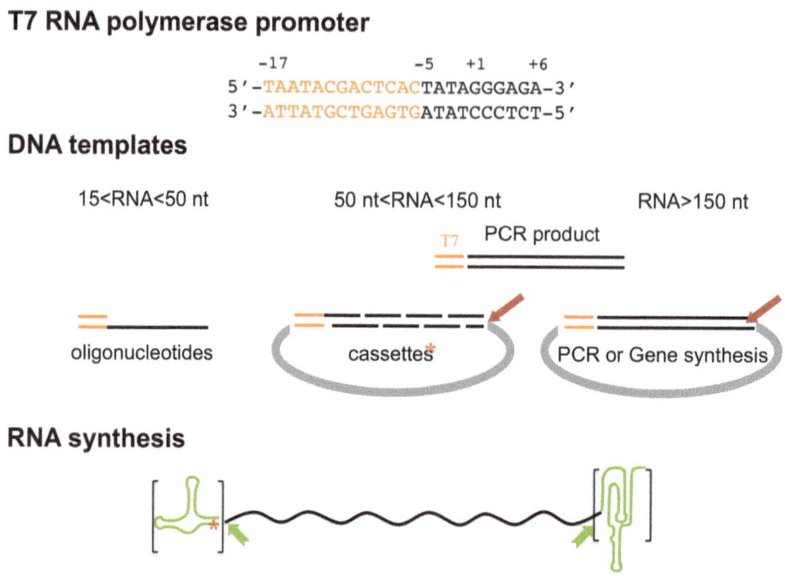

Fig. 2 RNA sample preparation by in vitro T7 transcription. The T7 RNA polymerase promoter encompasses 23 nucleotides (nt) and contains the recognition domain (orange, −17 through −5) and the initiation domain (black, −5 through +6); transcription initiates at position +1 with 3 Gs giving rise to an RNA transcript complementary to the template strand. The nature of the template may differ depending on the size of the desired product: hybridized oligonucleotides, PCR products, or linearized (red arrows) plasmids. In order to produce transcripts with homogeneous 5′ and 3′ extremities, ribozymes (green) may be inserted. Red asterisks indicate protocols detailed in this chapter

Readers eager to learn more about theoretical aspects of the method will find reviews on the topic [25–27], some of which specifically dedicated to SAXS and RNA [8, 20, 28, 29].

2 Materials

2.1 Cloning of Cassettes

1. Desalted DNA oligonucleotides.
2. Restriction enzymes: *Bam*H1, *Hind*III, *Bst*NI, or *Nsi*I.
3. Other enzymes: T4 polynucleotide kinase, T4 DNA ligase, FastAP.
4. Dry heat block.
5. LB/ampicillin agar plates.

2.2 Purification of RNA on Polyacrylamide Gels and Band Visualization

1. 33 × 40 cm^2 glass plates.
2. 2-mm-thick spacers.
3. 2-mm-thick combs.
4. 10× TBE: 0.89 M Tris–borate pH 8.0, 20 mM EDTA.

5. 300-mL denaturing polyacrylamide gel (acrylamide/bisacrylamide 19:1 ratio, 8 M urea, 1× TBE) or native polyacrylamide gel (acrylamide/bisacrylamide 37.5:1 ratio, 1× TBE).
6. 2× RNA denaturing gel loading dye solution: 95% (v/v) formamide, 10 mM EDTA, 25 mM Tris–HCl pH 8.0, 0.025% (w/v) xylene cyanol, and 0.025% (w/v) bromophenol blue.
7. 2× RNA native gel loading dye solution: 50% (w/v) glycerol, 25 mM Tris–HCl pH 8.0, 0.025% (w/v) xylene cyanol and 0.025% (w/v) bromophenol blue.
8. Plastic wrap.
9. Scalpel blade.
10. TLC plate with fluorescent dye (Polygram Gel 400 UV 254, Macherey-Nagel).
11. UV light source.

2.3 Autocatalytic Cleavage of RNA with Ribozyme

1. Dry heat block.
2. Transcription buffer (5×): 200 mM Tris–HCl pH 8.1, 150 mM $MgCl_2$, 5 mM spermidine, 25 mM DTE, 0.05% (v/v) Triton X-100, 20 mM of ATP, CTP, GTP, and UTP.
3. Dilution buffer: 40 mM Tris–HCl pH 8.1 and 30 mM $MgCl_2$.
4. Incubate 1 hour at 60°C.

2.4 RNA Electroelution

1. BioRad Sub-Cell GT or Mini-Sub Cell GT apparatus (BioRad).
2. Sterile 1× TBE.
3. Power supply.
4. Elutrap system and Biotrap membranes BT1 and BT2 (Schleicher & Schuell).
5. Alternatively to 4: Dialysis tube Spectra/Por of appropriate cutoff (Spectrum Laboratories, Inc.) and clips to seal dialysis tubes.

2.5 RNA Purification by Chromatography

1. Bio-Scale Q5 anion exchange column (BioRad).
2. Superdex 75 increase 10/300 column (GE Healthcare Life Sciences).
3. Superdex 200 increase 10/300 column (GE Healthcare Life Sciences).
4. Superose 6 increase 3.2/300 column for longer transcripts (GE Healthcare Life Sciences).
5. Anion exchange buffer QA: 50 mM Tris–HCl pH 7.5, 50 mM KCl, 5 mM $MgCl_2$.

Table 1
HPLC-SEC columns

Column	Manufacturer	Exclusion limits/volume/flow rate/matrix
Bio SEC3-100	Agilent	0.1–100 kDa/4.6 mL/0.3 mL/min/silica
Bio SEC3-150	Agilent	0.5–150 kDa/4.6 mL/0.3 mL/min/silica
Bio SEC3-300	Agilent	5–1250 kDa/4.6 mL/0.3 mL/min/silica
Superdex 200 increase	GE Healthcare	10–600 kDa/2.4 mL/0.15 mL/min/agarose

Table 2
SAXS structural parameters of selected RNA molecules

Name	Hairpin tetraloop	Hairpin T-loop	$tRNA^{Phe}$	gly-IRES
Nucleotides	32	35	76	237
Rg Guinier (Å)	15.59 ± 0.02	18.25 ± 0.04	22.04 ± 0.03	50.58 ± 0.06
Rg $p(r)$ (Å)	15.74 ± 0.01	18.51 ± 0.01	22.86 ± 0.02	50.51 ± 0.02
D_{max} (Å)	51	62	74	157
Vp (Å^3)	11,600	14,600	28,100	169,000
MM_{theor} (kDa)	10.2	11.4	23.5	71.6
MM_{Vc} (kDa)[a]	10.1	11.0	21.8	106

[a]Molecular mass estimate according to [33]

6. Anion exchange buffer QB: 50 mM Tris–HCl pH 7.5, 2 M KCl, 5 mM $MgCl_2$.
7. Size-exclusion buffer: 50 mM Tris–HCl pH 7.5, 150 mM KCl, 5 mM $MgCl_2$.
8. Amicon® Ultra Centrifugal Filters (Merck) with adapted volume and molecular mass cut-off.

2.6 HPLC-SEC Chromatography

1. Size-exclusion columns: *see* examples in Table 1.
2. HPLC system (Agilent LC System 1200–1260 series).
3. 2-mL Vials and conical glass inserts with polymer feet for the HPLC injectors (Agilent).
4. Size-exclusion buffer: 50 mM HEPES-Na pH 7.5, 150 mM KCl, 5 mM $MgCl_2$, filtered and degassed on filtration unit with 0.22-μm filter.

2.7 Data Processing and Analysis

1. Data processing is performed with *Foxtrot* [19] as implemented on SWING beamline.

2. Data analysis is performed with the *ATSAS* suite [24], a widely used package in the bioSAXS community and freely available to academics at https://www.embl-hamburg.de/biosaxs/software.html. Other packages are available and may be used as well, such as *BIOXTAS-RAW* and *SCATTER* [30, 31].
3. Model building and inspection are carried out using *PyMOL* (Schrödinger LTD).

3 Methods

3.1 RNA Production by In Vitro Transcription (See Fig. 2 and Note 1)

3.1.1 RNA Cloning Using the Cassette Technique

1. Reconstitute the gene with overlapping oligonucleotides as shown for a synthetic tRNA gene in Fig. 3.
2. Design 18- to 21-mers that overlap by 7–10 nucleotides.
3. Include in the construct (1) the upstream T7 RNA polymerase promoter sequence (−17 to −1), (2) *Hind*III and *Bam*HI restriction sites for cloning in pUC118 vector, and (3) the downstream *Bst*NI restriction site coincidental with the 3′-end of the sequence to ensure the correct CCA end of the tRNA.
4. If *Bst*NI is present in the tRNA sequence, then it may be replaced by *Nsi*I.
5. By substituting two oligonucleotides (one on each strand), point variants of the RNA gene can be easily cloned in parallel.

3.1.2 Phosphorylation of Oligonucleotides

1. Phosphorylate individually each oligonucleotide (400 pmol) with 10 U (1 μL) T4 polynucleotide kinase in 50 μL.
2. Incubate for 15 min at 37 °C.
3. Inactivate the kinase by rising the temperature to 65 °C for 10 min.

3.1.3 Hybridization of Phosphorylated Oligonucleotides

1. Mix 2 μL (16 pmol) of each phosphorylated oligonucleotide ($2 \times 10 = 20$ μL total volume).
2. Add 20 μL of buffer 150 mM Tris-HCl pH 7.5, 30 mM $MgCl_2$, 15 mM dithioerythritol.
3. Incubate the mixture at 85 °C for 10 min.
4. Switch off the heater block and leave the tube to slowly cool down to 30 °C (2–3 h).

3.1.4 Ligation of the RNA Gene into pUC 118 Vector

1. Digest successively pUC118 with *Bam*HI and *Hind*III.
2. Perform a dephosphorylation step of the digested vector with FastAP, stop the reaction with phenol extraction, precipitate and adjust the concentration at 1 μg/μL.

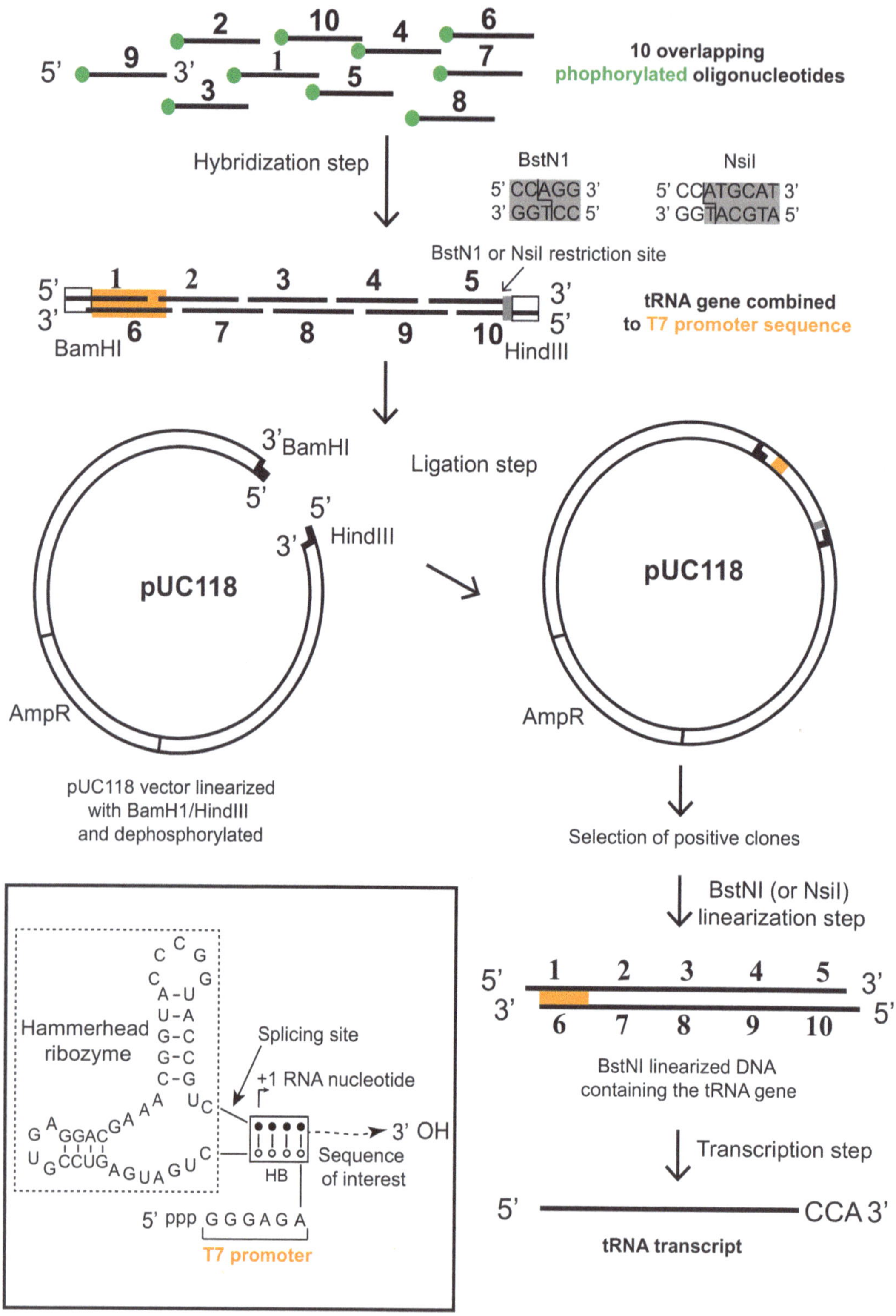

Fig. 3 Production of RNA transcripts using the cassette technique. The RNA synthetic gene (here a tRNA gene) in combination with the T7 RNA polymerase promoter and BstNI site is assembled before its ligation into the

3. Incubate 10 μL of the hybridization mixture (obtained in Subheading 3.1.3) with 1 μg pUC118 (digested and dephosphorylated) in the presence of 5 μL ATP 10 mM and 5 U (5 μL) T4 DNA ligase (Thermo Scientific) overnight at 16 °C.

3.1.5 Bacterial Transformation and Search for Positive Clones

1. Transform 200 μL of $CaCl_2$ chimio-competent *E. coli* TG1 cells with the ligation mixture.
2. Spread on LB/ampicillin plates and incubate overnight at 37 °C.
3. Positive clones are verified by sequencing.
4. Perform DNA maxi-preparation of the positive plasmid.

3.1.6 In Vitro T7 Transcription

1. Digest 50 μg of plasmid with *Bst*NI.
2. Prepare a 500 μL transcription mix containing 100 μL of transcription buffer (5×), 25 U T7 RNA polymerase, 50 μg of linearized plasmid and 4 mM of each nucleotide triphosphate (pH adjusted to 7.5).
3. Incubate at 37 °C for 3 h in a dry heat block.
4. Stop the reaction by ethanol precipitation (2.5 volumes) in the presence of 0.3 M acetate-Na pH 5.2.
5. Centrifuge the solution for 15 min at 15,000 *g* and store the dry RNA pellet at −20 °C.

3.1.7 RNA Cloning Using the Transzyme Strategy

1. Assemble the sequence of the synthetic transzyme gene with cassettes (*see* Subheading 3.1.1) as illustrated in Fig. 3.
2. Oligonucleotide#2, coding for the hammerhead ribozyme, remains identical for all sequences (boxed in dashed lines in Fig. 3).
3. The four nucleotides on the 5′ side of the hybridization box (HB), downstream of the strong T7 internal promoter, must be complementary to the first four nucleotides of the sequence of interest that form the 3′ side of HB.

Fig. 3 (continued) pUC118 plasmid. The tRNA gene-containing plasmid is then linearized with *Bst*NI and T7 RNA transcription is performed. Note that *Bst*NI enables the production of RNA molecules ending with the 3′-CCA sequence universally conserved in tRNA. In the transzyme strategy (insert), the oligonucleotide#2 incorporates the hammerhead ribozyme. The transzyme construct, i.e., a transcript integrating the self-cleavable ribozyme, contains, from 5′ to 3′, the strong internal promoter sequence for the T7 RNA polymerase, the hammerhead ribozyme sequence (boxed in dashed lines), and the domain called hybridization box (HB). The splicing of the RNA occurs between the hammerhead ribozyme and the HB. The open circles correspond to the four variable nucleotides immediately downstream of the T7 promoter. They are chosen to be complementary to the starting 5′ sequence of the RNA of interest (filled circle) allowing the formation of the base-paired structure (HB) that directs the specific cleavage of the ribozyme. The autocatalytic cleavage releases the desired RNA sequence with a 5′OH terminus

3.1.8 Transcription and Autocatalytic Cleavage

1. Prepare a 500 μL transcription mix containing 100 μL of transcription buffer (5×), 25 U T7 RNA polymerase, 50 μg of linearized plasmid in a 1.5-mL Eppendorf tube and 4 mM of each nucleotide triphosphate (pH adjusted to 7.5).
2. Incubate at 37 °C for 3 h in a dry heat block.
3. Dilute the transcription mix with 2 mL of dilution buffer (*see* **Note 2**).
4. Incubate at 60 °C for 1 h in a dry heat block.
5. Stop the reaction by ethanol precipitation (*see* Subheading 3.1.6).

3.2 RNA Purification (Fig. 4)

*3.2.1 RNA Purification on Gel (See **Note 3**)*

1. Resolubilize the precipitated transcript in water and RNA denaturing loading buffer (1:1 mix).
2. Clean glass plates, spacers, and combs.
3. Assemble gel plates using tape to seal the bottom carefully and put the plates upright.
4. Prepare 25 mL of gel and pour the "bottom" layer (5 cm high) of the gel; let it polymerize.
5. Pour the rest of the gel slowly, remove air bubbles, insert combs, and let it polymerize horizontally for at least 15 min.
6. Remove both the tape and the comb and rinse thoroughly the well with 1× TBE.
7. Perform the electrophoresis overnight at 650 V/55 mA at room temperature.
8. To insure an optimal separation, run the electrophoresis as long as possible being careful not to let the RNA migrate outside the gel. For example, to purify a 76-nt-long tRNA transcript, the transcription reaction is loaded on a 12% gel and the electrophoresis is stopped 2 h after the xylene cyanol dye has left the bottom of the gel.
9. Remove the gel from the plates at the end of the electrophoresis process, and cover it with a clean plastic wrapping film.
10. Place the gel on the TLC plate and illuminate with the UV source (254 nm). DNA and RNA molecules will appear as dark purple bands (*see* **Note 3**).
11. Cut the gel piece containing the RNA of interest, store it at −20 °C until needed, or use it immediately for electroelution.

3.2.2 RNA Extraction by Electroelution Using Biotrap Device

1. Before use, sterilize the Biotrap device wrapped in aluminum foil.
2. Insert a BT2 membrane in the second trap insert and two BT1 membranes at each end of the device (Fig. 5a).

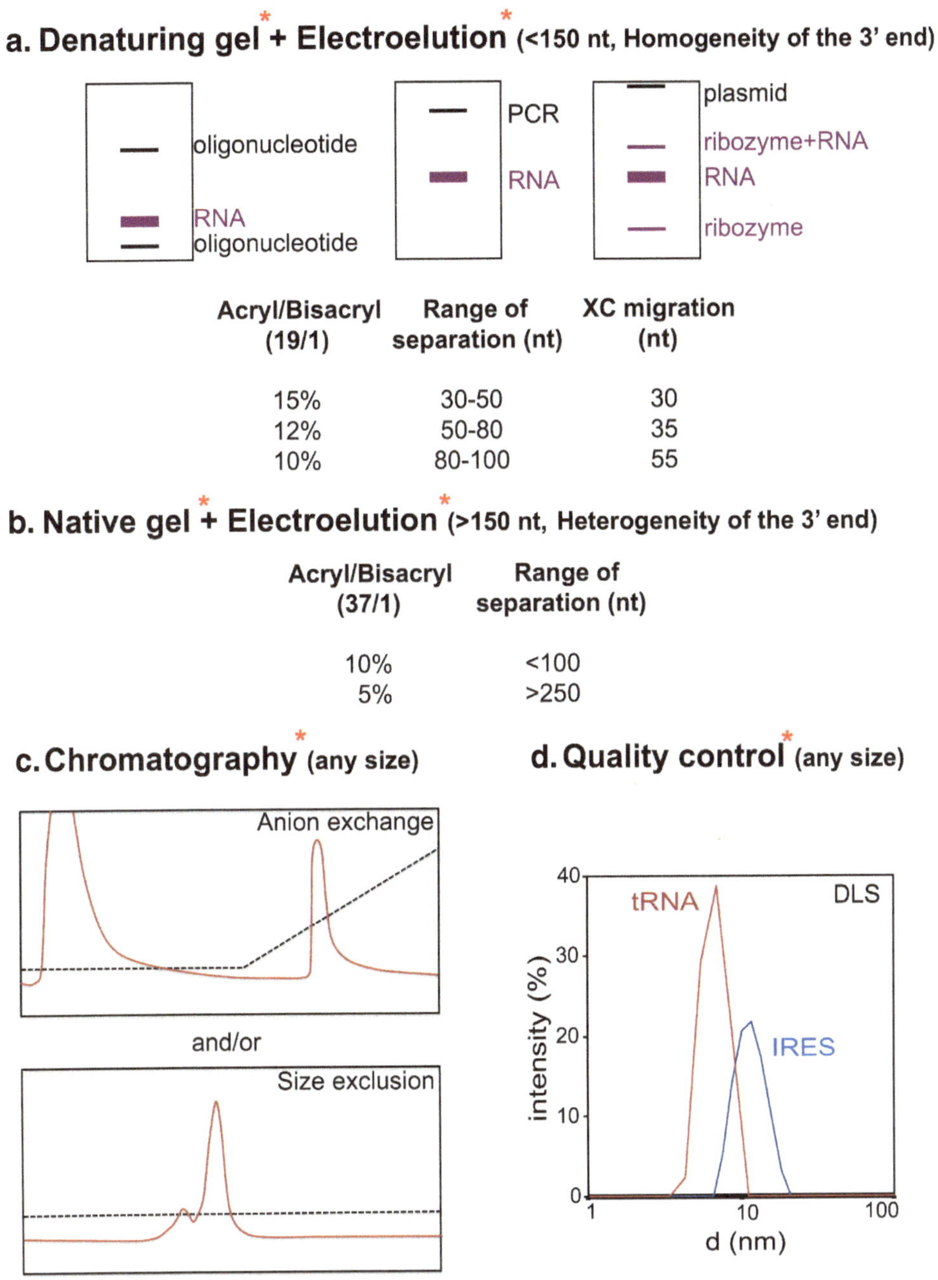

Fig. 4 Large-scale purification of RNA for SAXS. The purification procedure varies with the size of the RNA. (**a**) Molecules of less than 150 nt can be purified on denaturing polyacrylamide gels. Several profiles can be observed depending on the chosen DNA template (black) and the produced RNA molecules (purple). The percentage of the gel has to be adapted to the size (denaturing gels) or the shape (native) of the RNA. For longer RNAs (>150 nt), it is recommended to use (**b**) native gels and/or (**c**) chromatography (anion exchange or size exclusion) to remove unincorporated nucleotides and incomplete or unfolded transcripts. In all cases, native and denaturing gels are used to assess the purity and homogeneity of the sample, along with SEC and DLS, before SAXS analysis. (**d**) DLS is a simple, rapid, and noninvasive method ideal for quality control [44]. In the example two samples—yeast $tRNA^{Phe}$ and the IRES of human GlyRS mRNA—show a monomodal size distribution that confirms their homogeneity. Red asterisks indicate protocols detailed in this chapter

3. Put RNA-containing gel pieces in the device and immerse them in sterile 1× TBE buffer.
4. Perform a first electroelution during 1 h at 150 V in a cold room (4 °C).
5. Recover the eluate from the elution chamber (800 μL) and divide it in two times 400 μL.
6. Perform a second elution with fresh 1 × TBE and split again the recovered eluate.
7. Add 40 μL of sodium acetate 3 M pH 5.2 and 1 mL ethanol to each sample; let it precipitate overnight at −20 °C.
8. Centrifuge for 45 min at 15,000 *g*, wash, and dry the pellet.
9. The recovery yield of RNA from the gel with the Biotrap is about 80–90%.

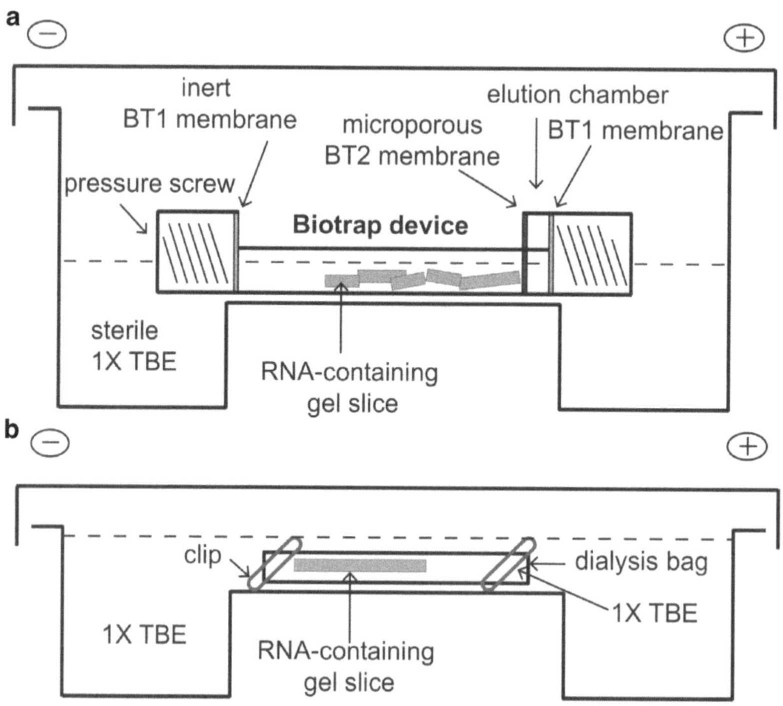

Fig. 5 Electroelution of RNA transcript from polyacrylamide gel. (**a**) Using the Biotrap apparatus: transcript-containing gel slices are introduced in a Biotrap device closed by BT1 membranes on each side and are submerged with 10-mL sterile 1× TBE. A BT2 membrane is inserted to form a chamber of adjustable volume in which the transcript will be trapped. The Biotrap is placed in the electrophoresis Sub-Cell GT system filled with 1× TBE buffer (buffer levels in and out of the Biotrap device should be identical). The electric field permits the migration of the transcript out of the gel pieces into the elution chamber formed between BT1 and BT2 membranes. (**b**) Using a dialysis bag: the transcript-containing gel slice is introduced in a dialysis tube filled with sterile 1× TBE and sealed on each end with a dialysis clip. The bag is placed on a Mini-Sub Cell GT or in a Sub-Cell GT apparatus (depending on the size of the bag) so that the gel piece is perpendicular to the electrodes. After immersion with 1× TBE buffer, the electric field is applied to permit the migration of the transcript out of the gel slice inside the bag

3.2.3 RNA Extraction by Electroelution in a Dialysis Bag

1. Use this alternative as a less sophisticated and less expensive option (Fig. 5b).
2. Insert RNA-containing gel pieces in the dialysis tube (its cutoff depends on the molecular weight of the RNA to be extracted, e.g., for tRNA transcripts, use a cutoff of 12–14 kDa to keep the eluted molecules inside the tube).
3. Place the tube in a migration device (Mini-Sub Cell GT or Sub-Cell GT) and immerse it in 1× TBE buffer.
4. Perform two successive electroelutions at 150 V in a cold room (4 °C) for 20 min.
5. Before recovering the transcript, invert the current during 30 s to unstick the eluted transcript from the inner bag membrane.
6. Pipet the buffer from the dialysis bag.
7. Reduce the volume (which is usually comprised between 2 and 10 mL) by a concentration step in appropriate centrifugal filter units.
8. If necessary, perform an additional precipitation step with sodium acetate and ethanol.
9. The recovery yield of the transcript from the gel with the dialysis tube is estimated at about 50–60%.

3.2.4 RNA Purification by Chromatography (See ***Note 4****)*

1. Centrifuge the transcription mix for 5 min at 15,000 *g* to remove the white precipitate that contains pyrophosphate (PPi).
2. Add 9 volumes of water to the supernatant.
3. Equilibrate the anion exchange column with five volumes of buffer QA.
4. Load the sample on the column using a reduced flow rate (no more than 1 mL/min).
5. Wash the column with 20 column volumes of buffer QA. The flow rate can be increased up to 2 mL/min.
6. Run a salt gradient from 50 mM KCl (buffer QA) to 1.5 M (buffer QB) over 15 column volumes.
7. Monitor the elution using the absorbance at 254 nm.
8. Analyze the fractions on a native gel.
9. If additional bands are present (corresponding to unfolded RNA or RNA aggregates), then a SEC step may be added to remove unwanted conformers. Perform an ethanol precipitation of the RNA sample or concentrate it in a Centricon centrifugal unit and run a SEC in the corresponding buffer. The main peak contains the RNA of interest, freed of aggregates.

10. Collect the corresponding fractions and ethanol precipitate or concentrate the RNA in a Centricon centrifugal unit with the adapted cutoff.

3.2.5 RNA Quantification and Quality Control (See **Note 5**)

1. Dissolve the RNA pellet in distilled water and determine the concentration of the solution with a spectrophotometer at 260 nm (1 absorbance unit corresponds to 40 μg/mL).
2. Denaturing conditions can alter the proper folding of the transcript. Therefore, before use, a denaturation/renaturation step of the transcript is recommended. It consists of incubating the water-solved transcript for 2 min at 65 °C followed by a 10-min cooling at 25 °C and subsequent addition of $MgCl_2$ (at the final concentration needed for functional assays).
3. Assess the purity and quality of the eluted transcript by performing an analytical polyacrylamide gel electrophoresis on an aliquot.
4. Run an analytical SEC with the column used for SEC-SAXS analysis (*see* Subheading 3.3.2) to assess the homogeneity of the sample in solution.
5. When available, use multiangle (MALS) or dynamic light scattering (DLS) to confirm sample homogeneity before running a SAXS experiment (*see* Fig. 4d).

3.3 SEC-SAXS Experiment

3.3.1 Column and Sample Preparation (See **Note 6**)

1. Rinse the SEC column with water (three column volumes).
2. Equilibrate with the elution buffer (five column volumes recommended) at the flow rate used for sample elution (0.3 mL/min with BioSEC columns f.ex.; *see* Table 1).
3. Centrifuge the sample (10 min, 15,000 *g*, 4 °C).
4. Transfer an aliquot of 20–80 μL in an adapted HPLC vial, and place the vial in the injector of the HPLC system.

3.3.2 Test Elution (Optional)

1. This step should be performed before the synchrotron session as part of sample quality control.
2. Inject 5 μL of sample and run a test elution.
3. Follow the elution based on UV absorbance signal (Fig. 6a).
4. Define the SAXS acquisition windows for the solvent (at the beginning of the run) and for the sample (starting at the end of the void volume V0).

3.3.3 Data Acquisition

1. Inject 10–70 μL of sample and run the SEC elution.
2. Follow the elution based on UV absorbance (*see* **Note 6** and Fig. 6a).
3. Collect scattering patterns for the solvent and the solvent plus sample in appropriate time windows (Fig. 6a).

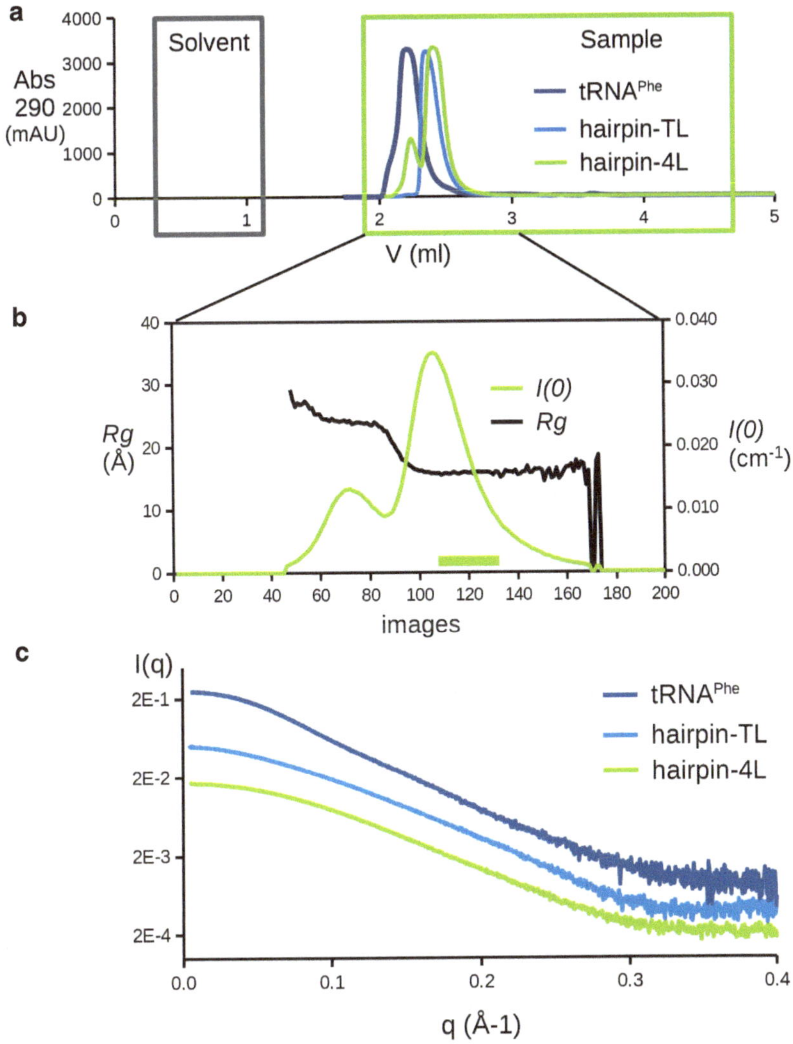

Fig. 6 Example of SEC-SAXS data collection and processing. (**a**) SEC chromatogram recorded at 290 nm to avoid signal saturation. The elution profiles performed on a BioSEC3-100 column at a flow of 0.3 mL/min are depicted for three samples: the native yeast $tRNA^{Phe}$ (in dark blue) and two hairpins corresponding to the acceptor stem of yeast $tRNA^{Phe}$ with its natural 7-nucleotide T-loop (hairpin TL in light blue) or with a tetraloop (hairpin 4L in green). The RNA quantity injected for these analyses was ~250 μg. The three RNAs elute in the expected order, from the largest (23.5 kDa) to the smallest (10.2 kDa). SAXS signal is recorded for the solvent in the grey window and then for the sample in the green window. The averaged solvent signal (*see* Fig. 1) is then subtracted from the curves obtained for each image in the green window to retrieve the sample signal from which X-ray scattering intensity *I*(0) and radius of gyration *Rg* can be estimated. (**b**) Evolution of *I*(0) and *Rg* values along the SEC elution of the hairpin 4L sample. A dataset of 640 SAXS images was collected in the green window, but the analysis is restricted to the region of the major elution peak (images 0–200). As in the chromatogram, two peaks are observed, the first characterized by *Rg* values of ~24 Å and likely containing a dimer of hairpin, the second with *Rg* values of ~16 Å corresponding to the monomeric RNA. SAXS profiles from images 112–130 (green bar) showing a stable *Rg* are averaged leading to the green profile for the hairpin 4L shown in (**c**). The same procedure was applied to the two other RNAs leading to the profiles displayed in (**c**). Resulting SAXS curves are represented on a relative scale with an offset to facilitate their comparison

3.3.4 Data Processing

1. Process the set of images collected with the solvent, i.e., create a unidimensional SAXS profile for each image by radial averaging of 2D patterns (Fig. 1e).
2. Superimpose and compare the solvent profiles to check that there is no drift in the signal.
3. Average these profiles to produce a low-noise solvent curve.
4. Process the set of images collected during the elution of the sample and create corresponding unidimensional SAXS profiles (*see* **Note 7**).
5. Subtract the averaged solvent profile to each profile to retrieve the characteristic SAXS signal of the solute(s) (Fig. 1f).
6. Run a global Guinier analysis to estimate the Rg and $I(0)$ of each subtracted frame (Fig. 6b).
7. Identify a region where the Rg is stable and average the profiles to obtain the SAXS curve of the sample (Fig. 6c).

3.4 Data Analysis

1. Display the SAXS curve of the sample in *PRIMUS* [24] (Fig. 7a).
2. In the analysis panel, select "radius of gyration" (*see* Fig. 7b and **Note 8**) and plot the data as $\ln(I)$ vs. q^2.
3. Adjust the high q value to match Guinier conditions ($qRg < 1.2$).
4. Adjust high and low q values to get a linear fit of the data and the low residual.
5. Record the Rg and associated standard deviation.
6. If the curve has been determined on an absolute scale (in cm^{-1}) with respect to a standard (water scattering, for example), then $I(0)$ should also be recorded for further molecular mass determination.
7. In the analysis panel, select "distance distribution" (*see* Fig. 7c and **Note 9**).
8. Starting from *AUTOGNOM* parameters, adjust the high and low q values to get a good fit of the SAXS curve (left panel).
9. Adjust the D_{max} to get a pair-distance distribution function (PDDF, right panel) or $p(r)$ that falls off smoothly at $p(D_{max}) = 0$.
10. For ab initio reconstructions, q_{max} should not exceed 0.25–0.3 ($qRg < 7–8$).
11. Record the Rg from $p(r)$ and D_{max}, and save the output (extension .out) file.
12. The panel "molecular weight" provides an estimate of the particle molecular mass based on various approaches [32, 33].

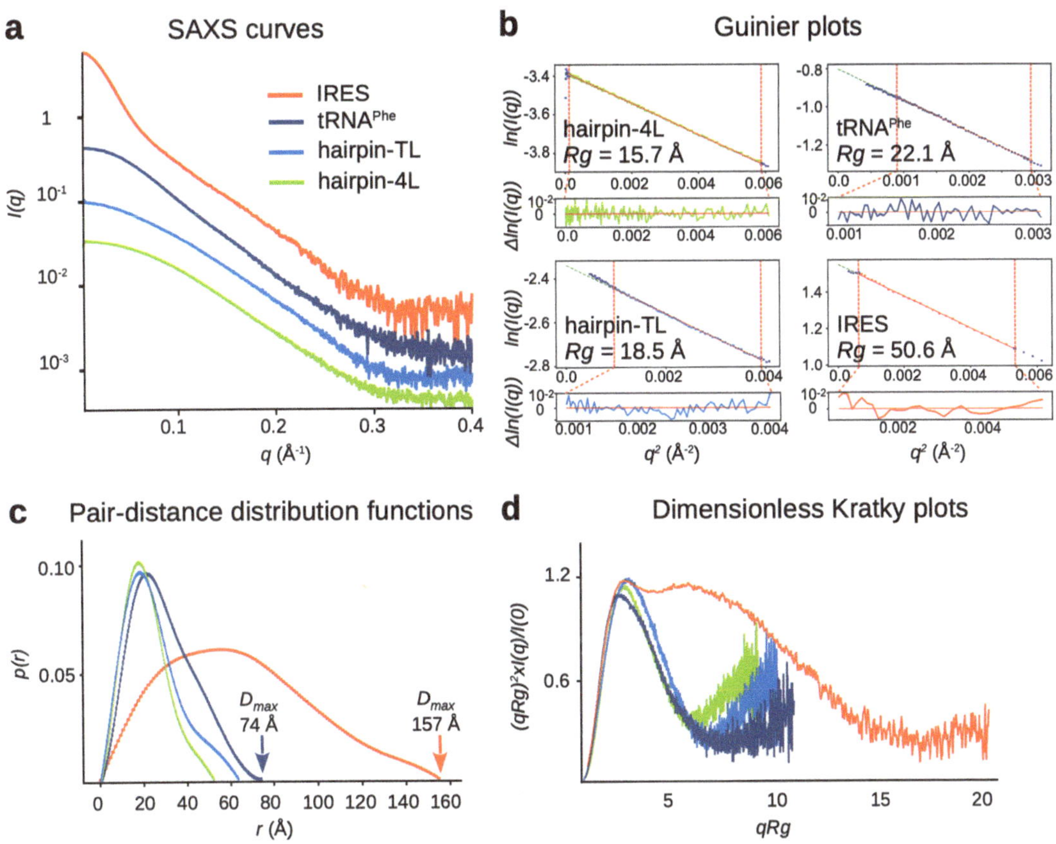

Fig. 7 SAXS data analysis on various RNAs. (**a**) SAXS curves of native yeast tRNA^{Phe} (in dark blue) and two derived acceptor arms, hairpin TL (light blue) and hairpin 4 L (green), and the IRES (red) present in the human GlyRS mRNA (*see* details in Table 2). Corresponding Guinier plots are shown in (**b**). In the Guinier approximation at low angles (typically $q_{max} < 1.2/Rg$ Å^{-1}), the data plotted as ln$I(q)$ vs. q^2 should be linear and the slope provides an estimation of *Rg*. *Rg values are indicated for the four RNAs.* The residual Δln($I(q)$) vs. q^2 shown below each Guinier plot indicates the quality of fit in the selected *q* interval (i.e., between red dotted lines). If the data are on absolute scale, the intercept at $q = 0$ gives the $I(0)$. (**c**) An indirect Fourier transform of the SAXS curves gives access to the pair-distance distribution function *p(r)* of the sample, which allows to determine the maximal distance D_{max}, as well as a more accurate estimate of *Rg* and $I(0)$. The shape of the $p(r)$ is correlated to the shape of the particle. Hairpins and tRNA^{Phe} show similar $p(r)$ profiles typical of elongate objects, whereas the IRES looks more as a large multidomain particle. (**d**) The dimensionless Kratky representation $I(q)/I(0) * (qRg)^2$ vs. *qRg* (plotted using *BioXtas-RAW* [31]) is used to estimate the "globularity" of the particles and to detect flexible or unfolded regions. Unfolded particles produce a hyperbolic Kratky plot that plateaus at high *q*, whereas folded particles show a return to the baseline. Perfectly globular objects show a peak at about 1.1 for $qRg = \sqrt{3}$. It is the case with the four studied RNAs, the IRES displaying a second peak indicative of a multidomain particle

13. The panel "Porod volume" provides an estimate of the particle volume from the Porod invariant *Q* (*see* **Note 10**).
14. Plot the data as $I(q)q^4$ vs. q (Porod plot) with data up to $0.25 < q < 0.3$ to derive the *Vp* of the particle.

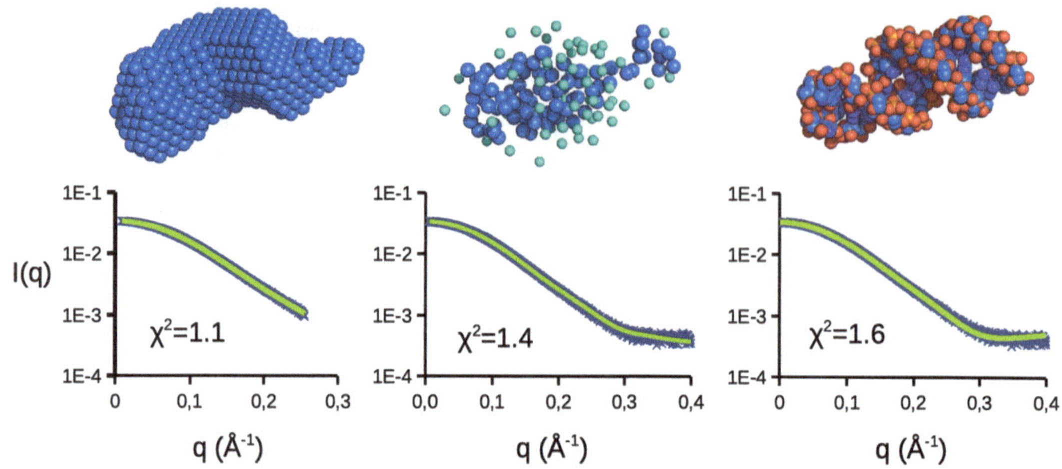

Fig. 8 SAXS-based RNA modeling. In this example dealing with the hairpin 4L (32 nt), the SAXS curve is used to test an atomic model or to generate ab initio models in combination with the *p*(*r*) (*see* Fig. 6). (Left) *DAMMIF* is run at low resolution to reconstruct molecular shapes made of dummy atoms (blue spheres) that reproduce (green curve) the SAXS curve (blue crosses). Twenty models were generated with a normalized spatial discrepancy (NSD) of 0.5 and the model showing the best goodness of fit (χ^2) is displayed. (Middle) Modeling against the full range of data can be performed with *GASBOR* creating chains made of dummy residues (blue spheres) folded by simulated annealing and surrounded by solvent molecules (light blue). The best model of an ensemble showing an NDS of 0.7 is displayed. (Right) When an atomic model is available, its theoretical SAXS curve (green curve) can be computed with *CRYSOL* and compared with experimental data (blue crosses). This model is derived from the acceptor arm of $tRNA^{Phe}$ (PDBid 1EHZ, ref. 56) in which the T-loop is replaced by a UUCG tetraloop. The minimized model is in excellent agreement with the SAXS data ($\chi^2 = 1.6$). Ab initio models—low-resolution envelopes and chains of dummy residues—provide a good idea of the overall shape of the hairpin. All models of hairpin 4L are aligned and displayed at the same scale for comparison

15. Plot the data as $I(q)q^2$ vs. q (Kratky in plot menu) to assess the compactness of the particle (*see* Fig. 7d and **Note 10**).

3.5 SAXS-Based RNA Modeling (See Note 11)

1. Use the output file of GNOM (**step 10** in Subheading 3.5) as input for *DAMMIF* [34] to generate a series of models made of dummy atoms (Fig. 8).
2. Run *DAMMIF* either from the *PRIMUS* interface or from the command line for a better control of the parameters.
3. Superimpose dummy atom models and compare them (*DAMAVER* procedure) to determine their normalized spatial discrepancy (NSD) using different tools of *ATSAS* [24]. The mean NSD should be lower than 0.7–0.9 for a unique solution.
4. Visualize a series of 10–30 ab initio models generated by *DAMMIF* in *PyMOL*.
5. Use the command “set grid_mode, on” in *PyMOL* to display the models on a grid.

6. Apply similar procedures with *GASBOR* [35] to data at higher angles ($0.4 < q < 0.6$ Å^{-1}) to create dummy residue models. The chain of dummy residues is folded by simulated annealing to fit the experimental data.

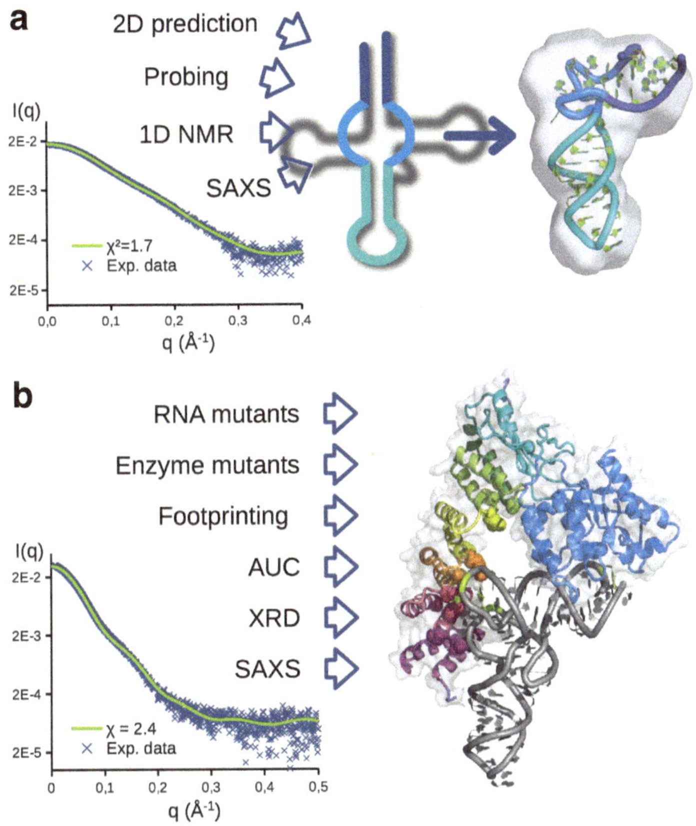

Fig. 9 SAXS, a powerful tool for integrative structural studies. (**a**) This example illustrates such an approach in the characterization of the world's smallest tRNAs, which are armless tRNAs found in the mitochondria of the nematode *Romanomermis culicivorax* [21]. A 2D structure prediction indicated that these tRNAs lacking both the D- and T-arms would fold as an elongated hairpin with a central bulge, as displayed at the center for tRNA^{Leu} with a shadow suggesting a classical tRNA cloverleaf fold in background. The 2D fold was confirmed by in-line RNA probing; NMR only detected Watson-Crick interactions but no tertiary interactions. The SAXS analysis (the fit to the experimental data of the final model is displayed) helped fold the RNA in 3D and confirmed its boomerang shape, in spite of its severely shortened sequence (45 nt instead of 75–80). (**b**) The integrative approach is also very helpful to analyze RNA/protein complexes. For example, in the study of the protein-only RNase P (PRORP2) from *Arabidopsis thaliana* in association with a pre-tRNA substrate [57], the model of the tRNA 5′ maturation complex was built based on a set of distance constrains – interacting domains were determined by footprinting and by binding tests in ultracentrifugation (AUC) using RNA and enzyme mutants—and structural information on individual partners from X-ray diffraction (XRD), in combination with SAXS (the fit to the experimental data of the final model is displayed)

7. When an atomic model is available, calculate its theoretical SAXS curve with *CRYSOL* [36] and compare it to experimental data.
8. Inspect atomic models in *PyMOL*, modify them, and compare them to experimental data using the *SASPY* plugin [37] which provides a direct link to the *ATSAS* package and the possibility to iteratively adjust the model.
9. The *ATSAS* suite may be further exploited to model flexible regions of the molecule, to perform rigid-body or multicomponent fitting to improve the goodness of fit (i.e., minimize the χ^2) of the model to the data.
10. Use complementary biochemical and biophysical methods to derive additional structural constrains and help refine the model against SAXS data (Fig. 9).
11. Follow recommendations from [38] for the presentation of SAXS data in publications.

4 Notes

1. When designing an RNA molecule to be produced by T7 RNA polymerase, several constraints have to be considered, which are the overall length of the RNA, the nature of its 5′ sequence, and the desired homogeneity in 3′.

   For short RNAs (20–50 nt), the template can be reduced to a double-stranded promoter region followed by a single-stranded template sequence [39]. For medium RNA molecules (50–150 nt), synthetic genes can be assembled from oligonucleotides (cassettes) and cloned in a standard vector (Subheading 3.1.1). Finally, for longer sequences, transcription can be performed directly from the product of a PCR reaction. However, it is recommended to clone the DNA template into a plasmid in order to improve the transcription yield. Since the transcription termination occurs at the extremity of the DNA template, it is important to insert a restriction site, which (1) will be used to linearize the plasmid and (2) will generate the proper sequence at the end of the DNA template that matches the RNA sequence. For example, the *Bst*NI restriction enzyme is used to linearize the DNA template when producing tRNA transcripts with their characteristic 3′ CCA end. These considerations are detailed in [23].

   Despite the presence of a strong promoter, the T7 RNA polymerase has intrinsic limitations related to the nature of its promoter which starts at nucleotide −17 and extends to the nucleotide +6 of the sequence to be transcribed [40, 41]. RNA transcripts starting at position +1 with one or more G residues are better transcribed by the T7 RNA polymerase and the presence of Cs and Ts in the +1 to +6 region of a transcript may be responsible for low transcription efficiency. However, the sequence can be modified to meet the T7 promoter wild-type sequence (GGGAGA or AGGGAG) if the 5′ extremity is not essential for RNA folding and function. Alternatively, a hammerhead ribozyme sequence can be introduced between the T7 RNA polymerase promoter and the sequence to be transcribed. Transcription of this construct gives rise to a "transzyme" molecule (Subheading 3.1.7). The autocatalytic activity of the ribozyme releases an RNA transcript starting with the proper 5′-OH nucleotide [42].

Importantly, the T7 RNA polymerase adds one or two nucleotides in a DNA-independent manner at the 3′ extremity of the transcript. The longer the transcript, the more difficult the elimination of this 3′ heterogeneity. For short RNA sequences (<100 nt), one can isolate the specific transcript from $n + 1$ and $n + 2$ contaminating transcripts by electrophoresis on thick denaturing gels (Subheading 3.2.1). However, for longer transcripts, the resolution of such gels does not allow an efficient purification. When the homogeneity of the 3′ end is essential, another autocatalytic ribozyme can be inserted in 3′ [43]. Longer transcripts can be purified either on native gels or by chromatography if the presence of one or two supplementary nucleotides at the 3′ extremity does not interfere with RNA folding or function (Subheadings 3.2.1 and 3.2.5).

2. Transcription conditions are adapted from a protocol described previously [23], with a $MgCl_2$ concentration of 30 mM instead of 22 mM and the final dilution and incubation time adapted to enhance the yield of autocatalytic cleavage. The latter may occur already during the transcription step. However, the yield of cleavage generally does not exceed 40% of the molecules. To increase the amount of cleavage, it is recommended to dilute the mix and incubate longer.

3. The percentage and concentration of acrylamide used in a purification gel has to be adapted to the size of the RNA of interest (Fig. 4) and the size of the wells should also be adapted to the amount of RNA to purify. For example, the product of an efficient 1 mL transcription reaction is phenol extracted, precipitated, and solubilized into 250 μL water and 250 μL loading dye and loaded in a $2 \times 10 \times 100\ mm^3$ well (contains up to 2 mL sample). It is important to get a thin and intense band that can easily be visualized and separated from the contaminating products. When purifying RNA on a native gel, it is necessary to test the migration conditions on an aliquot of the transcription reaction before; indeed, the migration behavior of

an RNA in native conditions will depend essentially on the shape of the molecule and cannot be predicted.

When looking at the result of denaturing gel under UV, three scenarios are possible depending on the DNA template nature and the cloning strategy. Suppose one is purifying an *n*-nucleotide (nt)-long RNA: (1) DNA bands of 17 nt and $n + 17$ nt will be detected when using an oligonucleotide template and (2) a unique band at $n + 17$ nt when using a PCR template, while (3) a linearized plasmid should not enter efficiently the gel and appear as a smear at the top of the gel.

4. A chromatography on a strong anion exchanger can be used to purify the desired RNA transcript from unincorporated nucleotides, DNA template, T7 RNA polymerase, and abortive RNA fragments. The proposed protocol was designed for the Bio-Scale Q5 Anion Exchange Column (BioRad) and can be adapted to other anion exchange resins like Mono Q or DEAE sepharose [44]. At least two peaks should be observed at 254 nm along the elution gradient (Fig. 4). The first one contains free nucleotides that do not bind to the resin and the second one contains the RNA of interest. Additional peaks corresponding to short abortive transcripts may appear depending on the efficiency of the transcription reaction. Importantly, the main peak containing the full-length RNA transcript has to be analyzed by loading a sample of each collected fraction on an analytical denaturing polyacrylamide gel. The product should be identified as a strong band on the gel. Multiple additional bands indicate that the RNA transcript is not fully separated from the DNA template. Then a phenol: chloroform (pH 5.5) extraction step followed by ethanol precipitation at the end of the transcription reaction and before loading the column will remove both the DNA template and the T7 RNA polymerase from the sample.
5. As indicated in the introduction, it is essential to guarantee sample purity and monodispersity before performing a SAXS analysis because the scattered signal contains information about all entities present in the solution. If the solution contains even a low amount of large particles (oligomers, aggregates), structural interpretations may be compromised. A rigorous quality control should therefore include at least an analysis of the sample on native and denaturing polyacrylamide gels, a DLS [45], or MALS analysis and a SEC separation identical to that used during the SAXS experiment.
6. The choice of HPLC-SEC column depends on the size of the RNA and its protein partner(s) and should allow a good separation of the molecule of interest from aggregates or potential oligomers. Some proteins tend to adsorb on silica-based Bio SEC3 matrices and agarose-based columns may be a better

option. It is advisable to use the same column in the laboratory as on the beamline to optimize the elution conditions prior to synchrotron shifts. During the SEC-SAXS analysis, sample elution is followed by UV absorbance, but the signal at 260 nm is often saturated in the major peak due to the high amount of RNA injected to ensure a good SAXS signal. One can record the elution profile at a higher wavelength, such as 280 nm in the case of a complex with a protein, or 290 nm where RNA absorption is significantly lower to record the entire profile of the main peak (Fig. 6a).

7. Data processing is generally performed on the fly during data acquisition and softwares may vary from beamline to beamline but the principle remains the same. The most critical step is the subtraction of solvent and setup contribution (Fig. 1f). It is important to determine the so-called solvent curve (which also accounts for the background from the SAXS capillary) with a buffer solution that is as close as possible to the environment of the macromolecule. An advantage of the SEC-SAXS analysis over an analysis performed in batch mode is indeed to place the RNA of interest in a well-defined mobile phase (buffer solution), the signal of which is measured at the beginning of the elution after extensive equilibration of the column.

8. Examples of data analysis in Fig. 7 use the *ATSAS* suite (version 2.8.4) [24]. The first step is the Guinier analysis [2] at very small angles (Fig. 7b) to derive an estimate of the radius of gyration Rg, which characterizes the mass distribution in the particle, and the forward scattered intensity $I(0)$ if the curve is on absolute scale. $I(0)$ is proportional to particle concentration and to the square of their volume and can be used to determine the molecular mass of the particle. The Guinier plot should be as linear as possible with a low residual. Lack of linearity is an indication of either repulsive (downturn of the plot) or attractive (aggregation, upturn of the plot) interactions (*see* Fig. 3 in [26]). Repulsion may arise from the high negative charge of the RNA backbone at low counterion concentrations and can be fixed by increasing the salt content. Attractive interactions leading to aggregation may appear at high concentrations used for SAXS analysis or may be due to inappropriate buffer conditions causing particle instability. This can be solved by adjusting the ionic strength and the valence of the cations.

9. The PDDF or $p(r)$ is the indirect Fourier transform of the SAXS curve (Fig. 7c) and represents the distribution of distance between scattering elements in the particle. The $p(r)$ profile contains information on the shape and volume of the particle (*see* Fig. 3 in [27]) and provides an estimate of the maximum interatomic distance D_{max}. $I(0)$ and Rg can also be determined from the $p(r)$ taking into account the entire scattering profile.

This step is performed with *GNOM* [46] and the output file is used for ab initio reconstruction (*see* Subheading 3.5).

10. In the Porod representation ($I(q)q^2$ vs. q), the Porod invariant Q corresponds to the integrated area under the curve and is related to the particle (Porod) volume by $Vp = 2\pi \cdot I(0)/Q$ [3, 47, 48]. The Kratky representation ($I(q)q^4$ vs. q) or its dimensionless version ($I(q)/I(0) * (qRg)^2$ vs. qRg [49] detects particles with a compact core which display one peak and a rising tail at higher angles due to the presence of disordered and flexible regions attached to globular domains, while totally unfolded molecules display no peak but a continuously rising curve (Fig. 7d).

11. SAXS data provide an experimental basis for RNA modeling (Fig. 8). The first possibility is the ab initio reconstruction of the particle shape at low resolution ($q_{max} < 8/Rg$ Å^{-1}) and does not require any a priori knowledge of the structure. It is carried out with *DAMMIF* [34] using dummy-atom ensembles which are adjusted to fit the $p(r)$ function and reproduce the SAXS curve with a low discrepancy measured by a χ^2. Another approach makes use of all available data (high resolution) and of *GASBOR* [35], which folds a chain of dummy residues to reproduce the SAXS profile. This procedure was developed for proteins made of amino acid chains, and owing to the difference in size and density with nucleotides, the number of dummy residues should be multiplied by three. In both approaches, one will seek models with χ^2 close to 1 and NSD of the series below 0.7–0.9.

    If a crystal, NMR, or EM structure is available, its theoretical SAXS profile can be readily computed with *CRYSOL* [36] and compared to the experimental SAXS profile, thus providing a means to adjust/improve 3D structure predictions using normal mode analysis, for instance [50]. When no structure is known, the 2D structure of an RNA may be inferred from sequence alignment and covariations and assembled into 3D models with programs like *ASSEMBLE* or *RNACOMPOSER* [51, 52]. Resulting models can then be evaluated with SAXS data. Other approaches based on coarse-grained representation or on RNA modular modeling have also been applied to RNA reconstruction filtered by SAXS information [53–55].

    Finally, while SAXS alone hardly provides an unambiguous and unique answer, it is a very powerful tool in integrative structural biology as illustrated in Fig. 9. The combination of complementary information from biochemical and biophysical methods is useful to sort a series of ab initio models and to support a final proposition.

Acknowledgments

The authors thank the team of the SWING beamline (SOLEIL synchrotron, Saint-Aubin, France), in particular Pierre Roblin and Javier Pérez, for beam time provision and assistance during data collection and processing, as well as José Teixeira (LLB, CEA Saclay, France) for his critical reading of the manuscript and fruitful discussion. The work described in this chapter was supported by the French Centre National de la Recherche Scientifique, the University of Strasbourg, the LabEx consortium "NetRNA" (ANR-10-LABX-0036_NETRNA).

References

1. Putnam CD, Hammel M, Hura GL et al (2007) X-ray solution scattering (SAXS) combined with crystallography and computation: defining accurate macromolecular structures, conformations and assemblies in solution. Q Rev Biophys 40:191
2. Guinier A (1939) La diffraction des rayons X aux très petits angles: application à l'étude de phénomènes ultramicroscopiques. Ann Phys 11:161–237
3. Glatter O, Kratky O (1982) Small angle X-ray scattering. Academic Press Inc Ltd., London
4. Witz J (2003) 1964: The first model for the shape of a transfer RNA molecule. An account of an unpublished small-angle X-ray scattering study. Biochimie 85:1265–1268
5. Lake JA, Beeman WW (1967) Yeast transfer RNA: a small-angle X-ray study. Science 156:1371–1373
6. Ninio J, Luzzati V, Yaniv M (1972) Comparative small-angle X-ray scattering studies on unacylated, acylated and cross-linked Escherichia coli transfer RNAIVal. J Mol Biol 71:217–229
7. Svergun DI, Nierhaus KH (2000) A map of protein-rRNA distribution in the 70 S *Escherichia coli* ribosome. J Biol Chem 275:14432–14439
8. Chen Y, Pollack L (2016) SAXS studies of RNA: structures, dynamics, and interactions with partners. Wiley Interdiscip Rev RNA 7:512–526
9. Fang X-W, Yang X-J, Littrell K et al (2001) The Bacillus subtilis RNase P holoenzyme contains two RNase P RNA and two RNase P protein subunits. RNA 7:233–241
10. Russell R, Millett IS, Tate MW et al (2002) Rapid compaction during RNA folding. Proc Natl Acad Sci 99:4266–4271
11. Lipfert J, Ouellet J, Norman DG et al (2008) The complete vs ribozyme in solution studied by small-angle X-ray scattering. Structure 16:1357–1367
12. Hammond JA, Rambo RP, Filbin ME et al (2009) Comparison and functional implications of the 3D architectures of viral tRNA-like structures. RNA 15:294–307
13. Schmitt E, Panvert M, Lazennec-Schurdevin C et al (2012) Structure of the ternary initiation complex aIF2–GDPNP–methionylated initiator tRNA. Nat Struct Mol Biol 19:450–454
14. Jones CP, Cantara WA, Olson ED et al (2014) Small-angle X-ray scattering-derived structure of the HIV-1 5′ UTR reveals 3D tRNA mimicry. Proc Natl Acad Sci 111:3395–3400
15. Colussi TM, Costantino DA, Hammond JA et al (2014) The structural basis of transfer RNA mimicry and conformational plasticity by a viral RNA. Nature 511:366–369
16. Tuukkanen AT, Spilotros A, Svergun DI (2017) Progress in small-angle scattering from biological solutions at high-brilliance synchrotrons. IUCrJ 4:518–528
17. Mathew E, Mirza A, Menhart N (2004) Liquid-chromatography-coupled SAXS for accurate sizing of aggregating proteins. J Synchrotron Radiat 11:314–318
18. Watanabe Y, Inoko Y (2009) Size-exclusion chromatography combined with small-angle X-ray scattering optics. J Chromatogr A 1216:7461–7465
19. David G, Pérez J (2009) Combined sampler robot and high-performance liquid chroma-

tography: a fully automated system for biological small-angle X-ray scattering experiments at the Synchrotron SOLEIL SWING beamline. J Appl Crystallogr 42:892–900

20. Rambo RP, Tainer JA (2010) Improving small-angle X-ray scattering data for structural analyses of the RNA world. RNA 16:638–646
21. Jühling T, Duchardt-Ferner E, Bonin S et al (2018) Small but large enough: structural properties of armless mitochondrial tRNAs from the nematode Romanomermis culicivorax. Nucleic Acids Res 46:9170–9180
22. Baronti L, Karlsson H, Marušič M et al (2018) A guide to large-scale RNA sample preparation. Anal Bioanal Chem 410:3239–3252
23. Beckert B, Masquida B (2011) Synthesis of RNA by in vitro transcription. In: Nielsen H (ed) RNA. Humana Press, Totowa, pp 29–41
24. Franke D, Petoukhov MV, Konarev PV et al (2017) *ATSAS 2.8*: a comprehensive data analysis suite for small-angle scattering from macromolecular solutions. J Appl Crystallogr 50:1212–1225
25. Lipfert J, Doniach S (2007) Small-angle X-ray scattering from RNA, proteins, and protein complexes. Annu Rev Biophys Biomol Struct 36:307–327
26. Jacques DA, Trewhella J (2010) Small-angle scattering for structural biology—expanding the frontier while avoiding the pitfalls. Protein Sci 19:642–657
27. Mertens HDT, Svergun DI (2010) Structural characterization of proteins and complexes using small-angle X-ray solution scattering. J Struct Biol 172:128–141
28. Cantara WA, Olson ED, Musier-Forsyth K (2017) Analysis of RNA structure using small-angle X-ray scattering. Methods 113:46–55
29. Pollack L (2011) Time resolved SAXS and RNA folding. Biopolymers 95:543–549
30. Förster S, Apostol L, Bras W (2010) Scatter: software for the analysis of nano- and mesoscale small-angle scattering. J Appl Crystallogr 43:639–646
31. Hopkins JB, Gillilan RE, Skou S (2017) BioXTAS RAW: improvements to a free open-source program for small-angle X-ray scattering data reduction and analysis. J Appl Crystallogr 50:1545–1553
32. Fischer H, Neto MDO, Napolitano HB et al (2010) Determination of the molecular weight of proteins in solution from a single small-angle X-ray scattering measurement on a relative scale. J Appl Crystallogr 43:101–109
33. Rambo RP, Tainer JA (2013) Accurate assessment of mass, models and resolution by small-angle scattering. Nature 496:477–481
34. Franke D, Svergun DI (2009) *DAMMIF*, a program for rapid *ab-initio* shape determination in small-angle scattering. J Appl Crystallogr 42:342–346
35. Svergun DI, Petoukhov MV, Koch MH (2001) Determination of domain structure of proteins from X-ray solution scattering. Biophys J 80:2946–2953
36. Svergun D, Barberato C, Koch MHJ (1995) *CRYSOL*—a program to evaluate X-ray solution scattering of biological macromolecules from atomic coordinates. J Appl Crystallogr 28:768–773
37. Panjkovich A, Svergun DI (2016) SASpy: a PyMOL plugin for manipulation and refinement of hybrid models against small angle X-ray scattering data. Bioinformatics 32 (13):2062–2064
38. Trewhella J, Duff AP, Durand D et al (2017) 2017 publication guidelines for structural modelling of small-angle scattering data from biomolecules in solution: an update. Acta Crystallogr Sect Struct Biol 73:710–728
39. Francklyn C, Schimmel P (1989) Aminoacylation of RNA minihelices with alanine. Nature 337:478
40. Oakley JL, Strothkamp RE, Sarris AH et al (1979) T7 RNA polymerase: promoter structure and polymerase binding. Biochemistry 18:528–537
41. Rong M, He B, McAllister WT et al (1998) Promoter specificity determinants of T7 RNA polymerase. Proc Natl Acad Sci 95:515–519
42. Fechter P, Rudinger J, Giegé R, Théobald-Dietrich A (1998) Ribozyme processed tRNA transcripts with unfriendly internal promoter for T7 RNA polymerase: production and activity. FEBS Lett 436:99–103
43. Price SR, Ito N, Oubridge C et al (1995) Crystallization of RNA-protein complexes. I. Methods for the large-scale preparation of RNA suitable for crystallographic studies. J Mol Biol 249:398–408
44. Koubek J, Lin KF, Chen YR et al (2013) Strong anion-exchange fast performance liquid chromatography as a versatile tool for preparation and purification of RNA produced by in vitro transcription. RNA 19:1449–1459
45. Lorber B, Fischer F, Bailly M et al (2012) Protein analysis by dynamic light scattering: methods and techniques for students. Biochem Mol Biol Educ 40:372–382
46. Svergun DI (1992) Determination of the regularization parameter in indirect-transform methods using perceptual criteria. J Appl Crystallogr 25:495–503

47. Rambo RP, Tainer JA (2011) Characterizing flexible and intrinsically unstructured biological macromolecules by SAS using the Porod-Debye law. Biopolymers 95:559–571
48. Reyes FE, Schwartz CR, Tainer JA et al (2014) Methods for using new conceptual tools and parameters to assess RNA structure by small-angle X-ray scattering. Methods Enzymol 549:235–263
49. Durand D, Vivès C, Cannella D et al (2010) NADPH oxidase activator p67phox behaves in solution as a multidomain protein with semi-flexible linkers. J Struct Biol 169:45–53
50. Suhre K, Sanejouand Y-H (2004) ElNemo: a normal mode web server for protein movement analysis and the generation of templates for molecular replacement. Nucleic Acids Res 32: W610–W614
51. Jossinet F, Ludwig TE, Westhof E (2010) Assemble: an interactive graphical tool to analyze and build RNA architectures at the 2D and 3D levels. Bioinformatics 26:2057–2059
52. Biesiada M, Pachulska-Wieczorek K, Adamiak RW et al (2016) RNAComposer and RNA 3D structure prediction for nanotechnology. Methods 103:120–127
53. Yang S, Parisien M, Major F et al (2010) RNA structure determination using SAXS data. J Phys Chem B 114:10039–10048
54. Gajda MJ, Martinez Zapien D, Uchikawa E et al (2013) Modeling the structure of RNA molecules with small-angle X-ray scattering data. PLoS One 8:e78007
55. Bhandari YR, Fan L, Fang X et al (2017) Topological structure determination of RNA using small-angle X-ray scattering. J Mol Biol 429:3635–3649
56. Shi H, Moore PB (2000) The crystal structure of yeast phenylalanine tRNA at 1.93 A resolution: a classic structure revisited. RNA 6:1091–1105
57. Pinker F, Schelcher C, Fernandez-Millan P et al (2017) Biophysical analysis of *Arabidopsis* protein-only RNase P alone and in complex with tRNA provides a refined model of tRNA binding. J Biol Chem 292:13904–13913

Chapter 15

Site-Specific Spin Labeling of RNA for NMR and EPR Structural Studies

Bertrand Vileno and Isabelle Lebars

Abstract

Many RNA architectures were discovered to be involved in essential biological pathways acting as catalysts and/or regulators of gene expression, transcription, translation, splicing, or viral infection. The key to understand their diverse biological functions is to investigate their structure and dynamic. Nuclear Magnetic Resonance (NMR) is a powerful method to gain insight into these properties. However, the study of high-molecular-weight RNAs by NMR remains challenging. Advances in biochemical and NMR methods over the recent years allow to overcome the limitation of NMR. In particular, the incorporation of paramagnetic probes, coupled to the measurement of the induced effects on nuclear spins, has become an efficient tool providing long-range distance restraints and information on dynamic in solution. At the same time, the use of spin label enabled the application of Electron Paramagnetic Resonance (EPR) to study biological macromolecules. Combining NMR and EPR is emerging as a new approach to investigate the architecture of biological systems.

Here, we describe an efficient protocol to introduce a paramagnetic probe into a RNA at a specific position. This method enables various combinations of isotopic labeling for NMR and is also of interest for EPR studies.

Key words RNA, NMR, EPR, Paramagnetic, Spin-label

1 Introduction

RNAs play a crucial role in a wide range of biological processes in all organisms. The classical vision of RNA as a genetic code carrier has completely evolved since the discovery of ribozymes [1–5], riboswitches [5, 6], and small regulatory RNAs such as small interfering RNAs (siRNAs), microRNAs, or PiwiRNAs [7, 8]. RNAs achieve their biological functions through their remarkable ability to adopt various architectures and to undergo structural rearrangements upon binding to proteins, nucleic acids, or small molecules [9–14]. The key to understand their diverse biological functions is to investigate their structure and dynamic. Nuclear Magnetic Resonance (NMR) is a powerful method to gain insight into these properties [15–18]. However, the study of high-molecular-weight

Véronique Arluison and Frank Wien (eds.), *RNA Spectroscopy: Methods and Protocols*, Methods in Molecular Biology, vol. 2113, https://doi.org/10.1007/978-1-0716-0278-2_15,

RNAs by NMR remains challenging due to spectral crowding and broadened linewidths. Advances in biochemical and NMR methods over the recent years allow to overcome some limitations of NMR [16, 18–23]. In particular, the incorporation of paramagnetic probes, coupled to the measurement of paramagnetic relaxation enhancement (PRE) has become an efficient tool providing long-range distance restraints and information on dynamic in solution [24, 25]. The use of PREs allows the structural characterization of RNA and RNA-protein complexes [26–28]. The commonly used spin labels to study RNAs are commercially available nitroxides, such as 3-(2-iodoacetamido)-proxyl (proxyl) and (1-oxyl-2,2,5,5-tetramethyl-Δ3-pyrroline-3-methyl) methanethiosulfonate (MTSL).

Several techniques known as site-directed spin labeling (SDSL) have been developed to incorporate paramagnetic probes into RNA. The first main approach of SDSL relies on noncovalent spin labeling strategies. Helmling et al. described the use of a short chemically synthesized spin-labeled RNA subsequently hybridized to a large in vitro transcribed RNA [29]. Noncovalent spin labeling using abasic sites in DNA and RNA duplexes was also reported for introducing a nitroxide group [30]. More recently, a first example of SDSL of unmodified RNA based on an aptamer approach has been described [31]. The second main method to incorporate spin labels into RNA uses covalent approaches. Several techniques were developed to attach a probe at the RNA 5′-end. The 2,2,6,6-tetramethylpiperidine 1-oxyl (TEMPO) radical phosphoramidite building block was grafted to the 5′-terminus of RNA, allowing PRE measurement by NMR [27]. Enzymatic methods to attach a spin label to a thio-modified nucleotide were also reported. A guanosine monophosphorothioate was introduced at the 5′-end of a RNA using in vitro transcription by T7 RNA polymerase, enabling the linkage with the paramagnetic probe [32]. Grant and Qin used the T4 polynucleotide kinase to attach a phosphorothioate group at the 5′-terminus of a RNA, allowing its reaction with the iodomethyl derivative of nitroxide [33]. A variety of methods to link covalently a spin label at a chosen position into RNAs are also available. Spin-labeled phosphoramidites were synthesized and incorporated into the RNA during the solid-phase synthesis [34, 35]. The post-synthetic spin labeling of convertible nucleosides introduced into RNAs was also described [36]. One approach using convertible nucleosides combined to DNA-catalyzed RNA ligation was reported to site-directed spin label a long riboswitch [37, 38]. In the case of large RNA, an enzymatic method combining SDSL and segmental labeling has been developed [39]. Another protocol using enzymatic ligation of 4-thiouridine-modified small RNAs led to obtain large doubly spin-labeled RNA providing EPR long-distance measurement [40]. More recently, a novel approach combining in vitro transcription and click chemistry enabled the measurement of distances and corresponding distance distributions by PELDOR between two nitroxide groups in RNA duplex [41]. Combining NMR and EPR is emerging as a new approach to investigate the architecture of biological systems [42, 43].

In this chapter, we provide a detailed protocol for site-directed spin labeling RNA using a fully enzymatic method with yields compatible with NMR and EPR studies [39] . By contrast with solid-phase chemical synthesis methods, the enzymatic approach can be applied to large RNA molecules while maintaining the advantage of selective and segmental isotopic labeling with nitrogen-15 and carbon-13. In addition, this technique is of particular interest for EPR studies where two specific sites have to be labeled in order to monitor the dipolar interaction between two electronic spins. This method therefore enables various combinations of isotopic labeling and spin labeling for both EPR and NMR structural studies.

2 Materials

Prepare all solutions using ultrapure water (water purified on resin filters followed by deionization to achieve a resistivity of 18.2 MΩ cm at 25 °C). This will be indicated as "water" in the text. Work with specific precautions recommended for RNA manipulation [44].

2.1 Synthesis of RNAs by In Vitro Transcription

1. DNA templates resuspended in water at a concentration of 100 μM stored at −20 °C (*see* **Note 1**).
2. Unlabeled and $^{13}C/^{15}N$-labeled nucleotide triphosphates (rNTPs) supplied in aqueous solution titrated to a pH at 7.3–7.5 with NaOH stored at −20 °C at a concentration of 100 mM (*see* **Note 2**).
3. 6-Thioguanosine-5′-O-monophosphate (6-T-GMP, Biolog, Germany) dissolved in water at a concentration of 100 mM stored at −20 °C.
4. T7 RNA polymerase ("in-house") stored in aliquots at −20 °C, in 20 mM sodium phosphate pH 7.7, 1 mM DTT, 1 mM EDTA, 100 mM NaCl, and 50% glycerol.
5. Stock solutions stored at −20 °C in single-use aliquots: 50% (w/v) polyethylene glycol (PEG-8000), 1 M $MgCl_2$, and 0.28 M $MgCl_2$.
6. Triton X-100 stored at room temperature.
7. 10× Transcription buffer stored in single-use aliquots at −20 °C: 400 mM Tris–HCl pH 8.1, 10 mM spermidine, 0.1% (v/v) Triton X-100, 50 mM DTT [45, 46].
8. EDTA 0.5 M, pH 8.0, stored at 4 °C.

9. Phenol:chloroform saturated solution at pH 4.7 stored at 4 °C.
10. Ammonium acetate 3 M, pH 5.5, stored at 4 °C.
11. Absolute ethanol stored at −20 °C.

2.2 Ligation of RNA Segments

1. DNA splint resuspended in water at a concentration of 100 μM stored at −20 °C.
2. RNA fragments purified as described in Subheading 3.1.3.
3. T4 RNA ligase 2 supplied in 10 mM Tris–HCl pH 7.5, 50 mM KCl, 35 mM $(NH_4)_2SO_4$, 0.1 mM DTT, 0.1 mM EDTA, and 50% glycerol.
4. 10× reaction buffer: 500 mM Tris–HCl pH 7.5, 20 mM $MgCl_2$, 10 mM DTT, and 4 mM ATP.

2.3 Analytical Electrophoresis of RNA

1. 10× TBE buffer: 0.9 M Tris–borate, 10 mM EDTA, pH 8.3, stored at room temperature.
2. 19:1 (w/w) Acrylamide/bisacrylamide 40% solution.
3. Analytical polyacrylamide gel electrophoresis under denaturing conditions (7 M urea) in 0.5× TBE (*see* **Note 3**).
4. Analytical polyacrylamide gel electrophoresis under native conditions in 0.5× TBE (*see* **Note 3**).
5. *N*,*N*,*N*,*N*′-tetramethyl-ethylenediamine (TEMED).
6. Ammonium persulfate: prepare a 10% solution (w/v) in water and freeze in single-use aliquots at −20 °C.
7. 7 M Urea solution.
8. 6× Bromophenol blue/xylene cyanol loading dye: 0.25% (w/v) bromophenol blue, 0.25% (w/v) xylene cyanol blue, and 30% glycerol.
9. Staining toluidine blue solution: dissolve 0.6 g of toluidine blue in 240 mL ethanol, and 6 mL acetic acid, and complete with water to 600 mL.

2.4 Purifications of RNA

1. Gel-running buffer: 45 mM Tris–borate, 1 mM EDTA, pH 8.3, stored at room temperature.
2. PAGE equipment.
3. Electroelution apparatus (Elutrap, Schleicher & Schuell).
4. Elution buffer: 45 mM Tris–borate, 1 mM EDTA, pH 8.3, stored at room temperature.
5. Ammonium acetate 3 M, pH 5.5, stored at 4 °C.
6. Absolute ethanol stored at −20 °C.

2.5 Spin Labeling of RNA

1. 3-(2-Iodoacetamido)-proxyl.
2. RNA from Subheading 2.3.
3. Reaction buffer: 100 mM sodium phosphate, pH 8.0.
4. Absolute ethanol stored at −20 °C.

2.6 NMR Studies of RNA

1. Dialysis tubes (MW cutoff 3500).
2. Dialysis buffer: 50 mM sodium phosphate, pH 6.4.
3. Lyophilization apparatus.
4. 3-mm NMR microtube.
5. Deuterium oxide 100%.

2.7 EPR Studies of RNA

1. Glass capillaries (Hirschmann ring caps 20 up to 50 μL).
2. Hirschmann tube-sealing wax.
3. EPR tubes (Wilmad [Clear Fused Quartz] 2.97/1.98 mm OD/ID).
4. 3-(2-Iodoacetamido)-proxyl.
5. Dialysis buffer: 50 mM sodium phosphate, pH 6.4, supplemented by 10% v/v D_2O.

3 Methods

The method described here is based on a segmental approach using ligation technique with T4 RNA ligase 2 [47–49]. Using a DNA splint, one transcribed RNA is ligated to a second RNA fragment in which the thio-modified nucleotide (6-T-GMP) is introduced at its 5′-terminus by in vitro transcription. These two RNA segments correspond each to a part of the studied full-length RNA. The procedure is described in Fig. 1.

The first important step is the choice of an appropriate attachment site of the spin label. The targeted nucleotide needs to be accessible and the probe should be placed in a region where the structure will not be modified upon SDSL. It is therefore recommended to have as precise as possible a NMR characterization of the RNA. In addition, the attachment site that corresponds to the segmentation position has also to be chosen such as to optimize the transcription yield by the T7 RNA polymerase and therefore starting preferentially with a G nucleotide [45, 50].

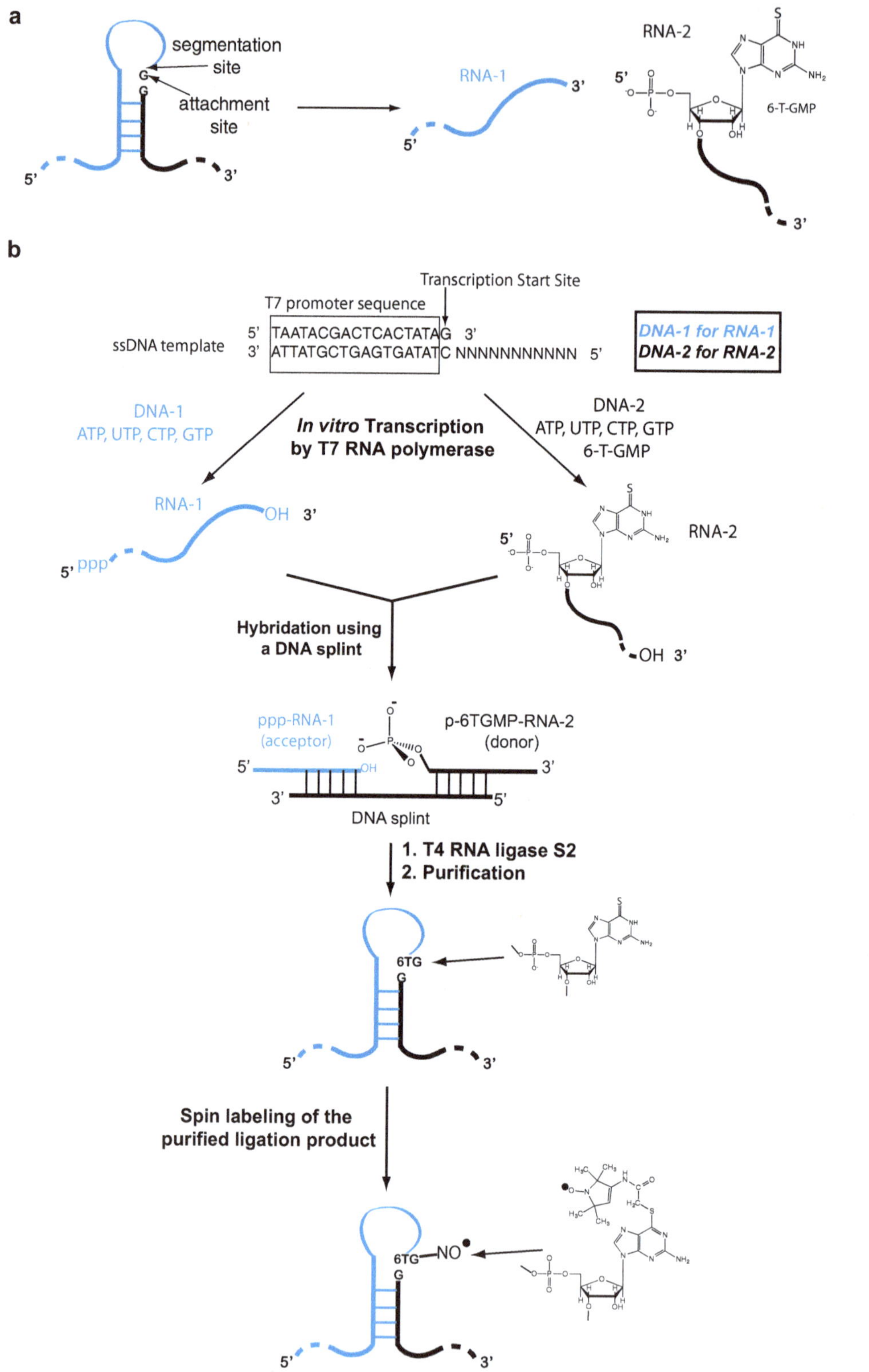

Fig. 1 Description of the method for site-specific spin labeling of RNA. (**a**) Choice of the attachment site. (**b**) Schematic representation of the approach to introduce a spin label at a chosen position into a RNA

3.1 Synthesis of RNA Segments by In Vitro Transcription Using T7 RNA Polymerase

3.1.1 Optimization of In Vitro Transcription Conditions

Transcription conditions are optimized with varying [$MgCl_2$]/[NTP] ratio from 0.8 to 2.0 in 25 μL reaction mixtures. The protocol below is detailed for [$MgCl_2$]/[NTP] ratio equal to 1. All stock solutions, except the T7 RNA polymerase, are brought to room temperature prior to mixing to avoid precipitation.

1. Mix the appropriate volume of water and 5 μL of 50% PEG 8000 in 1.5-mL tubes (*see* **Note 4**).
2. Add 2.5 μL of 10× transcription buffer and 1.4 μL of $MgCl_2$ (0.28 M).
3. Add appropriate volumes of each NTP and 6-T-GMP (only for RNA-2) (*see* **Note 4**).
4. Shake slowly the reaction mixture.
5. Add 2.5 μL of each DNA strand (*see* **Note 5**).
6. Add 1 μL of T7 RNA polymerase (final concentration: 0.04 mg/mL).
7. Incubate the reaction mixture at 37 °C for 4 h 30 min.
8. Add 5 μL of 0.5 M EDTA pH 8.0.
9. Vortex.
10. Centrifuge at room temperature, 2 min, at 5443 × *g*.
11. Mix 5 μL of the transcription reaction to 10 μL of 7 M urea solution and 3 μL of the 6× loading dye.
12. Pre-run the polyacrylamide denaturing gel for 15 min at 400 V in 0.5× TBE running buffer (*see* **Note 3**).
13. Rinse the wells thoroughly with 0.5× TBE solution prior to gel loading.
14. Load the sample.
15. Run the gel at 400 V in 0.5× TBE running buffer until the bromophenol blue reaches to the bottom of the gel.
16. Stain the gel with the toluidine blue solution for 15 min.
17. Remove the toluidine blue solution.
18. Destain the gel with water.
19. Evaluate the transcription yield (Fig. 2).

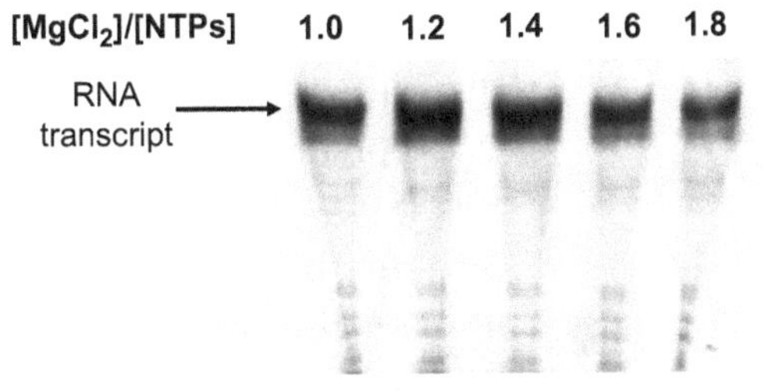

Fig. 2 Optimization of in vitro transcription yield. The best transcription yield is obtained at 1.2 [$MgCl_2$]/[NTP] ratio

3.1.2 Preparative Transcriptions

The protocol below is described at a 1.0 [$MgCl_2$]/[NTPs] ratio in 5-mL reaction volume.

1. Mix the appropriate volume of water and 1 mL of 50% PEG 8000 in 15-mL Falcon tube (*see* **Note 4**).
2. Add 500 μL of 10× transcription buffer and 80 μL of 1 M $MgCl_2$.
3. Add appropriate volumes of each NTP and 6-T-GMP (only for RNA-2) (*see* **Note 6**).
4. Shake slowly the reaction mixture.
5. Add 15 μL of each DNA strand (stock: 100 μM).
6. Add 20 μL of T7 RNA polymerase (final concentration: 0.04 mg/mL).
7. Incubate the reaction mixture at 37 °C for 4 h 30 min.
8. Add 0.5 mL of 0.5 M EDTA pH 8.0.
9. Mix.
10. Add 2.5 mL of phenol/chloroform pH 4.7.
11. Vortex.
12. Centrifuge for 10 min at room temperature at 8063 × *g*.
13. Transfer the aqueous phase in a 50-mL Falcon tube.
14. Add 2 mL of water to the phenol fraction.
15. Vortex.
16. Centrifuge for 10 min at room temperature at 8063 × *g*.
17. Combine the aqueous phase to the first one in the 50-mL Falcon tube.
18. Add 0.1 volume of ammonium acetate.
19. Mix.
20. Add 2.5 volumes of cold absolute ethanol.
21. Vortex.
22. Place at −20 °C overnight.

3.1.3 Purification of RNA Transcripts

1. Centrifugation at 8063 × *g* for 30 min at 4 °C.
2. Remove the liquid gently.
3. Resuspend the pellet in 1 mL of a 7 M urea solution.
4. Prepare a polyacrylamide denaturing (7 M urea) gel at the appropriate percentage of acrylamide with a two-well comb (*see* **Note 7**).
5. Pre-run the gel at a constant power of 40 W at about 50 °C in 0.5× TBE running buffer for 30 min.
6. Rinse the wells thoroughly with 0.5× TBE solution prior to gel loading.

7. Mix 200 μL of 7 M urea solution and 40 μL of 6× loading dye.
8. Load the mixture in the smaller well comb.
9. Load the pellet of RNA resuspended in 7 M urea solution in the second well comb.
10. Run the gel at a constant power of 40 W in 0.5× TBE running buffer until the bromophenol blue reaches to the bottom of the gel.
11. Remove the plates from the gel.
12. Place the gel on clean and clear cellophane over a silica chromatography plate.
13. Visualize RNA by UV shadowing at 254 nM.
14. Excise the band corresponding to the expected transcript RNA with a clean scalpel.
15. Elute the RNA from the gel slices using an Elutrap Electroelution System: recover the RNA in the sample chamber every hour, until the absorbance decreases.
16. Combine all fractions containing the desired RNA.
17. Add 0.1 volume of 3 M ammonium acetate pH 5.5.
18. Mix.
19. Add 2.5 volumes of cold absolute ethanol.
20. Place at −20 °C overnight.
21. Centrifuge at 8063 × *g* for 30 min at 4 °C.
22. Wash the pellet with 70% ethanol.
23. Resuspend the pellet in water.

3.2 Ligation of RNA Segments

The ligation by T4 RNA ligase 2 enables to link an acceptor RNA to a donor RNA modified at its 5′-end with 6-T-GMP using a DNA splint [49]. Prior ligation, the correct hybridization of the two RNA segments to the complementary DNA splint has to be verified to avoid the formation of side products.

The procedure is described in Fig. 1.

3.2.1 Hybridization of RNAs to the DNA Splint

1. Mix equimolar amounts of the two RNA fragments and the DNA splint at a final concentration of 10 μM each, in the 1× T4 RNA ligase 2 buffer, in a 10 μL reaction volume.
2. Heat the mixture at 95 °C (2 min) and snap-cool at 4 °C (5 min).
3. Centrifuge at 6720 × *g* for 5 min at room temperature.
4. Pre-run a native polyacrylamide gel for 15 min at 150 V in 0.5× TBE running buffer at 4 °C (*see* **Note 3**).
5. Rinse the wells thoroughly with 0.5× TBE solution prior to gel loading.

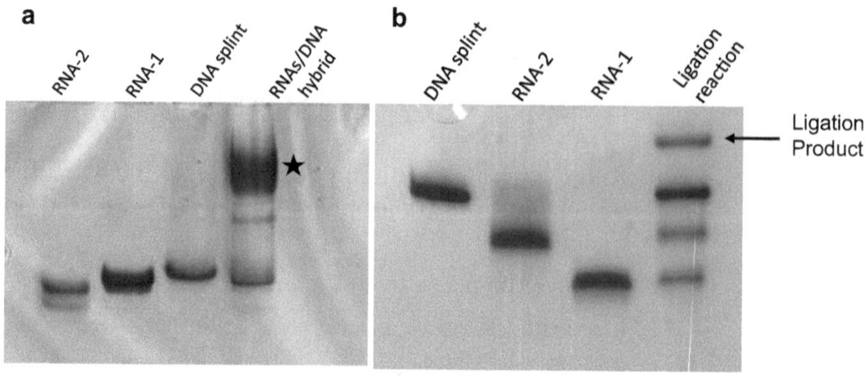

Fig. 3 (**a**) Free RNAs, DNA splint, and RNA/DNA hybrid are visualized on a native polyacrylamide gel. The hybrid is indicated by a star. (**b**) Free RNAs, DNA splint, and ligation reaction are visualized on a denaturing polyacrylamide gel

6. Load the sample.
7. Run the gel at 400 V in 0.5× TBE running buffer until the bromophenol blue reaches to the bottom of the gel.
8. Stain the gel with the toluidine blue solution for 15 min.
9. Remove the toluidine blue solution.
10. Destain the gel with water.
11. Free oligonucleotides and RNA/DNA hybrid are visualized (Fig. 3).

3.2.2 Ligation of RNA Fragments

1. Mix equimolar amounts of the two RNA fragments and the DNA splint at a final concentration of 30 μM each, in 1× T4 RNA ligase 2 buffer, in a 5 mL reaction volume (*see* **Note 8**).
2. Heat the mixture at 95 °C (5 min) and snap-cool at 4 °C (15 min).
3. Centrifuge at 8063 × *g* for 10 min at room temperature.
4. Add 2.25 kU of T4 RNA ligase 2.
5. Incubate at 37 °C overnight.
6. Add 0.1 volume of 3 M ammonium acetate pH 5.5 and 2.5 volumes of cold absolute ethanol.
7. Mix.
8. Place at −20 °C overnight.
9. Purify the ligation product as described in Subheading 3.1.3.
10. Precipitate by addition of 0.1 volume of 3 M ammonium acetate pH 5.5 and 2.5 volumes of cold absolute ethanol.
11. Place at −20 °C overnight.

3.3 Site-Specific Spin Labeling of RNA

All the following steps have to be performed in the dark to prevent light-induced degradation.

1. Centrifuge at 8063 × *g* for 30 min at 4 °C.
2. Wash the pellet with cold 70% ethanol solution.
3. Dry the pellet.
4. Resuspend the pellet in 400 μL of 100 mM sodium phosphate buffer pH 8.
5. Evaluate the concentration of the ligation product by UV absorption using a Nanodrop Spectrophotometer.
6. Dissolve 3-(2-iodoacetamido)-proxyl in absolute ethanol at a concentration of 0.15 M.
7. Add a 100-fold excess of 3-(2-iodoacetamido)-proxyl solution to the RNA.
8. Incubate at 40 °C for 7 h under stirring at 300 rpm.
9. Add again a 100-fold excess of 3-(2-iodoacetamido)-proxyl solution to the RNA.
10. Incubate at 40 °C overnight under stirring at 300 rpm.
11. Precipitate by addition of 0.1 volume of 3 M ammonium acetate pH 5.5 and 2.5 volumes of ethanol.
12. Place at −20 °C overnight.
13. Centrifuge at 8063 × *g* at 4 °C for 30 min.
14. Remove the supernatant.
15. Resuspend the pellet in water.
16. Dialyze the spin-labeled RNA against the buffer used for NMR and EPR experiments.
17. Change the dialysis buffer four times.

3.4 NMR Experiments

All the following steps have to be performed in the dark to prevent light-induced degradation.

1. Concentrate the sample by lyophilization.
2. Resuspend in 90/10 H_2O/D_2O in 150 μL.
3. Transfer the sample in a 3-mm NMR tube.
4. Record one-dimensional NMR spectra (spectral width of 20 ppm) at several temperatures to check the sample quality (Fig. 4; *see* **Note 9**).
5. Record all classical experiments for RNA structural studies by NMR and PRE measurements [16, 51, 52] (*see* **Note 9**).

3.5 EPR Experiments

EPR investigation is here performed to: (1) assess whether the spin label (i.e., the proxyl) is covalently bound to the RNA molecule and if yes how the motion of the rotamer is restricted (Fig. 5; *see* **Note 10**), (2) detect the presence of potential free spin label and

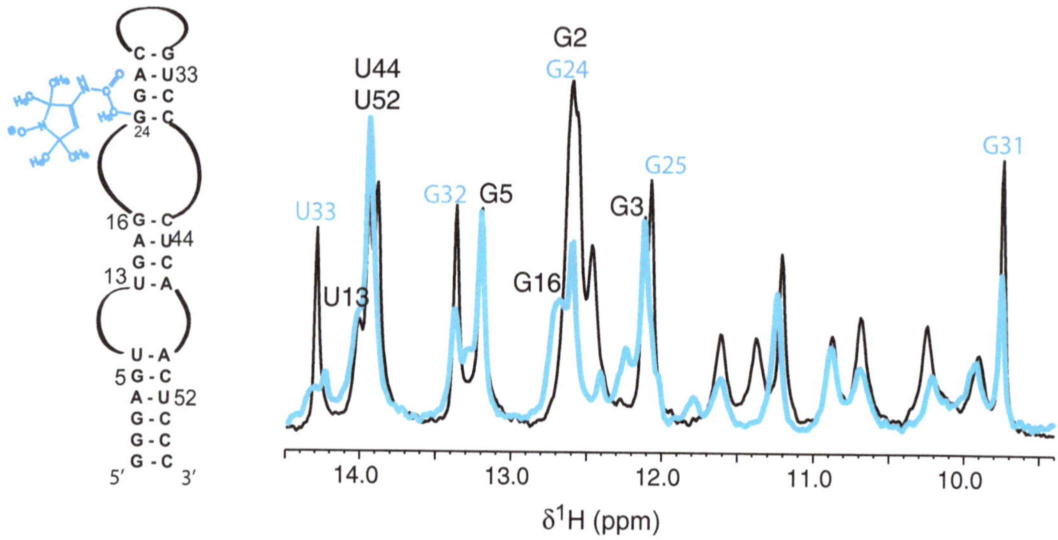

Fig. 4 NMR of a site-specific spin-labeled RNA [39]. Imino proton region of 1D spectra, recorded at 20 °C, of the free RNA (black) and of the SL-RNA (blue; *see* **Note 9**)

define the proxyl-RNA coupling efficiency. All the following steps must be performed in the dark to prevent potential light-induced degradation.

3.5.1 Preparation of RNA Samples

1. Load resuspended sample (from Subheading 3.4) into 20 μL glass capillaries sealed at both ends and place it into an EPR quartz tube (3/2 mm OD/ID).
2. Place it into the EPR cavity so that the whole length of the cavity is filled by the sample.
3. Record conventional X-band field swept EPR spectra centered around g of ca. 2.0055 (Fig. 6; *see* **Note 11**).
4. After potential baseline subtraction, perform a double integration of the spectra in order to obtain the EPR intensity (arb. unit), which is directly proportional to the number of paramagnetic centers detected within the cavity.

3.5.2 Spin Labeling Efficiency via Nitroxide Standards

Proceed following an identical protocol as for RNA sample measurements.

1. Prepare several 3-(2-iodoacetamido)-proxyl solutions into the buffer used for EPR and NMR investigations, with known concentrations around the expected RNA sample one.
2. Load each sample into 20 μL glass capillaries sealed at both ends.
3. Transfer capillaries into EPR quartz tubes, placed into the EPR cavity.

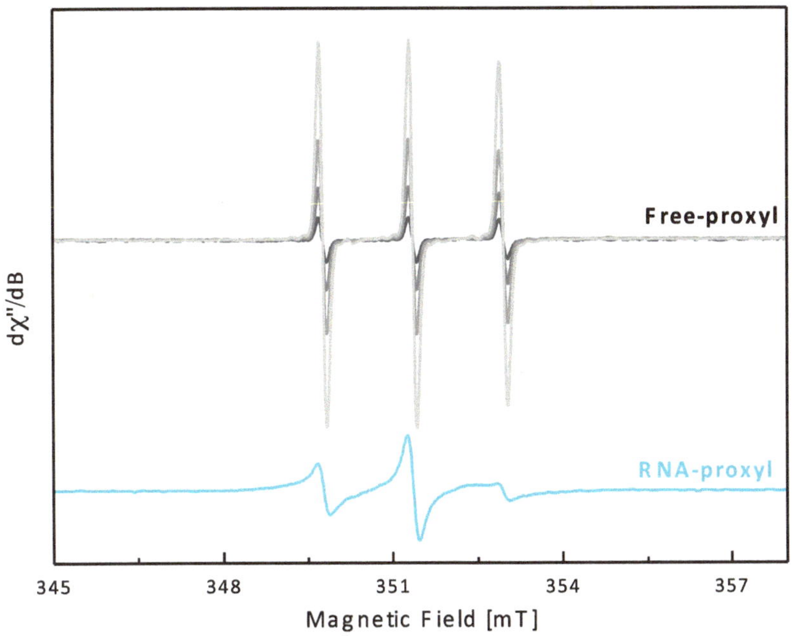

Fig. 5 EPR spectra of free proxyl taken at various concentrations ranging from 10 (black curve) to 200 μM (light gray curve) and compared to proxyl-labeled RNA (light blue curve). All spectra were acquired in phosphate buffer (50 mM, pH 6.4) supplemented by 10% v/v D_2O with identical experimental parameters. In this case, the concentration of the spin-labeled RNA was found to be 150 ± 15 μM

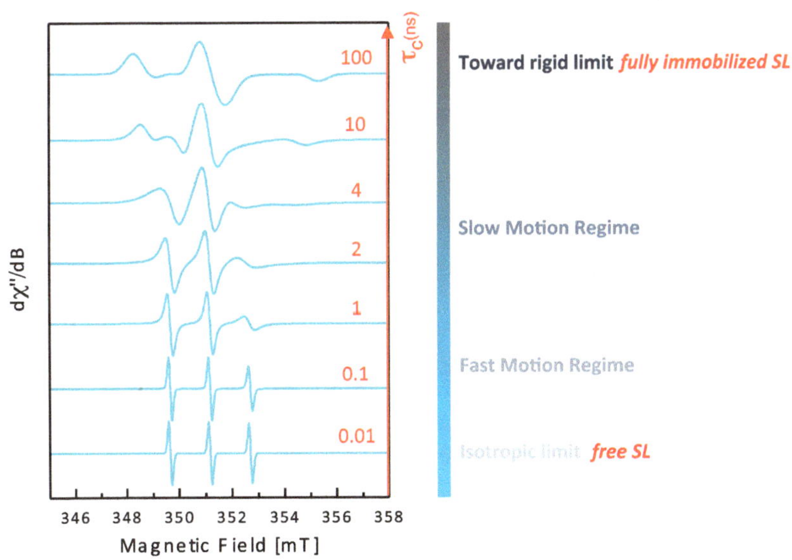

Fig. 6 Evolution of the EPR line shape for a nitroxide radical as a function of isotropic rotational correlation time (τ_r) ranging from 10 ps to 100 ns. This schematic representation highlights the four dynamical regime accessible by continuous-wave X-band EPR ranging from the isotropic limit (e.g., free spin-label) and fast motion regime to the slow motion regime and rigid limit (corresponding to fully immobilized nitroxide). Simulations were achieved using Easyspin toolbox [54]

4. Record conventional X-band field swept EPR spectra with the identical experimental parameters as for the RNA samples (Fig. 6; *see* **Note 11**).
5. After baseline subtraction (if attended for the RNA sample), perform a double integration of the spectra in order to obtain EPR intensity (arb. units), which can be compared to the RNA sample intensities to assess their spin concentrations and therefore the coupling efficiency.

4 Notes

1. DNA templates can be prepared according to three options: linearized plasmid, single-stranded, or double-stranded synthetic DNA templates. Here, the protocol describes the use of single-stranded or double-stranded synthetic DNA templates. DNA templates are purchased from SIGMA, IDT, or Eurogentec. Milligan et al. first used synthetic DNA templates and T7 RNA polymerase for the synthesis of short oligonucleotides [45, 50]. The efficiency of the transcription depends on the specific sequence at the 5′-end of the RNA: one or more guanine residues are required to improve the transcription yield. Care should be therefore taken when designing DNA templates.
2. The full-length RNA can be uniformly labeled by using $^{13}C/^{15}N$ rNTPs. Combinations of unlabeled and $^{13}C/^{15}N$-labeled rNTPs can also be used to produce selectively labeled A or U or C or G residues or any combinations. In addition, the ligation step in our protocol allows the segmental labeling of the RNA. This offers various possibilities of isotopic labeling schemes.
3. Polyacrylamide gels have to be prepared at an appropriate percentage according to the size of the RNA [53]. Analytical gel electrophoresis under denaturing conditions contains the appropriate percentage of acrylamide/bisacrylamide, 7 M urea in 0.5× TBE. Analytical gel electrophoresis under native conditions contains the appropriate percentage of acrylamide/bisacrylamide in 0.5× TBE.
4. Nucleotide triphosphates (rNTPs) and 6-T-GMP stock solutions are stored at a concentration of 100 mM. The first RNA segment (RNA-1) is transcribed in the presence of each NTP at a concentration of 4 mM: add 1 μL of each NTP. The appropriate volume of water is 6.1 μL. In the case of the 5′-thio-modified (6-T-GMP) RNA fragment (RNA-2), ATP, UTP, and CTP are added at a concentration of 4 mM, while the GTP and 6-T-GMP concentrations are, respectively, 3 mM and 9.1 mM:

add 1 μL of ATP, 1 μL of UTP, 1 μL of CTP, 0.8 μL of GTP, and 2.3 μL of 6-T-GMP. The appropriate volume of water is 4 μL.

5. For transcription tests, the DNA strand stocks at 100 μM are diluted to a concentration of 3 μM in order to add 2.5 μL of DNA. The final DNA template concentrations are 300 nM in 25 μL reaction volume.
6. Nucleotide triphosphates (rNTPs) and 6-T-GMP stock solutions are stored at a concentration of 100 mM. The first RNA (RNA-1) segment is transcribed with each NTP at a concentration of 4 mM: add 200 μL of each NTP. The appropriate volume of water is 2.57 mL. In the case of the 5′-thio-modified (6-T-GMP) RNA fragment (RNA-2), ATP, UTP, and CTP are added at a concentration of 4 mM, while the GTP and 6-T-GMP concentrations are, respectively, 3 mM and 9.1 mM: add 200 μL of ATP, 200 μL of UTP, 200 μL of CTP, 152 μL of GTP, and 456 μL of 6-T-GMP. The appropriate volume of water is 2.16 mL.
7. For preparative gels, the percentage of acrylamide/bisacrylamide is adjusted according to the size of the RNA. The size and the thickness of the gel are, respectively, 20 cm * 42 cm and 3 mm.
8. To optimize the ligation yield, a series of test experiments are performed in 10 μL with varying reaction times and RNA and enzyme concentrations.
9. Before PRE analysis, it is necessary to have as complete as possible the NMR chemical shift assignments of the free RNA. The spectra recorded for the spin-labeled RNA can next be compared to the free RNA to measure and analyze PRE effects. One-dimensional NMR spectra recorded over a spectral width of 20 ppm allow the analysis of imino proton found between 9 and 15 ppm. They constitute a good probe of the overall secondary structure of the RNA molecule. Their observation points out their protection from exchange with the solvent and therefore their involvement in hydrogen bonds. The resonances corresponding to those protons found between 12 and 15 ppm are characteristic of imino protons involved in Watson-Crick base pairs, whereas those found upfield (between 9 and 12 ppm) indicate the formation of noncanonical base-pairing. Assignment of these resonances provides thus precious information about the folding of the RNA. The comparison of the spectra of the wild-type RNA and the spin-labeled RNA allows to evaluate: (1) the efficiency of the reaction, (2) the quality of the sample, and (3) the impact of the paramagnetic probe on the global fold of the RNA. Then, all classical experiments for RNA structural studies by NMR and PRE

measurements can be performed and compared to the RNA-wt [16, 51, 52]. As exemplified in Fig. 4, the introduction of a spin label at position G24 induces reduced intensities of G24, G25, G31, G32, and U33, while G3 and G5 remain unaffected.

10. In solution the dynamic of the spin label, characterized by the tumbling of the molecule, is observable by EPR and characterized by the so-called rotational correlation time (τ_r) [54–56]. The line shape of the nitroxide derivative EPR spectrum gives a direct indication whether the spin probe is free or attached to a much more weighted molecule. In the first case the molecule undergoes a rapid tumbling and its corresponding spectrum illustrates the isotropic limit that consists in three sharp lines of nearly equal intensities ($\tau_r < 10^{-10}$ s; *see* Fig. 6). If covalently attached to a much larger molecule, the tumbling is slowed down accordingly as it is now restricted by the host molecule, potential steric hindrance, and/or restriction of the potential rotamer configurations. In such case, the spin probe goes from fast and slow motion regime towards the rigid limit corresponding to the fully immobilized spin label ($\tau_r > 10^{-7}$ s; *see* Fig. 6). Consequently, changes within the paramagnetic probe environment can be reflected by a change within the spectral nitroxide line shape. To evaluate the rotational correlation time, several methods are available ranging from the simulation of the experimental spectrum by using dedicated software/programs (e.g., the Easyspin toolbox under Matlab [Mathworks] environment [57, 58]) to various empirical methods [54, 55, 59]. It allows to define and deconvolute potential configuration of the spin label rotamer vs. the host molecule (*see* Fig. 7).

11. All experiments are performed at room temperature. An example of the principal experimental parameters is: X-band EPR spectrometer (e.g., EMXplus Bruker Biospin GmbH), equipped with a resonator (e.g., high-sensitivity resonator [4119HS-W1, Bruker]). The g factor is calibrated in the experimental conditions (e.g., using the Bruker strong pitch [g = 2.0028]): center field, ca. 3500 G; sweep width, ca. 15 mT; microwave power, ca. 2 mW; modulation amplitude, ca. 0.5 G (usually taken at least 30% inferior to the natural width of the free spin label to avoid overmodulation); conversion time, ca. 100 ms; and time constant, ca. 50 ms (as a safe rule of thumb the time constant is usually set up inferior to 50% of the conversion time to avoid spectrum distortion). Several scans can be accumulated to achieve a reasonable signal-to-noise ratio.

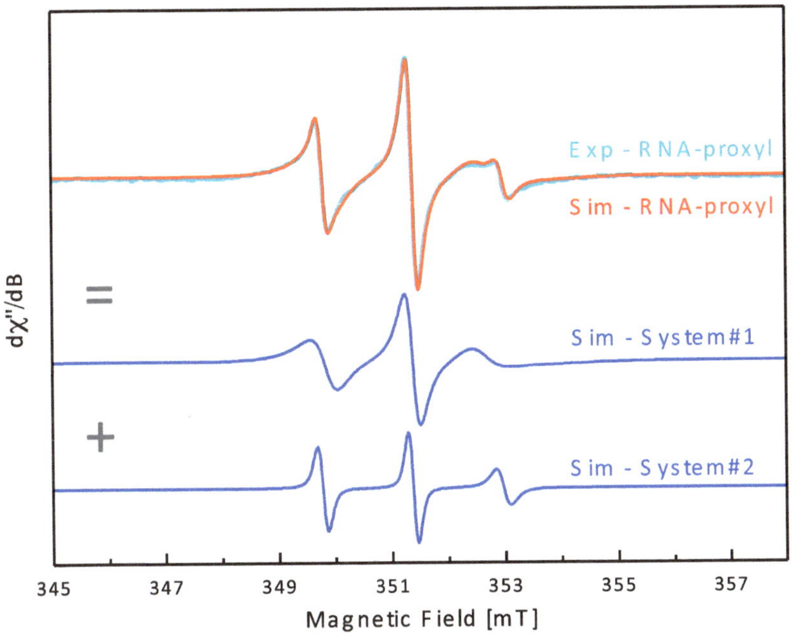

Fig. 7 Experimental EPR spectrum of proxyl-labeled RNA (light blue curve) and corresponding simulation (red curve). Simulation investigation allows to deconvolute the overall EPR fingerprint as the superposition of two spin systems at distinct dynamical regimes. It corresponds to two rotamer configurations of the proxyl, characterized by two rotational correlation times (τ_r) of ca. 4 $\pm$ 0.5 ns (slow motion regime, upper blue curve) and 0.8 $\pm$ 2 ns (fast motion regime, lower blue curve) with a ratio of 4:1, respectively. Noteworthy, no free spin label was detected, testimony of the spin labeling efficiency

References

1. Joyce GF (2002) The antiquity of RNA-based evolution. Nature 418:214–221
2. Kruger K, Grabowski PJ, Zaug AJ, Sands J, Gottschling DE, Cech TR (1982) Self-splicing RNA: autoexcision and autocyclization of the ribosomal RNA intervening sequence of Tetrahymena. Cell 31:147–157
3. Doudna JA, Cech TR (2002) The chemical repertoire of natural ribozymes. Nature 418:222–228
4. Walter NG (2007) Ribozyme catalysis revisited: is water involved? Mol Cell 28:923–929
5. Serganov A, Patel DJ (2007) Ribozymes, riboswitches and beyond: regulation of gene expression without proteins. Nat Rev Genet 8:776–790
6. Mandal M, Breaker RR (2004) Gene regulation by riboswitches. Nat Rev Mol Cell Biol 5:451–463
7. Grosshans H, Filipowicz W (2008) Molecular biology: the expanding world of small RNAs. Nature 451:414–416
8. Hannon GJ (2002) RNA interference. Nature 418:244–251
9. Bevilacqua PC, Blose JM (2008) Structures, kinetics, thermodynamics, and biological functions of RNA hairpins. Annu Rev Phys Chem 59:79–103
10. Brierley I, Digard P, Inglis SC (1989) Characterization of an efficient coronavirus ribosomal frameshifting signal: requirement for an RNA pseudoknot. Cell 57:537–547
11. Brunel C, Marquet R, Romby P, Ehresmann C (2002) RNA loop-loop interactions as dynamic functional motifs. Biochimie 84:925–944
12. Derrigo M, Cestelli A, Savettieri G, Di Liegro I (2000) RNA-protein interactions in the control of stability and localization of messenger RNA (review). Int J Mol Med 5:111–123
13. Kolb FA, Engdahl HM, Slagter-Jäger JG, Ehresmann B, Ehresmann C, Westhof E, Wagner EG, Romby P (2000) Progression of a loop-loop complex to a four-way junction is crucial for the activity of a regulatory antisense RNA. EMBO J 19:5905–5915

14. Tomizawa J, Som T (1984) Control of ColE1 plasmid replication: enhancement of binding of RNA I to the primer transcript by the Rom protein. Cell 38:871–878
15. Allen M, Varani L, Varani G (2001) Nuclear magnetic resonance methods to study structure and dynamics of RNA-protein complexes. Methods Enzymol 339:357–376
16. Latham MP, Brown DJ, McCallum SA, Pardi A (2005) NMR methods for studying the structure and dynamics of RNA. Chembiochem 6:1492–1505
17. Fürtig B, Buck J, Manoharan V, Bermel W, Jäschke A, Wenter P, Pitsch S, Schwalbe H (2007) Time-resolved NMR studies of RNA folding. Biopolymers 86:360–383
18. Getz M, Sun X, Casiano-Negroni A, Zhang Q, Al-Hashimi HM (2007) NMR studies of RNA dynamics and structural plasticity using NMR residual dipolar couplings. Biopolymers 86:384–402
19. Dominguez C, Schubert M, Duss O, Ravindranathan S, Allain FH-T (2011) Structure determination and dynamics of protein–RNA complexes by NMR spectroscopy. Prog Nucl Magn Reson Spectrosc 58:1–61
20. Duss O, Lukavsky PJ, Allain FH-T (2012) Isotope labeling and segmental labeling of larger RNAs for NMR structural studies. In: Atreya HS (ed) Isotope labeling in biomolecular NMR. Springer, Dordrecht, pp 121–144
21. Yadav DK, Lukavsky PJ (2016) NMR solution structure determination of large RNA-protein complexes. Prog Nucl Magn Reson Spectrosc 97:57–81
22. Schlundt A, Tants J-N, Sattler M (2017) Integrated structural biology to unravel molecular mechanisms of protein-RNA recognition. Methods 118–119:119–136
23. Barnwal RP, Yang F, Varani G (2017) Applications of NMR to structure determination of RNAs large and small. Arch Biochem Biophys 628:42–56
24. Cai S, Zhu L, Zhang Z, Chen Y (2007) Determination of the three-dimensional structure of the Mrf2–DNA complex using paramagnetic spin labeling. Biochemistry 46:4943–4950
25. Hennig J, Warner LR, Simon B, Geerlof A, Mackereth CD, Sattler M (2015) Structural analysis of protein–RNA complexes in solution using NMR paramagnetic relaxation enhancements. Methods Enzymol 558:333–362
26. Ramos A, Varani G (1998) A new method to detect long-range protein–RNA contacts: NMR detection of electron–proton relaxation induced by nitroxide spin-labeled RNA. J Am Chem Soc 120:10992–10993
27. Wunderlich CH, Huber RG, Spitzer R, Liedl KR, Kloiber K, Kreutz C (2013) A novel paramagnetic relaxation enhancement tag for nucleic acids: a tool to study structure and dynamics of RNA. ACS Chem Biol 8:2697–2706
28. Iwahara J, Clore GM (2006) Detecting transient intermediates in macromolecular binding by paramagnetic NMR. Nature 440:1227–1230
29. Helmling C, Bessi I, Wacker A, Schnorr KA, Jonker HRA, Richter C, Wagner D, Kreibich M, Schwalbe H (2014) Noncovalent spin labeling of riboswitch RNAs to obtain long-range structural NMR restraints. ACS Chem Biol 9:1330–1339
30. Shelke SA, Sandholt GB, Sigurdsson ST (2014) Nitroxide-labeled pyrimidines for non-covalent spin-labeling of abasic sites in DNA and RNA duplexes. Org Biomol Chem 12:7366–7374
31. Saha S, Hetzke T, Prisner TF, Sigurdsson ST (2018) Noncovalent spin-labeling of RNA: the aptamer approach. Chem Commun (Camb) 54:11749–11752
32. Macosko JC, Pio MS, Tinoco I, Shin YK (1999) A novel 5 displacement spin-labeling technique for electron paramagnetic resonance spectroscopy of RNA. RNA 5:1158–1166
33. Grant GPG, Qin PZ (2007) A facile method for attaching nitroxide spin labels at the 5′ terminus of nucleic acids. Nucleic Acids Res 35:e77
34. Höbartner C, Sicoli G, Wachowius F, Gophane DB, Sigurdsson ST (2012) Synthesis and characterization of RNA containing a rigid and nonperturbing cytidine-derived spin label. J Org Chem 77:7749–7754
35. Gophane DB, Endeward B, Prisner TF, Sigurdsson ST (2018) A semi-rigid isoindoline-derived nitroxide spin label for RNA. Org Biomol Chem 16:816–824
36. Shelke SA, Sigurdsson ST (2012) Site-directed spin labelling of nucleic acids. Eur J Org Chem 2012:2291–2301
37. Büttner L, Seikowski J, Wawrzyniak K, Ochmann A, Höbartner C (2013) Synthesis of spin-labeled riboswitch RNAs using convertible nucleosides and DNA-catalyzed RNA ligation. Bioorg Med Chem 21:6171–6180
38. Wawrzyniak-Turek K, Höbartner C (2014) Deoxyribozyme-mediated ligation for incorporating EPR spin labels and reporter groups into RNA. Methods Enzymol 549:85–104
39. Lebars I, Vileno B, Bourbigot S, Turek P, Wolff P, Kieffer B (2014) A fully enzymatic

method for site-directed spin labeling of long RNA. Nucleic Acids Res 42:e117

40. Duss O, Yulikov M, Jeschke G, Allain FH-T (2014) EPR-aided approach for solution structure determination of large RNAs or protein–RNA complexes. Nat Commun 5:3669
41. Domnick C, Hagelueken G, Eggert F, Schiemann O, Kath-Schorr S (2019) Posttranscriptional spin labeling of RNA by tetrazine-based cycloaddition. Org Biomol Chem 17:1805.
42. Qin PZ, Dieckmann T (2004) Application of NMR and EPR methods to the study of RNA. Curr Opin Struct Biol 14:350–359
43. Duss O, Yulikov M, Allain FHT, Jeschke G (2015) Combining NMR and EPR to determine structures of large RNAs and protein-RNA complexes in solution. Methods Enzymol 558:279–331
44. Nielsen H (2011) Working with RNA. In: Nielsen H (ed) RNA. Humana Press, Totowa, pp 15–28
45. Milligan JF, Groebe DR, Witherell GW, Uhlenbeck OC (1987) Oligoribonucleotide synthesis using T7 RNA polymerase and synthetic DNA templates. Nucleic Acids Res 15:8783–8798
46. Wyatt JR, Chastain M, Puglisi JD (1991) Synthesis and purification of large amounts of RNA oligonucleotides. BioTechniques 11:764–769
47. Nelissen FHT, van Gammeren AJ, Tessari M, Girard FC, Heus HA, Wijmenga SS (2008) Multiple segmental and selective isotope labeling of large RNA for NMR structural studies. Nucleic Acids Res 36:e89
48. Romaniuk PJ, Uhlenbeck OC (1983) Joining of RNA molecules with RNA ligase. Methods Enzymol 100:52–59
49. Bullard DR, Bowater RP (2006) Direct comparison of nick-joining activity of the nucleic acid ligases from bacteriophage T4. Biochem J 398:135–144
50. Milligan JF, Uhlenbeck OC (1989) Synthesis of small RNAs using T7 RNA polymerase. Methods Enzymol 180:51–62
51. Scott LG, Hennig M (2008) RNA structure determination by NMR. In: Keith JM (ed) Bioinformatics. Humana Press, Totowa, pp 29–61
52. Fürtig B, Richter C, Wöhnert J, Schwalbe H (2003) NMR spectroscopy of RNA. Chembiochem 4:936–962
53. Sambrook J, Fritsch E, Maniatis T (1989) Molecular cloning: a laboratory manual, vol vol. 2, 2nd edn. Cold Spring Harbor Laboratory Press, Cold Spring Harbor
54. Fajer PG (2006) Electron spin resonance spectroscopy labeling in peptide and protein analysis. In: Encyclopedia of analytical chemistry. https://doi.org/10.1002/9780470027318.a1609
55. Oppenheim SF, Buettner GR, Rodgers VGJ (1996) Relationship of rotational correlation time from EPR spectroscopy and protein-membrane interaction. J Membr Sci 118:133–139
56. Stoll S, Schweiger A (2007) EasySpin: simulating cw ESR spectra. Magn Reson 27:299–321
57. Stoll S, Schweiger A (2006) EasySpin, a comprehensive software package for spectral simulation and analysis in EPR. J Magn Reson 178:42–55
58. Etienne E, Le Breton N, Martinho M, Mileo E, Belle V (2017) SimLabel: a graphical user interface to simulate continuous wave EPR spectra from site-directed spin labeling experiments. Magn Reson Chem 55:714–719
59. Qin PZ, Butcher SE, Feigon J, Hubbell WL (2001) Quantitative analysis of the isolated GAAA tetraloop/receptor interaction in solution: a site-directed spin labeling study. Biochemistry 40:6929–6936

Chapter 16

Analysis of the HIV-1 Genomic RNA Dimerization Initiation Site Binding to Aminoglycoside Antibiotics Using Isothermal Titration Calorimetry

Serena Bernacchi and Eric Ennifar

Abstract

Isothermal titration calorimetry (ITC) provides a sensitive, powerful, and accurate tool to suitably analyze the thermodynamic of RNA binding events. This approach does not require any modification or labeling of the system under analysis and is performed in solution. ITC is a very convenient technique that provides an accurate determination of binding parameters, as well as a complete thermodynamic profile of the molecular interactions. Here we show how this approach can be used to characterize the interactions between the dimerization initiation site (DIS) RNA localized within the HIV-1 viral genome and aminoglycoside antibiotics. Our ITC study showed that the 4,5-disubstituted 2-desoxystreptamine (2-DOS) aminoglycosides can bind the DIS with a nanomolar affinity and a high specificity.

Key words HIV-1, Viral RNA, Aminoglycosides, Dimerization initiation site, ITC, Thermodynamics, RNA–drug interaction

1 Introduction

Interaction of biological macromolecules either with each other or with small ligands constitutes the basis for molecular recognition. The development of the approaches providing a detailed understanding of these interactions at the molecular level is required to facilitate the design and development of drugs. In this frame, the isothermal titration calorimetry (ITC) is an equilibrium solution technique that directly *characterizes* all parameters regulating the molecular interactions [1–4]. In a typical experiment, aliquots of a titrant in the syringe are injected into the cell containing the sample. Upon each titration, the amount of heat either released or absorbed during the binding event is measured, leading in one single experiment the determination of all the binding parameters: the equilibrium dissociation constant (K_d), the reaction stoichiometry (n), the binding enthalpy (ΔH), the binding entropy (ΔS), and the Gibbs free energy change (ΔG). Moreover, the analysis at

Véronique Arluison and Frank Wien (eds.), *RNA Spectroscopy: Methods and Protocols*, Methods in Molecular Biology, vol. 2113, https://doi.org/10.1007/978-1-0716-0278-2_16,

several temperatures gives also access to the heat capacity change (ΔCp). Importantly, thermodynamic analysis can give access to structural data and hence provides information on hydrogen bonding, hydrophobic interactions, and charge–charge interactions. A serious advantage of ITC over other alternative approaches is that ITC is a true solution technique which does not require any modification, labelling, or immobilization of any partner (truly label-free) and that there is no restriction in the molecular weight of the reagents or in the choice of the buffer composition. Since the measured enthalpy results not only from the heat of the binding events but also from other heat sources such as solvent effects and/or heat of dilution, one has to pay attention to prevent any buffer mismatch between the syringe and the sample cell (*see* **Note 1**). Importantly, the application field of ITC has been recently extended to kinetics, in addition to thermodynamics, by the development of kinITC [5–9]. This method leads to the determination of on- and off-rates close to those obtained by surface plasmon resonance [10, 11].

In this chapter, we provide a detailed protocol for the characterization of interactions between a fragment of viral RNA with aminoglycoside drugs. ITC is a well-adapted technique for the study of RNA/ligand interactions [12–17]. The RNA used in this chapter is a synthetic 23-nucleotide containing the HIV-1 genomic RNA dimerization initiation site (DIS). Dimerization of the genomic RNA is a crucial step of the HIV-1 retrovirus replication cycle. The DIS is a highly conserved stem–loop sequence located in the 5′-untranslated region of the viral genome which is essential for genome dimerization [18, 19]. A six-nucleotide palindromic sequence within the DIS promotes genome dimerization by forming a loop–loop complex (or kissing-loop complex) [20–23]. This initial kissing-loop complex is further stabilized in vitro into an extended duplex form by extension of interstrand Watson–Crick base pairs [24–28]. X-ray structures of the DIS as a loop–loop complex [29, 30] and as an extended duplex [31, 32] revealed striking resemblance with the bacterial 16 S ribosomal A site, which is the natural target of aminoglycoside antibiotics, thus opening the opportunity of targeting the HIV DIS using such compounds [33, 34]. In this frame, ITC analysis showed that the 4,5-disubstituted 2-desoxystreptamine (2-DOS) aminoglycosides, but not 4,6-disubstituted 2-DOS, can bind specifically DIS dimers. Binding parameters revealed that the specific interaction between aminoglycosides and DIS kissing loop is ensured by rings I, II, and III, while rings IV and V enhanced affinity through unspecific interaction with sugar phosphate backbone. The RNA determinants of this interaction were also investigated, and we could show that DIS hairpin monomer, as well as the HIV-1 subtype-B DIS sequence, is not specifically recognized by aminoglycosides. This exhaustive ITC analysis of the binding and thermodynamic parameters contributed to solve crystal structures of the DIS kissing loop and extended duplex in complex with several 4,5-disubstituted 2-DOS aminoglycosides [35–39].

2 Materials

2.1 Instrumentation

1. iTC_{200}, PEAQ-ITC (MicroCal, Malvern Panalytical, Malvern, UK) or equivalent highly sensitive isothermal titration calorimeter (*see* **Note 2**).
2. Refrigerated centrifuge for a rotor adapted for 50-mL Falcon tubes.
3. Nanodrop ND-1000 Spectrophotometer (Thermo Fisher Scientific, Waltham, MA, USA) or equivalent.
4. Dry bath for 90 °C incubation of Eppendorf tubes.

2.2 Instrument Accessories

1. Hamilton glass loading syringes for calorimeter.
2. BD Luer-Lok 1-mL syringe.

2.3 Sample Preparation

1. 23-mer HIV-1 RNA (sequence 5′ CUGCUUGAAGUGCACA CAGCAAG 3′) was chemically synthetized at the 1 μmol scale and purified (>98%) as described previously [32].
2. Lividomycin, neomycin, or paromomycin sulfate, powder. These aminoglycosides were purchased from Sigma-Aldrich and used without further purification. Paromomycin is shown in this example.
3. Amicon Ultra-4 centrifugal filter unit 10,000 MWCO (Millipore, Billerica, MA, USA).
4. 10× ITC buffer: 250 mM potassium chloride, 20 mM magnesium chloride, 200 mM sodium cacodylate, pH 7.0 (*see* **Note 3**). Prepare 10 mL.
5. 1× ITC buffer: 150 mM potassium chloride, 2 mM magnesium chloride, 20 mM sodium cacodylate, pH 7.0. Prepare 10 mL.

3 Methods

3.1 Sample Preparation

For each experiment, 300 μL of RNA at 10–30 μM should be prepared.

1. RNA is aliquoted in 3 Eppendorf tubes, each tube containing 1 mL of RNA at 2 μM in pure water (*see* **Note 4**).
2. Heat RNA sample for 5 min at 90 °C and snap cool in water on ice at 0 °C for 10 min.

3. Add 111 μL of 10× ITC buffer in each Eppendorf tube.
4. RNA sample is concentrated to a final volume of ~300 μL using an Amicon Ultra-4 concentrator. Centrifuge for ~10 min at 4500 g at 10 °C (recommended speed in a swinging rotor). Higher temperature should be avoided in order to prevent the conversion from the loop–loop complex RNA to the extended duplex form [40].
5. Check the absorbance at 260 nm using a spectrophotometer (take a full spectrum and not a single value). Here a final concentration of 16.8 μM (in strand) is obtained.
6. Prepare 1 mL of aminoglycoside solution at a 400 μM concentration in 1× ITC buffer. This chapter shows an example with the paromomycin aminoglycoside. The optimal aminoglycoside concentration may be different for another compound.

3.2 Designing a Typical ITC Experiment

3.2.1 Experimental Considerations

During an ITC experiment, aliquots of ligand are injected into the sample cell. The interaction between the ligand and the molecule in the cell induces the release (exothermic interaction) or absorption (endothermic interaction) of heat which is directly measured by the calorimeter. The titration curve of kcal mol^{-1} vs. molar ratio (ligand/sample) is generated by the integration of each heat pulse with time and normalized for concentration. After the fitting of the resulting isotherm with a binding model, the affinity constant ($K_a = 1/K_d$), the stoichiometry (n), and the enthalpy of interaction (ΔH) are obtained. The Gibbs free energy change (ΔG) and the entropy change (ΔS) are deduced via the thermodynamic relationships:

$$\Delta G = -RT \ln K_a = RT \ln K_d$$

$$\Delta G = \Delta H - T\Delta S$$

The determination of the appropriate concentration of the component placed in the sample cell (MT) depends on the binding affinity, the stoichiometry, and the heat of binding ΔH. If the Wiseman coefficient ($c = n \cdot K_a \cdot MT$) calculated is in the 1–1000 range (preferably between 10 and 500) [3], a sigmoid is then obtained, and this allows the determination of the accurate binding constant and ΔH.

In general the required concentrations of sample in the cell are ranging from 5 to 100 μM. Typical concentration for the ligand in the injection syringe is normally 10–20 times higher (and even higher in the case of low Wiseman c value experiments [41–43]). These conditions will ensure that the cell sample will become saturated at the end of the titration experiment. Therefore, the sample requirement in order to perform an ITC experiment under ideal conditions (c~100) might be significant for biological low affinity systems. A crucial point to obtain high-quality ITC experiments is to minimize the heat signal originated from mismatches between the solutions that are mixed (*see* **Note 1**). For this reason, it is imperative to prepare the two samples (from the syringe and the cell) rigorously in the same buffer.

3.2.2 Instrument Parameters

On iTC$_{200}$ and PEAQ-ITC microcalorimeters, the glass titration syringe holds up to 40 μL. Typical injections are ranging from 0.25 to 2 μL. The number of injections per experiment should be set in order to saturate the RNA at the end of the titration (a final molar ratio of 3–5 is acceptable in most situations). The injection speed should be 0.5 μL s^{-1}. The spacing between the injections is a frequently overlooked parameter, leading to truncated peaks and erroneous data if a complete return to the baseline is not achieved. We have shown that the time required for equilibration is strongly dependent on the rate constants k_{on} and k_{off} of each system [9]. The spacing value can be set to 120 s by default on the iTC$_{200}$, but it might be increased to 240 s or even longer in some situations [5, 10]. The filter value on iTC$_{200}$ instrument corresponds to the integration time that will be used to produce a single data point. Usually the filter period is set to 5 by default; however, setting 1 or 2 for a better sampling of each peak in view of performing a kinetic analysis of the thermodynamic data is also possible [6, 9]. This value is fixed to 1 on the PEAQ-ITC instrument and cannot be modified.

The feedback mode of the calorimeter setup should be set to "high gain," yielding the fastest response time. The "reference power" value corresponds to the baseline (in μCal s^{-1}) during the experiment. In the high gain mode, the instrument's reference power ranges from 0 to 12.25 μCal s^{-1} for the iTC$_{200}$. In most cases, the reference power can be fixed to 5 μCal s^{-1}. The stirring speed depends on the syringe geometry. It should be set to 1000 rpm for straight-bladed syringes on iTC$_{200}$, but this value can be reduced to 750 rpm for helical-bladed syringe available on both iTC$_{200}$ and PEAQ-ITC (*see* **Note 5**).

Most experiments are performed between 4 and 37 °C (the instrument's theoretical operating range is 2–80 °C, but extremes are not usable in practice). The choice of the experiment temperature is often overlooked, while it might be crucial to optimize results in some situations. Indeed, most biomolecular interactions are characterized by a temperature dependence of the ΔH due to a negative heat capacity change (*see* **Note 6**). In addition, the van't Hoff relationship

$$\partial \ln K_{\mathrm{d}}/\partial T = -\Delta H/\left(RT^2\right)$$

also shows that K_{d} is affected by temperature change. Changing the cell temperature is thus an easy way to play on the Wiseman c value.

3.3 Performing the ITC Experiment

Below is described the procedure for an iTC_{200} microcalorimeter. The experiment will be performed at 25 °C.

1. Fill the reference cell with ultrapure water using a 0.5-mL glass Hamilton syringe.
2. In the instrument controls tab, set the temperature at 25 °C in "Thermostat Control," and click "Set Jacket Temp."
3. Rinse the sample cell, two or three times, with 1× ITC buffer.
4. Warm up the RNA sample at 25 °C before loading in the sample cell.
5. Gently load the sample cell with 300 μL of RNA sample. The small excess should be removed. The final sample volume in the cell is ~280 μL. Care must be taken to avoid bubble formation in the sample cell.
6. To load the titration syringe, place a microcentrifuge tube with at least 70 μL of aminoglycoside sample at 400 μM. Place the syringe titration above the microcentrifuge tube in order that the tip of the syringe is immerged in ligand solution. Manual loading of the syringe is recommended since it is faster and safer than the semiautomatic procedure on the iTC_{200}, especially if the sample is viscous (i.e., with a buffer containing some glycerol or with highly concentrated protein). Connect the BD syringe to the titration syringe and load the sample manually by aspiration into the titration syringe (before aspiration, make sure that the plunger is in the open position). When the sample gets to the plunger, click "Close Fill Port" in the instrument controls tab; this will move the plunger tip down and block the fill port. Perform the "purge/refill" procedure to remove small air bubbles in the syringe. Unplug the BD syringe from the titration syringe.
7. Place the titration syringe into the sample cell.
8. Set the reference power to 5 μCal s^{-1}.
9. Set the initial delay to 100 s and the stirring speed to 1000 rpm or 750 rpm depending on the syringe geometry.
10. Set the number of injections to 28, the sample cell concentration to 0.0168 mM, and the concentration in the syringe to 0.4 mM. This ensures a final molar ratio of ~5, required in such RNA/ligand experiments.
11. Set the titration parameters (the number and volume of the injections as well as the spacing between two injections) in the "Injection Parameters" window into the "Advanced Experimental Design" tab. Set the first injection volume to 0.3 μL (*see* **Note 7**) and the next injection volume to 1.5 μL. Set the spacing between each injection to 150 s.

12. Press the "Start" button to begin the experiment. The experiment will start when both cells are thermostated at 25 °C.
13. A blank experiment should be also performed by injecting aminoglycoside into 1× ITC buffer without RNA, in order to subtract the heat of dilution from the reaction heat data.

3.4 ITC Data Analysis

Various software are available for ITC data processing: MicroCal Origin 7 (VP-ITC and iTC_{200} from MicroCal); NanoAnalyze (TA Instrument); AFFINImeter [8]; the NITPIC, SEDPHAT, and GUSSI package [44, 45], or the MicroCal PEAQ-ITC analysis software (Malvern PEAQ-ITC). ITC data analysis using the MicroCal Origin7 software was previously described in details in [46].

Here we will describe the procedure for data processing using the MicroCal PEAQ-ITC analysis software provided with the PEAQ-ITC instrument.

1. Open the software and select both the RNA–ligand titration and the blank experiment (use the "Ctrl" key to select both files). They are quickly processed and integrated by the software and integrated heats are displayed on the left tab (Fig. 1). Binding is detected for the RNA/ligand titration ("Binding" is displayed in green, Fig. 1a), whereas "No binding" is displayed in orange for the blank experiment (Fig. 1b).
2. On the right arrowhead of the blank experiment, change "No binding" for "Control" (now displayed in blue).
3. On the main "Fitting Model" window, change the fitting model from "One Set of Sites" to "Two Sets of Sites" (*see* **Note 8**).
4. On top of the main window, change from "Overview" to "Adjust Baseline." This tab allows to manually correct integration heat for each data point (Fig. 2).
5. The "Time Factor" parameter on the bottom of the window might be useful if the baseline is not stable (large oscillation of the baseline due to air-conditioning, for instance), but 5 is generally a good standard value.
6. The number of points per injection is set by default to 25, but this can be changed if necessary.
7. Baseline for the RNA/ligand experiment and the buffer/ligand control can be adjusted, if necessary. This is generally not required in good experimental conditions: well-maintained instrument, samples without aggregates and injections without bubbles, and no heat perturbation around the instrument.
8. On top of the main window, change from "Adjust Baseline" to "Assign Control." On "Control Parameter," change from "Fitted Offset" to "Single." The control experiment is then automatically selected. The "Line" Method is generally

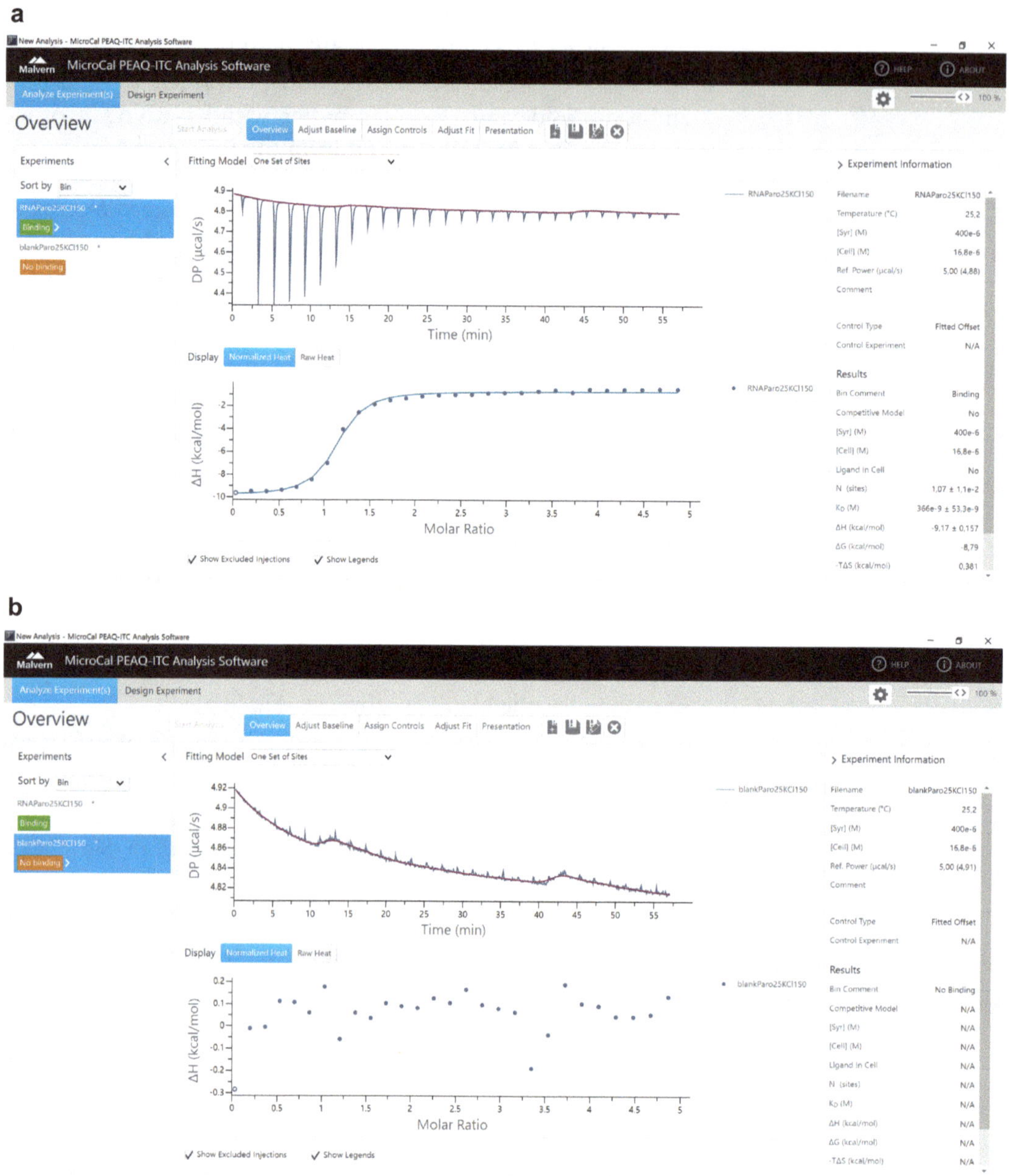

Fig. 1 Screenshot of the "Overview" tab after selection of the RNA/ligand (**a**) and the blank (**b**) experiment

preferable over the "Point-to-point" and "Mean" ones (Fig. 3). In the absence of control experiment, it is possible to use the heat of dilution from the last injections of the experiment, provided the RNA oversaturated with ligand. It is however highly recommended to perform a control experiment to obtain more accurate results.

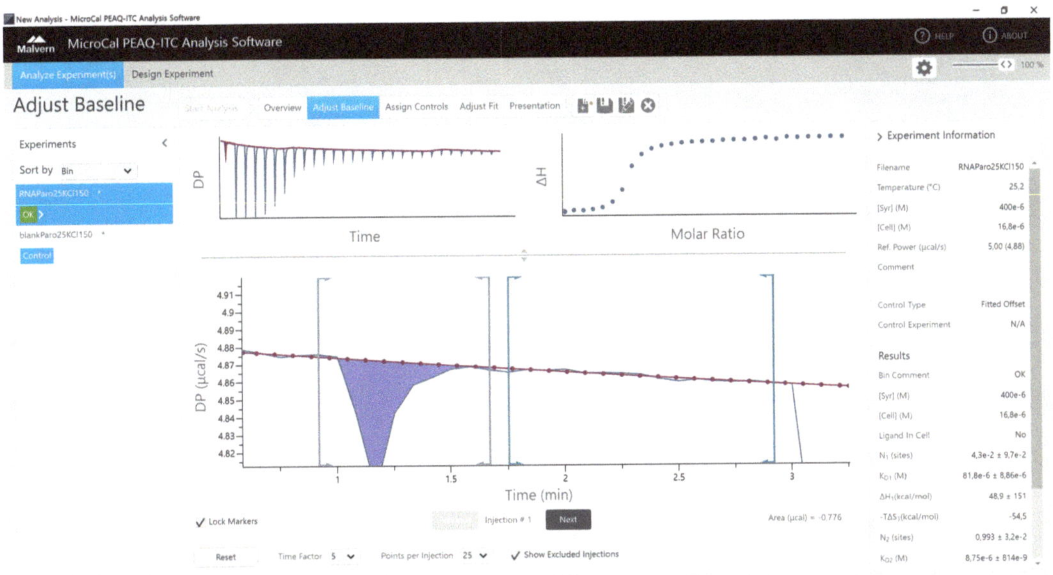

Fig. 2 Screenshot of the "Adjust Baseline" tab. The blue area corresponds to the integrated heat. It is possible to adjust the baseline by moving individually each point and the markers. Individual points can be excluded (or included) on the top-right part of the window

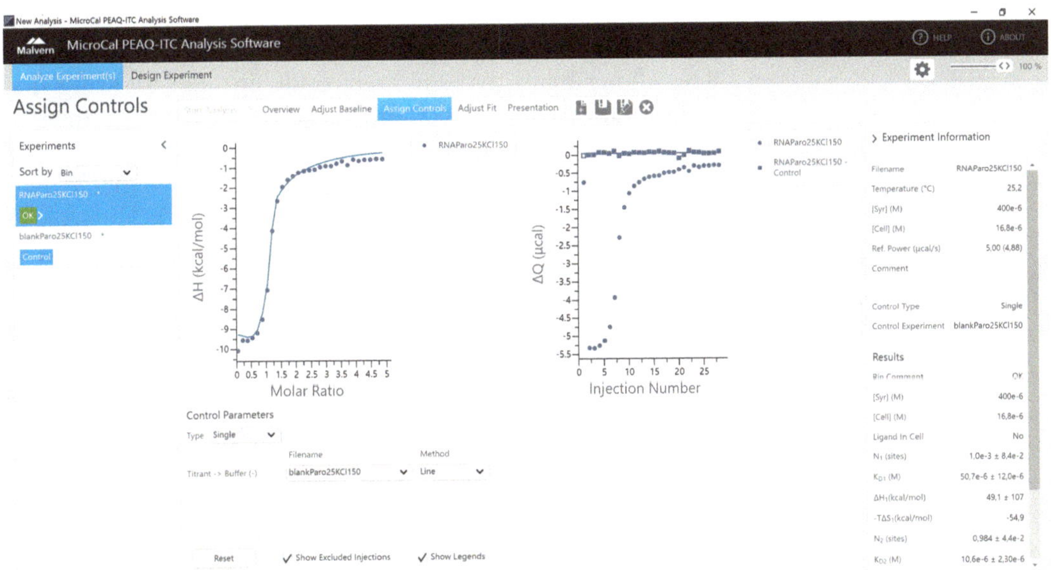

Fig. 3 Screenshot of the "Assign Control" tab. On the left part are shown the integrated heats for the experiment. On the right part are shown the integrated heats for the blank experiment and the corrected integrated heats for the experiment with RNA as well

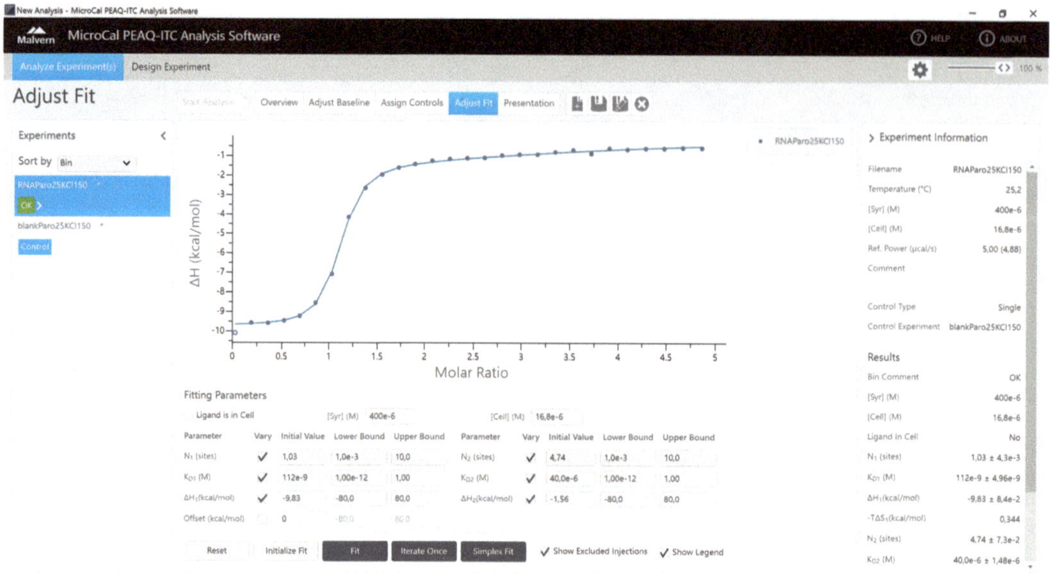

Fig. 4 Screenshot of the "Adjust Fit" tab

9. On top of the main window, change from "Assign Control" to "Adjust Fit." Fitted parameters can be adjusted here (Fig. 4). Here, N1 (which corresponds to the stoichiometry of the first class of binding sites) is very close to 1.0 (as expected since the RNA dimer binds two aminoglycoside ligands), showing that concentrations of the RNA and ligand were correctly estimated and that the RNA is properly folded. This binding site corresponds to the specific binding site and entails the highest affinity ($K_{d1} = 112$ nM, $\Delta H = -9.8$ kcal mol^{-1}). The stoichiometry of the second set of binding sites, N2, is relevant of unspecific binding due to electrostatic interactions with the lowest affinity ($K_{d2} = 40$ μM, $\Delta H = -1.6$ kcal mol^{-1}).
10. It is important to check errors on refined parameters! Indeed, in this model, the high correlation between the different fitted parameters could lead to multiple equivalent solutions.
11. On top of the main window, change from "Adjust Fit" to "Result Table." To visualize correlation between fitted parameters, click on "Show Errors" in the bottom of the window. The "dependency" value should be examined for all refined parameters (closer to 0 is better; closer to 1.0 is worst).
12. Select the "Final Figure" tab to display a publication quality figure (Fig. 5) which can be easily exported in various graphical formats.

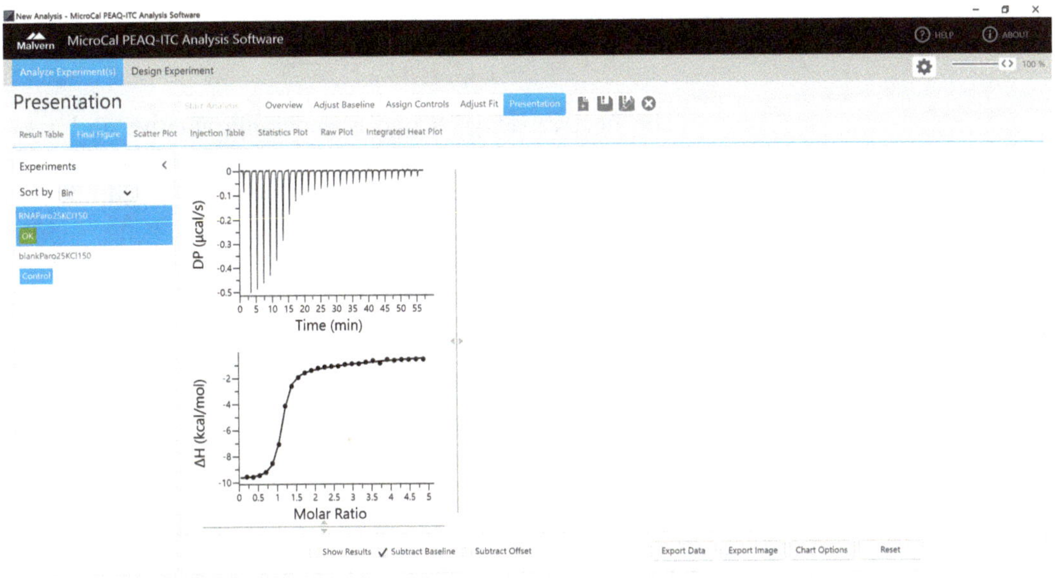

Fig. 5 Screenshot of the "Presentation" tab, showing the final figure ready to be exported for publication

4 Notes

1. Buffer mismatch is due to differences between the ligand and the sample solution, as well as any difference in pH, salt, glycerol concentration, buffer concentration, or additives. In order to minimize these differences, interactants must be prepared in identical buffers or dialyzed against the same buffer.
2. Since microcalorimeter (especially iTC_{200}) is very sensitive to temperature variations, it is highly recommended to place the instrument far away from any airflow, especially air-conditioning, and in a room exposed to minimal temperature changes (avoid heaters, large windows, or other instruments generating lots of heat such as freezers). The instrument should be regularly and meticulously cleaned.
3. Low-salt conditions were chosen at this temperature to prevent conversion of the loop–loop dimer into the extended duplex dimer [40].
4. RNA concentration at this step should be less than ~3 μM in order to prevent formation of extended duplex dimers as observed in [40].
5. Although the reduction of the stirring speed improves the baseline stability, it might introduce artifacts due to insufficient mixing of the injected ligand.

6. Heat capacity change is defined by the relationship:

$$\Delta C_p = (\delta \Delta H / \delta T)$$

The heat capacity change is generally negative and associated to a local folding following binding [47]. It is a valuable tool for optimizing the study of any binding event: for an exothermic reaction, ΔH will be more negative at 35 °C than 25 °C and, consequently, the signal will be increased at 35 °C. In some situations, no signal can be detected by ITC. This can result from athermal reactions (ΔH is null in the conditions of the experiment) where the interaction exists, but cannot be detected by ITC (an example is provided in [48]). In such situations, the change in temperature may allow the interaction to be observed due to the heat capacity change of the system.

7. A first small injection (0.3–0.5 μL) is performed at the beginning of the titration. It will be discarded during data analysis because it is generally affected by error due to sample dilution during the equilibration procedure and to the backlash of the screw mechanism of the syringe motor.

8. Since the aminoglycoside ligand is positively charged, the RNA/aminoglycoside interaction in this example is characterized by a highly specific interaction coexisting with unspecific interactions due to electrostatic interactions. These unspecific interactions can be strongly reduced by increasing salt concentrations (the specific interactionis also reduced, however). This situation holds true for most of RNA/ligand interactions.

References

1. Leavitt S, Freire E (2001) Direct measurement of protein binding energetics by isothermal titration calorimetry. Curr Opin Struct Biol 11:560–566
2. Privalov PL, Dragan AI (2007) Microcalorimetry of biological macromolecules. Biophys Chem 126:16–24
3. Velazquez Campoy A, Freire E (2005) ITC in the post-genomic era...? Priceless. Biophys Chem 115:115–124
4. Wiseman T, Williston S, Brandts JF, Lin LN (1989) Rapid measurement of binding constants and heats of binding using a new titration calorimeter. Anal Biochem 179:131–137
5. Bec G, Meyer B, Gerard MA, Steger J, Fauster K, Wolff P, Burnouf D, Micura R, Dumas P, Ennifar E (2013) Thermodynamics of HIV-1 reverse transcriptase in action elucidates the mechanism of action of non-nucleoside inhibitors. J Am Chem Soc 135:9743–9752
6. Dumas P, Ennifar E, Da Veiga C, Bec G, Palau W, Di Primo C, Pineiro A, Sabin J, Munoz E, Rial J (2016) Extending ITC to kinetics with kinITC. Methods Enzymol 567:157–180
7. Munoz E, Sabin J, Rial J, Perez D, Ennifar E, Dumas P, Pineiro A (2019) Thermodynamic and kinetic analysis of isothermal titration calorimetry experiments by using KinITC in AFFINImeter. Methods Mol Biol 1964:225–239
8. Pineiro A, Munoz E, Sabin J, Costas M, Bastos M, Velazquez-Campoy A, Garrido PF, Dumas P, Ennifar E, Garcia-Rio L, Rial J, Perez D, Fraga P, Rodriguez A, Cotelo C (2019) AFFINImeter: a software to analyze molecular recognition processes from experimental data. Anal Biochem 577:117–134
9. Burnouf D, Ennifar E, Guedich S, Puffer B, Hoffmann G, Bec G, Disdier F, Baltzinger M, Dumas P (2012) kinITC: a new method for obtaining joint thermodynamic and kinetic

data by isothermal titration calorimetry. J Am Chem Soc 134:559–565
10. Guedich S, Puffer-Enders B, Baltzinger M, Hoffmann G, Da Veiga C, Jossinet F, Thore S, Bec G, Ennifar E, Burnouf D, Dumas P (2016) Quantitative and predictive model of kinetic regulation by E. coli TPP riboswitches. RNA Biol 13:373–390
11. Zihlmann P, Silbermann M, Sharpe T, Jiang X, Muhlethaler T, Jakob RP, Rabbani S, Sager CP, Frei P, Pang L, Maier T, Ernst B (2018) KinITC-one method supports both thermodynamic and kinetic SARs as exemplified on FimH antagonists. Chemistry 24:13049–13057
12. Barbieri CM, Srinivasan AR, Pilch DS (2004) Deciphering the origins of observed heat capacity changes for aminoglycoside binding to prokaryotic and eukaryotic ribosomal RNA a-sites: a calorimetric, computational, and osmotic stress study. J Am Chem Soc 126:14380–14388
13. Feig AL (2007) Applications of isothermal titration calorimetry in RNA biochemistry and biophysics. Biopolymers 87:293–301
14. Feig AL (2009) Studying RNA-RNA and RNA-protein interactions by isothermal titration calorimetry. Methods Enzymol 468:409–422
15. Kaul M, Pilch DS (2002) Thermodynamics of aminoglycoside-rRNA recognition: the binding of neomycin-class aminoglycosides to the A site of 16S rRNA. Biochemistry 41:7695–7706
16. Pilch DS, Kaul M, Barbieri CM, Kerrigan JE (2003) Thermodynamics of aminoglycoside-rRNA recognition. Biopolymers 70:58–79
17. Salim NN, Feig AL (2009) Isothermal titration calorimetry of RNA. Methods 47:198–205
18. Haddrick M, Lear AL, Cann AJ, Heaphy S (1996) Evidence that a kissing loop structure facilitates genomic RNA dimerisation in HIV-1. J Mol Biol 259:58–68
19. Paillart J-C, Berthoux L, Ottmann M, Darlix J-L, Marquet R, Ehresmann C, Ehresmann B (1996) A dual role of the dimerization initiation site of HIV-1 in genomic RNA packaging and proviral DNA synthesis. J Virol 70:8348–8354
20. Laughrea M, Jetté L (1994) A 19-nucleotide sequence upstream of the 5′ major splice donor site is part of the dimerization domain of human immunodeficiency virus 1 genomic RNA. Biochemistry 33:13464–13474
21. Muriaux D, Girard PM, Bonnet-Mathonière B, Paoletti J (1995) Dimerization of HIV-1lai RNA at low ionic strength. J Biol Chem 270:8209–8216
22. Paillart JC, Skripkin E, Ehresmann B, Ehresmann C, Marquet R (1996) A loop-loop "kissing" complex is the essential part of the dimer linkage of genomic HIV-1 RNA. Proc Natl Acad Sci U S A 93:5572–5577
23. Skripkin E, Paillart JC, Marquet R, Ehresmann B, Ehresmann C (1994) Identification of the primary site of the human immunodeficiency virus type I RNA dimerization *in vitro*. Proc Natl Acad Sci U S A 91:4945–4949
24. Laughrea M, Jetté L (1996) Kissing-loop model of HIV-1 genome dimerization: HIV-1 RNA can assume alternative dimeric forms, and all sequences upstream or downstream of hairpin 248-271 are dispensable for dimer formation. Biochemistry 35:1589–1598
25. Muriaux D, Fossé P, Paoletti J (1996) A kissing complex together with a stable dimer is involved in the HIV-1_{Lai} RNA dimerization process *in vitro*. Biochemistry 35:5075–5082
26. Rist MJ, Marino JP (2002) Mechanism of nucleocapsid protein catalyzed structural isomerization of the dimerization initiation site of HIV-1. Biochemistry 41:14762–14770
27. Takahashi KI, Baba S, Chattopadhyay P, Koyanagi Y, Yamamoto N, Takaku H, Kawai G (2000) Structural requirement for the two-step dimerization of human immunodeficiency virus type 1 genome. RNA 6:96–102
28. Takahashi KI, Baba S, Koyanagi Y, Yamamoto N, Takaku H, Kawai G (2001) Two basic regions of NCp7 are sufficient for conformational conversion of HIV-1 dimerization initiation site from kissing-loop dimer to extended-duplex dimer. J Biol Chem 276:31274–31278
29. Ennifar E, Dumas P (2006) Polymorphism of bulged-out residues in HIV-1 RNA DIS kissing complex and structure comparison with solution studies. J Mol Biol 356:771–782
30. Ennifar E, Walter P, Ehresmann B, Ehresmann C, Dumas P (2001) Crystal structures of coaxially stacked kissing complexes of the HIV-1 RNA dimerization initiation site. Nat Struct Biol 8:1064–1068
31. Ennifar E, Walter P, Dumas P (2010) Cation-dependent cleavage of the duplex form of the subtype-B HIV-1 RNA dimerization initiation site. Nucleic Acids Res 38:5807–5816
32. Ennifar E, Yusupov M, Walter P, Marquet R, Ehresmann B, Ehresmann C, Dumas P (1999) The crystal structure of the dimerization initiation site of genomic HIV-1 RNA reveals an

extended duplex with two adenine bulges. Structure 7:1439–1449
33. Ennifar E, Paillart JC, Marquet R, Ehresmann B, Ehresmann C, Dumas P, Walter P (2003) HIV-1 RNA dimerization initiation site is structurally similar to the ribosomal A site and binds aminoglycoside antibiotics. J Biol Chem 278:2723–2730
34. Bernacchi S, Freisz S, Maechling C, Spiess B, Marquet R, Dumas P, Ennifar E (2007) Aminoglycoside binding to the HIV-1 RNA dimerization initiation site: thermodynamics and effect on the kissing-loop to duplex conversion. Nucleic Acids Res 35:7128
35. Ennifar E, Aslam MW, Strasser P, Hoffmann G, Dumas P, van Delft FL (2013) Structure-guided discovery of a novel aminoglycoside conjugate targeting HIV-1 RNA viral genome. ACS Chem Biol 8:2509–2517
36. Ennifar E, Paillart JC, Bernacchi S, Walter P, Pale P, Decout JL, Marquet R, Dumas P (2007) A structure-based approach for targeting the HIV-1 genomic RNA dimerization initiation site. Biochimie 89:1195–1203
37. Ennifar E, Paillart JC, Bodlenner A, Walter P, Weibel JM, Aubertin AM, Pale P, Dumas P, Marquet R (2006) Targeting the dimerization initiation site of HIV-1 RNA with aminoglycosides: from crystal to cell. Nucleic Acids Res 34:2328–2339
38. Freisz S, Lang K, Micura R, Dumas P, Ennifar E (2008) Binding of aminoglycoside antibiotics to the duplex form of the HIV-1 genomic RNA dimerization initiation site. Angew Chem Int Ed Engl 47:4110–4113
39. Bodlenner A, Alix A, Weibel JM, Pale P, Ennifar E, Paillart JC, Walter P, Marquet R, Dumas P (2007) Synthesis of a neamine dimer targeting the dimerization initiation site of HIV-1 RNA. Org Lett 9:4415–4418
40. Bernacchi S, Ennifar E, Toth K, Walter P, Langowski J, Dumas P (2005) Mechanism of hairpin-duplex conversion for the HIV-1 dimerization initiation site. J Biol Chem 280:40112–40121
41. Tellinghuisen J (2008) Isothermal titration calorimetry at very low c. Anal Biochem 373:395–397
42. Tellinghuisen J (2016) Analysis of multitemperature isothermal titration calorimetry data at very low c: global beats van't Hoff. Anal Biochem 513:43–46
43. Turnbull WB, Daranas AH (2003) On the value of c: can low affinity systems be studied by isothermal titration calorimetry? J Am Chem Soc 125:14859–14866
44. Brautigam CA, Zhao H, Vargas C, Keller S, Schuck P (2016) Integration and global analysis of isothermal titration calorimetry data for studying macromolecular interactions. Nat Protoc 11:882–894
45. Keller S, Vargas C, Zhao H, Piszczek G, Brautigam CA, Schuck P (2012) High-precision isothermal titration calorimetry with automated peak-shape analysis. Anal Chem 84:5066–5073
46. Da Veiga C, Mezher J, Dumas P, Ennifar E (2016) Isothermal titration calorimetry: assisted crystallization of RNA-ligand complexes. Methods Mol Biol 1320:127–143
47. Spolar RS, Record MT Jr (1994) Coupling of local folding to site-specific binding of proteins to DNA. Science 263:777–784
48. Ramirez J, Recht R, Charbonnier S, Ennifar E, Atkinson RA, Trave G, Nomine Y, Kieffer B (2015) Disorder-to-order transition of MAGI-1 PDZ1 C-terminal extension upon peptide binding: thermodynamic and dynamic insights. Biochemistry 54:1327–1337

Chapter 17

Use of Fluorescent Nucleotides to Map RNA-Binding Sites on Protein Surface

V. Balobanov, N. Lekontseva, A. Mikhaylina, and A. Nikulin

Abstract

Currently, studies of RNA/protein interactions occupy a prominent place in molecular biology and medicine. The structures of RNA–protein complexes may be determined by X-ray crystallography or NMR for further analyses. These methods are time-consuming and difficult due to the versatility and dynamics of the RNA structure. Furthermore, due to the need to solve the "phase problem" for each dataset in crystallography, crystallographic structures of RNA are still underrepresented. Structure determination of single ribonucleotide–protein complexes is a useful tool to identify the position of single-stranded RNA-binding sites in proteins. We describe here a structural approach that incorporates affinity measurement of a protein for various single ribonucleotides, ranking the RNA/protein complexes according to their stability. This chapter describes how to perform these measurements, including a perspective for the analysis of RNA-binding sites in protein and single-nucleotide crystal soaking.

Key words RNA-binding proteins, RNA–protein complexes, Nucleotide–protein interactions, Ribonucleotide, Fluorescence, Crystal soaking

1 Introduction

The structures of RNA–protein complexes obtained by X-ray crystallography or NMR are fundamental to the detailed analysis and understanding of RNA–protein interactions. At this time, the determination of such structures continues to be technically challenging. An important stage in the structural studies of RNA–protein complexes is the search for a compact and well-ordered RNA fragment, which should be specific for the target protein and bind it with a high affinity, comparable to the native RNA. The location of the protein-binding region in a specific RNA is commonly determined by footprinting methods [1, 2] and must be identified before commencing high-resolution 3D structural studies. Due to the widespread use of genomic methods, such as a PAR-CLIP [3, 4], it has become possible to quickly identify target RNAs in the cell for specific proteins; however, the results produce

Véronique Arluison and Frank Wien (eds.), *RNA Spectroscopy: Methods and Protocols*, Methods in Molecular Biology, vol. 2113, https://doi.org/10.1007/978-1-0716-0278-2_17,

a wide spectrum of RNAs, and it is not always straightforward to deduce a consensus binding site among these sequences. The next step after identification of the recognized RNA consists of an optimization of the length of the specifically bound RNA fragment. Large RNA molecules are structurally very flexible, which represents a barrier to the crystallization of RNA–protein complexes. Furthermore, the large size of such complexes does not usually allow the use of NMR to determine their structures. Therefore, it is important to determine the smallest RNA fragment that still has a near-native affinity for the protein. However, variations in the size and sequence of the RNA fragments (polydispersity) may significantly influence the resolution of the diffraction pattern of the RNA–protein crystal and, consequently, the quality of the solved structures [5–7].

Recently we demonstrated that single-stranded RNA-binding proteins, such as Hfq or archaeal LSm proteins, bind isolated ribonucleotides with high specificity [8, 9]. Hfq binds UTP and ATP in poly(U) and poly(A) RNA-binding sites, respectively. The structures of the ribonucleotide–protein complexes revealed that Hfq binds ribonucleotides in the RNA-binding sites, and their position and conformation correspond to the position of the RNA bases in known structures of Hfq ribonucleoprotein complexes. The structure of the single nucleotide–protein complex can thus be used to map the RNA-binding site(s) on the protein. Just recently a similar technique has been applied to study m^6A methyltransferase active sites using bisubstrate analogues [10]. This technique is comparable with widely used crystallographic fragment screening for efficient drug discovery (for review *see* [11, 12]). More extended data can be obtained using short oligo-RNA fragments, but in this case the exact sequence of the oligo-RNA should be known prior to solution of the complex, which has not always been the case [13]. Otherwise, identification of these sites may be carried out experimentally by measuring the affinity of the RNA for a range of proteins with single amino acid substitutions at putative RNA-binding sites. However, this mutagenic approach is a time-consuming process. We suggest our alternative approach, which consists in solving the structure of single ribonucleotide–protein complexes, which will determine the specificity of the protein to the RNA and reveal the exact position of the RNA-binding site(s) on the protein.

Measuring the affinity of a protein for ribonucleotides facilitates estimating the stability of the corresponding complexes and, consequently, improves success in the crystallization of the nucleotide–protein complex. Common methods used for determining the stability of ligand–protein complexes are isothermal calorimetry (described in Chap. 17) [13], thermophoresis [14], or surface plasmon resonance (described in Chap. 5) [15]. These techniques all have limitations associated with the small size of nucleotides, sometimes poor solubility of the protein, or nonspecific binding of reaction components to surfaces.

Alternatively, a cost-effective way to measure affinity of a protein for a nucleotide is fluorescence. Two main approaches can be applied: (1) measuring of anisotropy change of a fluorophore-labeled nucleotide or (2) measuring the change in the intrinsic fluorescence of a protein upon nucleotide binding. The first method is based on the differential mobility of small molecules in the free and bound states, namely, a change in the rotational diffusion coefficient D_r (in radians^2/s). This coefficient should not be confused with translational diffusion coefficient D_t (in m^2/s) that is directly related to the hydrodynamic or Stokes radius (Stokes/Einstein relation) and is measured using fluorescence correlation spectroscopy (FCS). This change in D_r is reflected by changes of the fluorescence polarization (or anisotropy). Depolarization of light by a rotational diffusion of spherical particles is described by Perrin's Eq. (1):

$$r_0/r = 1 + 6tD_r \quad (1)$$

where τ = fluorescence lifetime, r_0 is the fundamental anisotropy, and r is the measured anisotropy.

Since nucleotides are not naturally fluorescent, this approach requires the use of a fluorescent label graft on the nucleotide that may induce undesirable contacts with the protein and affect the measured values. Nevertheless, this technique has been successfully applied to analyze AMP and GMP affinity for LSm proteins (*see* **Note 1**).

As mentioned, another approach is based on the change in the intrinsic fluorescence of a protein upon nucleotide binding (*see* **Note 2**). The position of the fluorescence spectrum maximum and its intensity depends on the environment of the fluorophore. Ligand binding alters the fluorophore surroundings and consequently its emission. This effect is visible when the fluorophore is located at the nucleotide-binding site or in its close environment. Regardless of the signal detection used during titration, the concentrations of both free protein and free nucleotide change during the experiment, and these changes of ligand or protein must be considered. However, approximations can be made to factor out the concentration change of one of the components, namely, by using an excess of one of the compounds. Change in the fluorescent compound concentration should be avoided because this results in change in fluorescent intensity independently of complex formation.

Recently, we compared the affinity of ribonucleotides among several RNA-binding proteins, measured by fluorescence, with the presence of the ribonucleotides in the crystal structures of the complexes obtained from both co-crystallization and soaking

[16]. It was found that ribonucleotides bound tightly to the proteins and can be confidently identified in experimental electron density maps with equilibrium dissociation constants (K_D) <10 μM. For higher K_D (20–40 μM), longer incubation of the protein crystals in the nucleotide solution is required to obtain the structures of the complexes. The complexes with K_D value higher than 50 μM are usually not stable enough to survive in crystallization conditions. As the K_D may help to estimate the probability of the nucleotide presence in the crystal of the complexes, we describe how to use this approach for further identification of the protein's RNA-binding sites.

2 Materials

2.1 Enzymes, Buffers, and Chemicals

1. Protein solution (stock solution concentration depends on protein, but suggested final concentration is around 3–5 μM).
2. Nucleotide solution (UTP, UMP, CTP, or CMP, stock solution 100 mM).
3. MANT-AMP and MANT-GMP ribonucleotides (Jena Bioscience, Germany) (*see* **Notes 1** and **3**).
4. Interaction buffer (protein-specific); for instance, 0.1–0.3 M NaCl and 50 mM Tris–HCl, p 8.0, may be used as a first step. Ionic strength and pH may be adapted for each protein, in particular because high ionic strength may abolish nucleotide complex formation [17].

2.2 Fluorescence Spectroscopy

1. Fluorimeter, equipped with magnetic stirring and with anisotropy measurement setup (e.g., Cary Eclipse Fluorescence Spectrometer, Agilent) (Fig. 1).
2. Fluorescence quartz cuvettes (different from absorbance cuvette, L-format detection). Low-volume cuvettes adapted for magnetic stirring are recommended.
3. Data analysis and graphing software (e.g., Origin, Igor, QtiPlot, KaleidaGraph, or SigmaPlot).

3 Methods

3.1 Determination of Apparent Dissociation Constant K_D Using Intrinsic Protein Fluorescence

1. Place ~100 μL of a 0.95 μM protein solution in a quartz cuvette for measuring fluorescence.
2. Set excitation wavelength at 280 nm to measure fluorescence emission spectrum in the range of 290–350 nm. Note that excitation/emission wavelength may vary if the protein does not contain tryptophan.

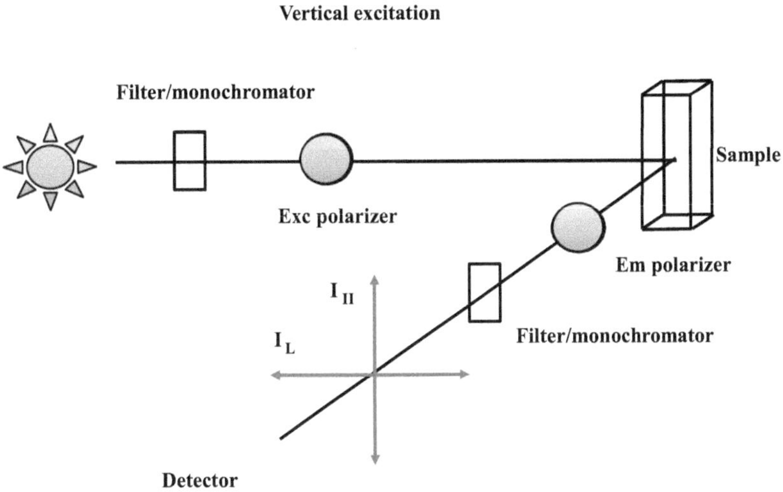

Fig. 1 Scheme of the fluorimeter, equipped with magnetic stirring and anisotropy measurement setup (e.g., Cary Eclipse Fluorescence Spectrometer, Agilent)

3. Add a few μL of concentrated nucleotide solution step by step. The concentration must be adapted depending on the K_D range and should cover protein/nucleotide ratios at least from 1:10 to 10:1. After each addition of nucleotide, measure the emission fluorescence spectrum. The volume change must be kept to a minimum to avoid dilution correction; therefore, the highest available concentration of the nucleotide should be used [18]. In the general case, changes of both concentrations must be considered (*see* **Note 4**).
4. Plot the fluorescence intensities vs. the concentrations of nucleotide solution (Fig. 2).
5. Saturation on the titration curve must be observed to measure K_D precisely by curve fits using software (*see* **Note 5**).
6. Apply the fitting procedure to the titration curves to obtain the value of the apparent dissociation constant K_D.
7. Repeat at least three independent measurements to calculate the average value of the K_D and obtain a standard deviation.
8. Analyze the titration curve.

Titration of a protein B by an excess of nucleotide A may be described as follows (assuming no cooperativity and only one binding site):

$$[AB] = \frac{[A]\,[B]_0}{K_D + [A]} \tag{2}$$

where K_D is the equilibrium dissociation constant, [AB] is the concentration of complex formed, and [A] and [B] are the free concentrations of nucleotide and protein, respectively. Note that if

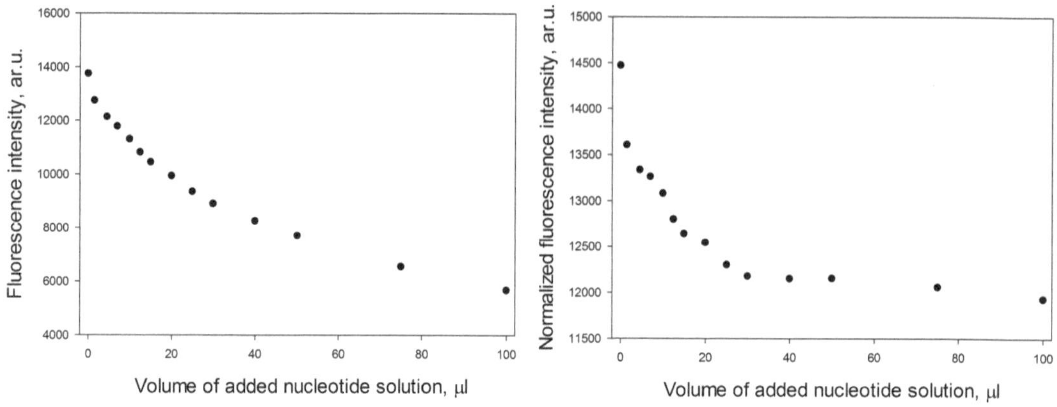

Fig. 2 Dependence of fluorescence (left panel) and fluorescence normalized to protein concentration (right panel) on the volume of the added nucleotide solution

the nucleotide is in excess, then the total concentration of nucleotide $[A]_0$ may be approximated to be equal to [A] (the concentration of free nucleotide).

Because at saturation (i.e., $[A] \gg K_D$), the maximum concentration of complex $[AB]_{max} = [B]_0$, we have

$$[AB] = \frac{[A]_0 [AB]_{max}}{K_D + [A]_0} \quad (3)$$

As fluorescence intensity I_f is proportional to [AB] (corrected from baseline), we can substitute

$$I_f = \frac{I_{f\,max} [A]_0}{K_D + [A]_0} \quad (4)$$

where I_f is the measured fluorescence intensity and $I_{f\,max}$ is the measured fluorescence intensity at saturation.

Using this approach, K_D data may be extracted from fitting the saturation curve.

Conversely, titration of a nucleotide by an excess of protein may also be performed; then [A] and [B] are the free concentrations of protein and nucleotide, respectively (in Eqs. (2) and (4)); where the protein is in excess, the total concentration of protein $[A]_0$ may be approximated to equal [A] (the concentration of free protein).

3.2 Determination of Apparent Dissociation Constant Using Fluorescence Anisotropy of the Labeled Nucleotide

1. Place a MANT-labeled nucleotide solution with a concentration of 1 μM in a quartz cuvette for measuring fluorescence (*see* **Note 6**).
2. Use an excitation wavelength of 355 nm, and measure the fluorescence anisotropy at 448 nm. The emission is recorded five times or averaged over time; the mean value of the

anisotropy and the standard deviation are calculated by the standard Cary Eclipse program.

3. Add a solution of protein step by step. The concentration must be adapted depending on the K_D range and should cover protein/nucleotide ratios at least from 1:10 to 10:1. After each addition of nucleotide, measure the emission fluorescence spectrum. The volume change must be kept to a minimum to avoid dilution correction; therefore, the highest available concentration of the nucleotide should be used [18]. In the general case, changes of both concentrations must also be considered (*see* **Note 4**).
4. Apply the curve fitting procedure to the titration curves to measure the value of the apparent dissociation constant K_D, as described in Subheading 3.1.

3.3 Obtaining Ribonucleotide–Protein Crystals by Co-crystallization

1. The K_D obtained using fluorescent titration allows one to estimate the stability of the nucleotide–protein complex and the chance of crystallization. Choose high-affinity complexes for initial trials.
2. Add the ribonucleotide stock solution to the crystallization drop to a final ribonucleotide concentration of 5–25 μM. Crystallize the protein with a suitable procedure [19, 20].

   For crystallization of Hfq in the presence of NTP, the crystallization drops were prepared by mixing 2 μL protein stock solution (30 mg/mL), 2 μL ribonucleotide stock solution (100 mM), and 2 μL reservoir solution. The final concentration of the nucleotide was 4.75 mM.
3. Before collecting the diffraction data, place the obtained crystals into a cryoprotective solution with an appropriate concentration of the ribonucleotide added (5–10 mM), for instance, 15% PEG 4000, 15% PEG 400, 200 mM ammonium chloride, and 50 mM Tris–HCl pH 8.5 [9].

3.4 Obtaining Ribonucleotide-Protein Crystals by Soaking of the Protein Crystals in Ribonucleotide Solution

1. Crystallize the protein by a suitable procedure [21, 22].
2. Incubate the obtained crystals in 2-μL drop of the mother solution with 20–50 μM ribonucleotide. The incubation time can vary from 10 min to several hours, depending on the affinity of the ribonucleotide for protein and the crystal packing (*see* **Notes 7** and **8**). For soaking, a protein may be placed in ADPNP, ATP, CTP, UTP, or GTP solutions. The protein crystals are placed into 2 mL of a 50 mM solution of the appropriate ribonucleotide, 20% PEG 400, and 50 mM Tris–HCl pH 8.5 [9].
3. Before collecting the diffraction data, place the obtained crystals into the cryoprotective solution with the appropriate con centration of the ribonucleotide added, as above.

3.5 Determination of the Ribonucleotide Position in the Experimental Electron Density

1. Determine the crystal structure of the ribonucleotide–protein complex by an appropriate method [19].
2. Place water molecules into the experimental electron density.
3. Check the unmodeled electron density maps and try to fit the ribonucleotides in these positions. The “ligand search” procedure implemented in Coot is extremely useful to locate the position and determine conformation of the ribonucleotides [23, 24].
4. Validate the position, geometry, and fit of the ribonucleotides by common procedures using wwPDB Validation Service [25].

4 Notes

1. Commercially available MANT-labeled ribonucleotides have been used. MANT is randomly attached to the 2′,3′-hydroxyl group of the ribose and usually does not hinder contacts of the nucleotide base and 2′-OH with the amino acids of the protein.
2. Alternatively, analysis with fluorescently labeled nucleotide and nonlabeled protein may also be done using protein-induced fluorescence enhancement (PIFE) [26].
3. Probes with a long lifetime are better probes for polarization measurements as their signal is low when the probe is free—the light is not polarized. Consequently, cyanines (Cy3, Cy5, etc.) are not the best adapted for polarization measurements because they have a short lifetime; their signal is already high even when the probe is free. Traditionally, used florescent-labelled nucleotides with ET3 (2′,3′-O-trinitrophenyl-), MANT (2′/3′-(*N*-methyl-anthraniloyl)-), or ANT (2′/3′-O-anthraniloyl-) chemical groups covalently attached to the ribose.
4. The total fluorescence intensity at each point is the sum of the fluorescence of the free protein and the protein with bounded nucleotide:

$$I = I_{\mathrm{P}} * [\mathrm{P}] + I_{\mathrm{PN}} * [\mathrm{PN}] \tag{5}$$

where I = measured fluorescence intensity, I_{P} = molar fluorescence of free protein, I_{PN} = molar fluorescence of protein with bounded nucleotide, and [P] and [PN] = concentration of free and bound proteins, respectively.

Knowing that

$$[\mathrm{P}] = [\mathrm{P_t}] - [\mathrm{PN}] \tag{6}$$

where $[\mathrm{P_t}]$ = total protein concentration, we can substitute

$$I = I_{\mathrm{P}} * [\mathrm{P_t}] + [\mathrm{PN}] * (I_{\mathrm{PN}} - I_{\mathrm{P}}) \tag{7}$$

In accordance with the definition of the dissociation constant, the concentration of the complex depends on the total concentration of the components as follows [27]:

$$[\mathrm{PN}] = \frac{[\mathrm{P_t}] + [\mathrm{N_t}] + K_\mathrm{d} - \sqrt{([\mathrm{P_t}] + [\mathrm{N_t}] + K_\mathrm{d})^2 - 4*[\mathrm{P_t}]*[\mathrm{N_t}]}}{2} \tag{8}$$

where K_D is the dissociation constant, and $[\mathrm{P_t}]$ and $[\mathrm{N_t}]$ are total protein and nucleotide concentrations, respectively.

Substituting [PN] from this equation into (7), we obtain the following equation for describing the titration curve in space:

$$I = I_\mathrm{P} * [\mathrm{P}_t] + (I_\mathrm{PN} - I_\mathrm{P}) * \frac{[\mathrm{P}_t] + [\mathrm{N_t}] + K_\mathrm{d} - \sqrt{([\mathrm{P_t}] + [\mathrm{N_t}] + K_\mathrm{d})^2 - 4 * [\mathrm{P_t}] * [\mathrm{N_t}]}}{2} \tag{9}$$

Thus, we obtain a general equation describing the dependence of the fluorescence intensity on the concentrations of both components. By applying the fitting procedure available in many programs, we can get the value of the apparent dissociation constant.

Fluorescence anisotropy depends only on the fraction of the bound nucleotide, which simplifies the calculations somewhat. The equations for describing the dependence of anisotropy on the concentration of components will be as follows:

$$A = A_\mathrm{N} + (A_\mathrm{PN} - A_\mathrm{N}) * \frac{[\mathrm{PN}]}{[\mathrm{N_t}]} \tag{10}$$

where A is measured fluorescence anisotropy, and A_N and A_PN are fluorescence anisotropy of free and bounded nucleotide, respectively:

$$A = A_\mathrm{N} + (A_\mathrm{PN} - A_\mathrm{N}) * \frac{[\mathrm{P_t}] + [\mathrm{N_t}] + K_\mathrm{d} - \sqrt{([\mathrm{P_t}] + [\mathrm{N_t}] + K_\mathrm{d})^2 - 4 * [\mathrm{P_t}] * [\mathrm{N_t}]}}{2[\mathrm{N_t}]} \tag{11}$$

Using this approach, we can carry out a titration by any of the components in a wide range of concentrations (Fig. 3).

5. Note that even if an apparent saturation is observed, it is always underestimated by eye (~30%). Appropriate software must be used to determine a precise K_D.

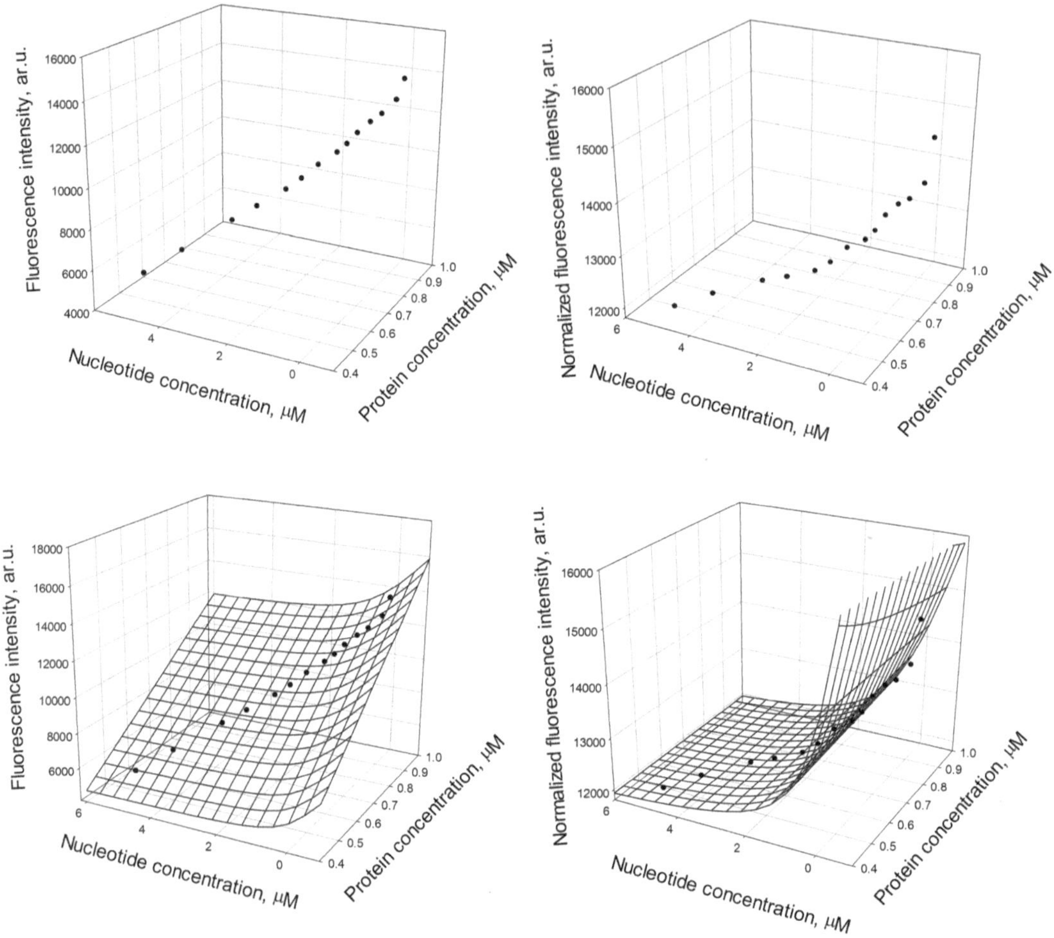

Fig. 3 Titration curves of a SmAP protein with UMP in coordination with fluorescence intensity, protein concentration, and nucleotide concentration (left panels) or normalized fluorescence intensity, protein concentration, and nucleotide concentration (right panels). The bottom panels superimpose the result of the fitting procedure

6. The concentration of fluorescent molecules must be kept low in order to avoid inner filter effect (reabsorption of emitted light) [18].
7. Additional incubation of the co-crystallized ribonucleotide–protein complexes in an appropriate solution of ribonucleotide with the higher concentration of the ligand usually increases the likelihood of detecting a bound ribonucleotide. Soaking time affects binding less than the concentration of ribonucleotide in solution.
8. Triphosphate, diphosphate, and monophosphate ribonucleotides may give different results for the equilibrium dissociation constant and in determining the structure of complexes

depending on the contacts of the phosphates with the protein atoms. Monophosphates usually mimic the RNA skeleton more accurately. Try all of them to get the best results.

Acknowledgments

This work was supported by RFBR grant #18-04-00222. We thank V. Arluison and R. Lease for their constructive comments on the paper.

References

1. Peng Y, Soper TJ, Woodson SA (2012) RNase Footprinting of protein binding sites on an mRNA target of small RNAs. In: Bacterial regulatory RNA. Humana Press, Totowa, pp 213–224
2. Nilsen TW (2014) RNase Footprinting to map sites of RNA-protein interactions. Cold Spring Harb Protoc 2014:pdb.prot080788
3. Ascano M, Hafner M, Cekan P, Gerstberger S, Tuschl T (2012) Identification of RNA-protein interaction networks using PAR-CLIP. Wiley Interdiscip Rev RNA 3:159–177
4. Garzia A, Morozov P, Sajek M, Meyer C, Tuschl T (2018) PAR-CLIP for discovering target sites of RNA-binding proteins. Methods Mol Biol 1720:55–75
5. Garber M, Gongadze G, Meshcheryakov V, Nikonov O, Nikulin A, Perederina A, Piendl W, Serganov A, Tishchenko S (2002) Crystallization of RNA/protein complexes. Acta Crystallogr D Biol Crystallogr 58:1664–1669
6. Ke A, Doudna JA (2004) Crystallization of RNA and RNA-protein complexes. Methods 34:408–414
7. Obayashi E, Oubridge C, Pomeranz Krummel D, Nagai K (2007) Crystallization of RNA-protein complexes. Methods Mol Biol 363:259–276
8. Nikulin A, Mikhailina A, Lekontseva N, Balobanov V, Nikonova E, Tishchenko S (2017) Characterization of RNA-binding properties of the archaeal Hfq-like protein from Methanococcus jannaschii. J Biomol Struct Dyn 35:1615–1628
9. Murina V, Lekontseva N, Nikulin A (2013) Hfq binds ribonucleotides in three different RNA-binding sites. Acta Crystallogr Sect D Biol Crystallogr 69:1504–1513
10. Oerum S, Catala M, Atdjian C, Brachet F, Ponchon L, Barraud P, Iannazzo L, Droogmans L, Braud E, Ethève-Quelquejeu M, Tisné C (2019) Bisubstrate analogues as structural tools to investigate m 6 a methyltransferase active sites. RNA Biol 16:798–808
11. Verlinde CLMJ, Fan E, Shibata S, Zhang Z, Sun Z, Deng W, Ross J, Kim J, Xiao L, Arakaki TL, Bosch J, Caruthers JM, Larson ET, Letrong I, Napuli A, Kelly A, Mueller N, Zucker F, Van Voorhis WC, Buckner FS, Merritt EA, Hol WGJ (2009) Fragment-based cocktail crystallography by the medical structural genomics of pathogenic protozoa consortium. Curr Top Med Chem 9:1678–1687
12. Patel D, Bauman JD, Arnold E (2014) Advantages of crystallographic fragment screening: functional and mechanistic insights from a powerful platform for efficient drug discovery. Prog Biophys Mol Biol 116:92–100
13. Schumacher MA, Pearson RF, Møller T, Valentin-Hansen P, Brennan RG (2002) Structures of the pleiotropic translational regulator Hfq and an Hfq-RNA complex: a bacterial Sm-like protein. EMBO J 21:3546–3556
14. Linke P, Amaning K, Maschberger M, Vallee F, Steier V, Baaske P, Duhr S, Breitsprecher D, Rak A (2016) An automated microscale thermophoresis screening approach for fragment-based lead discovery. J Biomol Screen 21:414–421
15. Wang S, Poon GMK, Wilson WD (2015) Quantitative investigation of protein-nucleic acid interactions by biosensor surface Plasmon resonance. Methods Mol Biol 1334:313–332
16. Nemchinova M, Balobanov V, Nikonova E, Lekontseva N, Mikhaylina A, Tishchenko S, Nikulin A (2017) An experimental tool to estimate the probability of a nucleotide presence in the crystal structures of the nucleotide–protein complexes. Protein J 36:157–165
17. Hwang W, Arluison V, Hohng S (2011) Dynamic competition of DsrA and rpoS

fragments for the proximal binding site of Hfq as a means for efficient annealing. Nucleic Acids Res 39:5131–5139
18. Hulme EC, Trevethick MA (2010) Ligand binding assays at equilibrium: validation and interpretation. Br J Pharmacol 161:1219–1237
19. Wlodawer A, Dauter Z, Jaskolski M (eds) (2017) Protein crystallography. Springer, New York
20. Ennifar E (ed) (2016) Nucleic acid crystallography. Springer, New York
21. Sauter C, Basquin J, Suck D (2003) Sm-like proteins in eubacteria: the crystal structure of the Hfq protein from Escherichia coli. Nucleic Acids Res 31:4091–4098
22. Nikulin A, Stolboushkina E, Perederina A, Vassilieva I, Blaesi U, Moll I, Kachalova G, Yokoyama S, Vassylyev D, Garber M, Nikonov S (2005) Structure of Pseudomonas aeruginosa Hfq protein. Acta Crystallogr D Biol Crystallogr 61:141–146
23. Emsley P, Cowtan K (2004) Coot: model-building tools for molecular graphics. Acta Crystallogr D Biol Crystallogr 60:2126–2132
24. Debreczeni JÉ, Emsley P (2012) Handling ligands with coot. Acta Crystallogr D Biol Crystallogr 68:425–430
25. Pozharski E, Deller MC, Rupp B (2017) Validation of protein-ligand crystal structure models: small molecule and peptide ligands. Methods Mol Biol 1607:611–625
26. Hwang H, Myong S (2014) Protein induced fluorescence enhancement (PIFE) for probing protein-nucleic acid interactions. Chem Soc Rev 43:1221–1229
27. Wang ZX, Kumar NR, Srivastava DK (1992) A novel spectroscopic titration method for determining the dissociation constant and stoichiometry of protein-ligand complex. Anal Biochem 206:376–381

Chapter 18

Fluorescent Oligonucleotide Probes for the Quantification of RNA by Real-Time qPCR

Florent Busi

Abstract

Quantitative real-time PCR (qPCR) is a widely adopted technique used for scientific, clinical, diagnostic, or quality control purposes. One of the main applications of qPCR is gene expression analysis, although mutation detection, genotyping, DNA detection, and quantification (from pathogens or genetically modified organisms) are also investigated using this technique.

Although nonspecific detection based on DNA-binding dyes (including SYBR Green I) offers versatility in qPCR assays, detection of the PCR product using fluorescent probes confers higher specificity and sensitivity to assays, justifying the use of fluorescent probes as a detection method.

This chapter seeks to propose a procedure for the design of qPCR assays using fluorescent hydrolysis probe technology. Particular attention will be paid to explaining the steps necessary to ensure the specificity of the oligonucleotides used as primers or fluorescent probes.

Key words qPCR, Hydrolysis probes, Fluorophore, Quencher

Abbreviations

bp	Base pair
BHQ(-1 or -2)	Black Hole Quencher (-1 or -2)
Cq	Quantification cycle
FAM	Carboxyfluorescein
gDNA	Genomic DNA; DABCYL: 4-4-dimethylamino-phenyl-azo-benzoic acid
HEX	6-Carboxy-2,4,4,5,7,7-hexachlorofluorescein
JOE	6-Carboxy4′,5′-dichloro-2′,7′-dimethoxyfluorescein
nt	Nucleotide
qPCR	Quantitative real-time PCR
ROX	Carboxy-X-rhodamine
RT	Reverse transcription reaction
ss/ds	Single/double stranded
TAMRA	Carboxytetramethylrhodamine
TET	6-Carboxy-2′,4,7,7′-tetrachlorofluorescein
Tm	Melting temperature

Véronique Arluison and Frank Wien (eds.), *RNA Spectroscopy: Methods and Protocols*, Methods in Molecular Biology, vol. 2113, https://doi.org/10.1007/978-1-0716-0278-2_18, © Springer Science+Business Media, LLC, part of Springer Nature 2020

1 Introduction

1.1 Context

The quantification of nucleic acids and in particular RNAs is a crucial issue in biological research. Indeed, whatever the biological field, researchers very often seek to determine the relative or absolute concentrations of specific nucleic acids to study the effects of different factors that are purpose of the studies (normal versus pathological conditions, treated sample versus control, etc.) and in particular their effects on gene expression measured by RNA level determination. For long, the golden standard technique for RNA and especially mRNA quantification (although all categories of RNA are eligible for this technique) has been northern blot (and southern blot as far as DNA quantification is concerned) [1, 2]. However, since the introduction of the quantitative real-time PCR (qPCR) technique in the late 1990s [3–6] and its democratization in research labs, thanks to the drop of the cost of qPCR reagents and real-time thermal cycler equipments, blotting techniques have often been relegated as the ultimate techniques when qPCR does not allow to answer researcher's specific biological questions. This situation owes much to the rather fastidious, numerous, and long-lasting steps of blotting methods that can advantageously be replaced by the straightforward qPCR techniques, allowing to generate results in less than 1 h for as many submicrogram-scale samples that are available in multi-well plates at once (96, 384, and even 1536 in the most recent apparatus) [7–9]. These elements need to be compared to a couple of days for the analysis of a few tens of samples available in the microgram range using northern blot, hence the replacement of blotting-based methods to the advantage of qPCR in nowadays labs.

qPCR has remained the technique of choice for studying variations in gene expression for the validation of data from DNA chip approaches. However, with the advent of high-throughput techniques such as RNAseq, the use of qPCR may be on the verge of suffering the same fate as the northern blot when qPCR itself began to be widely used. The cost of qPCR still allows it to do well while high-throughput sequencing techniques are still expensive. However, the frontier is tenuous because the cost increases with the number of target genes for which quantitative information is sought with qPCR, while the RNAseq, at constant cost, makes it possible to obtain quantitative information on all the transcripts in the different samples studied.

1.2 Real-Time qPCR Quantification Principle

Unlike the northern blot, where the quantities of RNA specifically detected using a probe are directly inferred from the intensity of the measured signals, qPCR quantification is carried out indirectly by analyzing a fluorescence signal measured at each PCR cycle, and representative of the quantity of PCR product at the time of measurement, given that for RNA, a reverse-transcription (RT) step is performed before qPCR. It is an extrapolation from the number of cycles required to cross an arbitrary fluorescence threshold called Cq (quantitative cycle) that allows quantification [6, 10]. Indeed, by assuming that when samples cross the fluorescence threshold (regardless of the number of cycles required), the fluorescence is equal to the threshold and is therefore proportional to the amount of DNA in the tube, the number of cycles Cq depends on the amount of DNA initially present in the tube monitored at each cycle for fluorescence. Thus, the larger the amount of (specific) matrix before starting PCR, the less cycles will have to be performed to reach the arbitrary fluorescence level.

The analysis will then take into account the difference between the Cqs of different samples to determine the factor existing between the quantities of specific RNA present in these different samples before the PCR takes place, while having determined or assumed the efficiency of the said PCR. This approach will be referred to as relative quantification and is the main way of dealing with research results in the literature [10]. Another way to analyze the data is to use serial dilutions of standards with known nucleic acid concentrations of targets to be amplified. The Cqs of the unknown samples analyzed are then transferred to the calibration curve plotted as Cq as a function of the concentration inferred from the standards with their known concentrations. A concentration can then be determined for each analyzed sample with unknown concentrations; these determinations correspond to experiment of absolute quantification by qPCR.

1.3 Real-Time qPCR Detection Chemistries

Two main strategies (summarized in Fig. 1) among many approaches [11] are used for the detection of nucleic acids throughout the PCR cycles to determine Cqs and thus quantify specific nucleic acids in biological samples.

1. The first is based on the use of DNA binding chemical compounds (DNA binding dyes or dyes) that fluoresce significantly more after binding to dsDNA and are referred to as nonspecific detection systems. Therefore, when used in large excess, the fluorescence level of these dyes measured along the cycles is correlated with the corresponding amount of nucleic acids in the tube at the time of measurement (at the end of the elongation step when DNA is double stranded). SYBR Green I was the first dye to be marketed in qPCR kits from several manufacturers. Several other dyes with improved fluorescence or binding with higher density are now available from competitors [11]. Although an intellectual property issue has certainly motivated the development of these competing dyes, these dyes really surpass the SYBR Green I in some specific applications, such as high-resolution melting (HRM) experiments for mutation/polymorphism detection. Dye-based reagents are

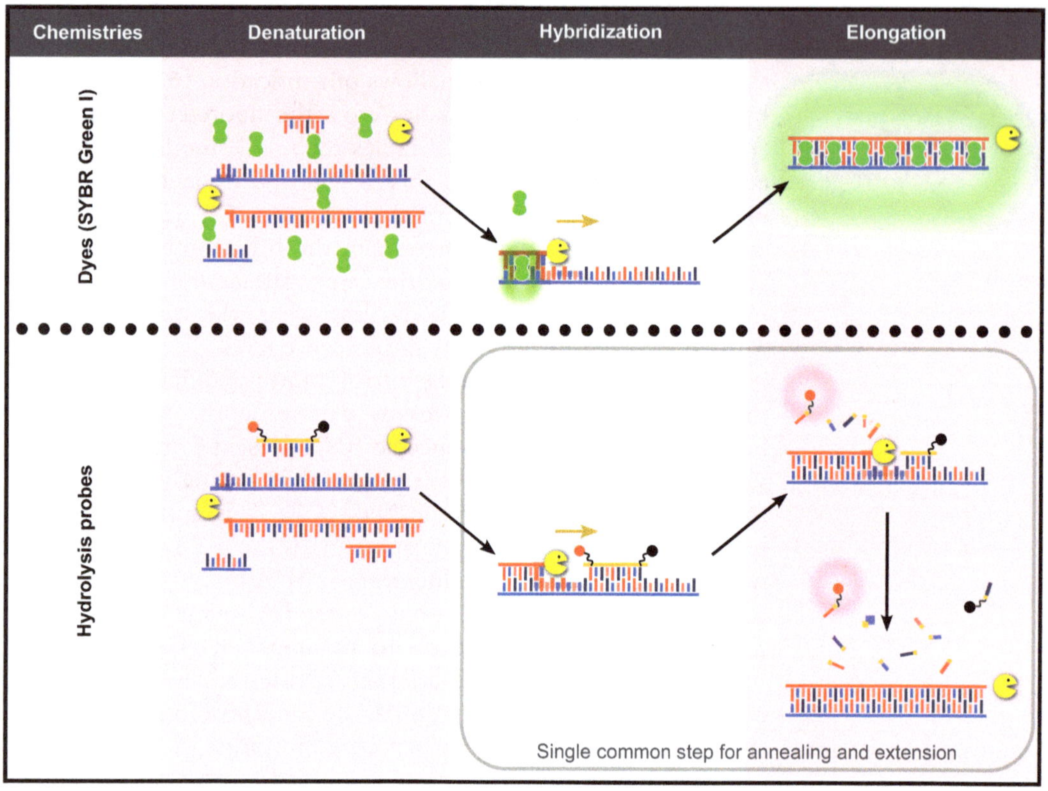

Fig. 1 Schematic representation of the two major chemistry families used for DNA quantity evaluation during qPCR cycles and their overall principle. dsDNA binding dyes acquire fluorescence upon DNA binding; therefore, the more dsDNA in the tube, the more fluorescence. Fluorescent probes (hydrolysis probes in the selected example) are capable to generate fluorescence only if a specific annealing of the probe occurs inside the amplified region

more affordable, reinforced by the fact that they offer researchers the possibility to use the same reagent for the study of the expression of different target genes as only two new oligonucleotides are needed to switch from quantification of one gene expression to another. It should be noted that in the case of dye-based qPCR, specificity is provided only by the two oligonucleotide primer sequences, which design needs to be particularly meticulous (*see* [12] for information on the design of such assays).

2. The second physical approach used to monitor the increase in the amount of PCR product in the tube along the cycles is based on the use of fluorescent oligonucleotide probes. Several strategies are currently being used although hydrolysis probes (*see* **Note 1**) are the most widely used technology, particularly because it is widely adopted in hospitals for diagnostic or genotyping applications and also in other environments for quality control and pathogen or genetically modified organism

DNA detection or quantification. As shown in Fig. 1, the hydrolysis probe is an oligonucleotide covalently linked to a fluorophore in its 5′ position and whose fluorescence is extinguished by a quencher attached to its 3′ position. After hybridization of the oligonucleotide and the probe on the matrix strand, the extension will be carried out by Taq using the free deoxyribose 3′ hydroxyl of the primer thanks to the 5′ > 3′ polymerase activity. As the polymerase reaches the probe, the 5′ > 3′ exonuclease activity will hydrolyze the probe nucleotides one after another so that elongation can continue using the matrix strand [13, 14]. This step will release the hydrolyzed nucleotides from the probe into the reaction medium and in particular the one coupled to the fluorophore, allowing it to recover its fluorescence by removing it from the vicinity of the quencher. Therefore, fluorescence increases over the cycles with amplification of the target amplicon. In this case, the increase in fluorescence is strictly dependent on the hybridization of the probe at its target site in the amplicon region: unspecific binding of the probe or unspecific amplification cannot lead to fluorescence release conversely to DNA binding dyes that detect nonspecific amplification and primer-dimer formation.

Alternative probe systems include fluorescent hybridization probes (also called FRET probes) [11, 15, 16], molecular beacons [11, 17–19], and scorpion probes [11, 16, 20]. In addition, modified versions of hydrolysis probes have also been developed to confer improved specificity, mismatch discrimination capability, or universal probes. These oligonucleotide modifications are mainly represented by locked nucleic acid (LNA) [11, 18, 21–24] or minor groove binder (MGB) [11, 18, 25, 26].

1.4 Design of Fluorescent Probes for the Quantification of RNAs by qPCR

Historically, hydrolysis probes have been introduced under the brand name TaqMan(R) by Applied Biosystems (now Thermo Fisher Scientific), the design of the assays (primers and probe) being performed by the proprietary software PrimerExpress(R). Various commercial software have since been implemented for the design of such assays, such as AlleleID(R) from Premier Biosoft and others. Most often, companies manufacturing oligonucleotides offer online tools to design assays (primers and probes). These software are based on tools such as Prime+ (available in GCG Wisconsin Package), Beacon Designer (TM) (Premier Biosoft), or Primer3 (among others) and which are customized and interfaced with the manufacturer's website.

The procedure described below will develop a design proposal, among many others, using a semi-manual method to design hydrolysis probes using bioinformatics tools available free of charge on the Internet. These probes can be used in qPCR experiments for the quantification of RNA.

Beware: This procedure works as described at the beginning of 2019. Minor and sometimes major adaptations will certainly be required when bioinformatic web tools are updated, possibly with changes in the options of the web interface or website. Nevertheless, most of the steps can be performed in several ways and also with alternative web interfaces; thus, the procedure will still be relevant as a step-by-step guideline but may require the use of other equivalent tools.

2 Materials

2.1 Bioinformatics Tools Needed Prior to Design

1. RefSeq sequences are available from the NCBI nucleotide database:
   https://www.ncbi.nlm.nih.gov/nucleotide/
2. NCBI Genome Data Viewer can be accessed here:
   https://www.ncbi.nlm.nih.gov/genome/gdv/
3. NCBI Blast [27] homepage is accessible using the following address:
   https://blast.ncbi.nlm.nih.gov/Blast.cgi

2.2 Bioinformatics Tools Needed for the Design

Primer3Web and Primer3plus are two web interfaces relying on the same Primer3 core program [28–30]. Both interfaces can be used for the design of the assay.

1. Primer3Web program is available here:
   http://primer3.ut.ee/
2. Primer3Plus Program can be accessed at the following URL:
   https://primer3plus.com/

2.3 Other Bioinformatics Tools

1. Multalin [31] is useful to generate multiple sequence alignments with easy-to-interpret formatted outputs:
   http://multalin.toulouse.inra.fr/multalin/
2. Mfold web server [32] or RNAfold [33] are used to predict the structure of the amplicon and are accessible respectively at the following URL:
   http://unafold.rna.albany.edu/?q=mfold
   http://rna.tbi.univie.ac.at/cgi-bin/RNAWebSuite/RNAfold.cgi

3 Methods

3.1 Example of Design in Eukaryotes: Hydrolysis Probe Assay for the Detection of a Specific Panel of Homo Sapiens CYP3A5 mRNA Transcript Variants

This example is chosen as it will illustrate the complex design of assays targeting genes belonging to superfamilies, and therefore the design should allow discrimination with other related cytochrome P450s (namely, CYP3A4, CYP3A7, and CYP3A43). In addition, CYP3A5 is subject to alternative splicing generating transcript variants of the CYP3A5 mRNA [34, 35], which also need to be taken into account during the design of the CYP3A5 mRNA hydrolysis probe qPCR assay.

A key point to ensure in eukaryotes the specific amplification of a transcribed mRNA of interest rather than a sequence present in the supposedly contaminating genomic DNA (in particular the corresponding gene, which will have strictly the same coding sequence) is to position the assay on an exon–exon junction. Therefore, genomic DNA is selectively excluded from the amplification process because of the following:

1. One of the oligonucleotides or the probe cannot bind to the gDNA because its position overlaps the exon–exon junction which significantly weakens its hybridization to the gDNA.
2. The size of the intron does not allow the amplification of the product derived from the gDNA due to insufficient elongation time, allowing only the amplification of the shortest product (i.e., the cDNA).
3. Lastly, the oligonucleotide primer (not the probe) is capable to bind both cDNA and gDNA, but it has a floating 3′ extremity after binding to genomic DNA, because of one (the most 3′ nucleotide) or ideally several nucleotides within the 3′ extremity of the primer, and this situation yield very poor priming efficiency during PCR; this last option can be used because Taq DNA polymerase lacks nucleotide misincorporation proofreading activity (3′ > 5′ exonuclease activity). This strategy is also used in PCR amplification of specific alleles (PASA)–allele-specific PCR (AS-PCR) [36, 37].

3.2 Elaboration of an Amplification Strategy to Achieve qPCR Assay Specificity to Quantify the RNA of Interest

This procedure is detailed step by step in the left part of Fig. 2.

1. A search for "homo sapiens cyp3A5" in the NCBI Nucleotide database retrieves five RefSeq sequences corresponding to five splice variants. Alternatively, searches can be completed with search field tags (*see* **Note 2**): "homo sapiens"[organism] cyp3A5[gene] or filters.
2. Among the results, the CYP3A5 gene summary indicates that five RefSeq transcripts are available for this gene (these can be accessed directly with the corresponding hyperlink).

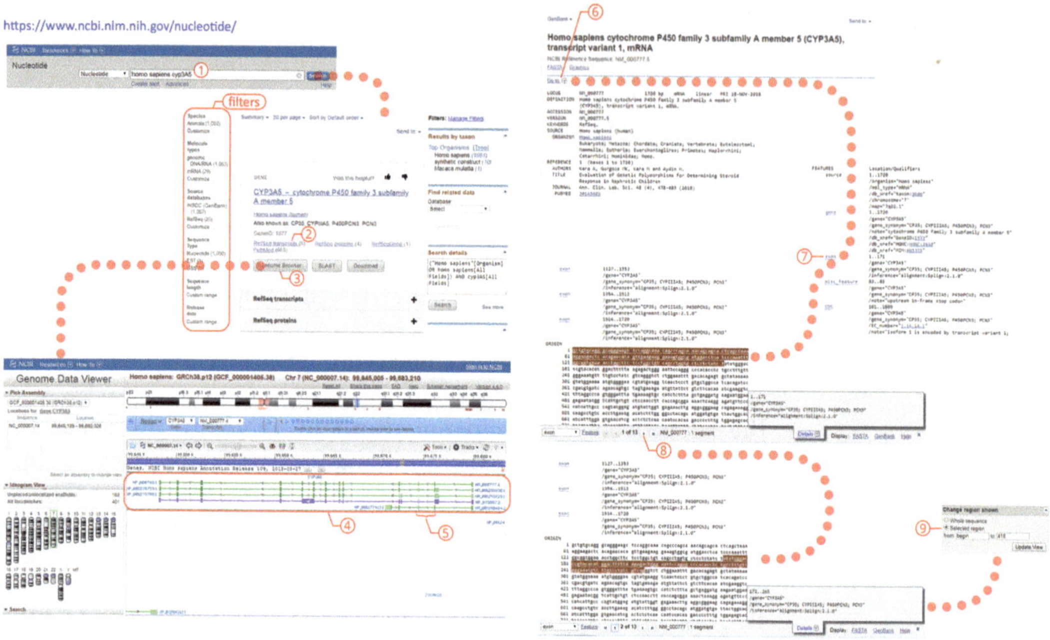

Fig. 2 Elaboration of the strategy to quantify the desired mRNA by visualizing (in the Genome Data Viewer) the different splice variants available in the nucleotide database (left, **steps 1–5**) and retrieval of the sequences of interest (right, **steps 6–9**)

3. Clicking on Genome Browser will allow visualizing all the sequences related to the gene of interest.
4. The five RefSeq transcripts are aligned with the latest human genome assembly.
5. With this example, if the purpose is to quantify all CYP3A5 mRNAs using a hydrolysis probe-based qPCR assay, a region common to all transcripts has to be found. Visually, the region encompassing exons 2 and 3 (indicated by label "(5)" in Fig. 2) is found in all RefSeq transcripts, and amplification of this region could therefore constitute a good strategy for the quantification of all transcripts.

3.3 Retrieval of RNA Sequence of Interest

The illustration of the procedure is continued in right part of Fig. 2.

1. From Nucleotide, display the sequence of interest in Genbank format (for instance NM_001291830.1 in the present example, but any sequence harboring the selected target region could also be used). Bellow the Genbank sequence title, directly select "features" from the "Go to" menu. This will directly display the features section of the Genbank sequence.
2. Then click on exon to highlight the exons features directly superimposed onto the sequence.

3. In the bottom of the internet browser window, use the feature navigation bar to navigate to the desired exon. In the Details sub-window, the start and end position numbers of each exon are indicated. Here are the information for the four first exons:

Exon number > start..end
Exon 1 > 1..171
Exon 2 > 172..265
Exon 3 > 266..318
Exon 4 > 319..418

4. The sequence of interest can then be copied after setting the sequence to display in the "Change region shown" box in the top right of the page (from exon 1 to exon 4 in the example).

3.4 Retrieval of RNA Sequences from Homologous Genes

The illustration of the procedure is continued in Fig. 3.

1. After displaying the selected region of the Genbank sequence, click on "Run Blast" in the "Analyze this sequence" box in the right column. This will fill the blast form with the sequence of interest.
2. Adjust the blast parameter to fill your needs and run "BLAST." In the example, the selected sequence from CYP3A5 mRNA is blasted against "human genomic + transcript" database as the aim is to identify homologous sequences in the same organism (Human).
3. On the "Descriptions" section of the result page, select the relevant sequences (duplicated entries are not needed).
4. Click on "Graphics" to obtain the corresponding multiple sequences alignment. Using the zoom and the scroll functions, the region of interest can be displayed, and nucleotides that will confer specificity during amplification are emphasized: mis matches, insertions, and deletion can easily be located.

3.5 Design of the Oligonucleotide Primers and Probe for the Assay

The strategy to achieve specificity of amplification of one mRNA (here CYP3A5) against related mRNAs that have been identified by Blast relies on two major principles (*see* Fig. 4). Knowing the mismatches between all related sequences and the sequence of interest, as highlighted with multiple sequence alignments, (1) the principle of allele-specific PCR [36, 37] can be used in order to position the 3′ extremity of primers on one nucleotide that is mismatched in all related cDNAs, or (2) one oligonucleotide used as a primer or the probe is positioned roughly centered with regard to the mismatch, which will ensure maximal efficiency of hybridization of the oligonucleotide to its target sequence and maximal destabilization with the related unspecific sequence.

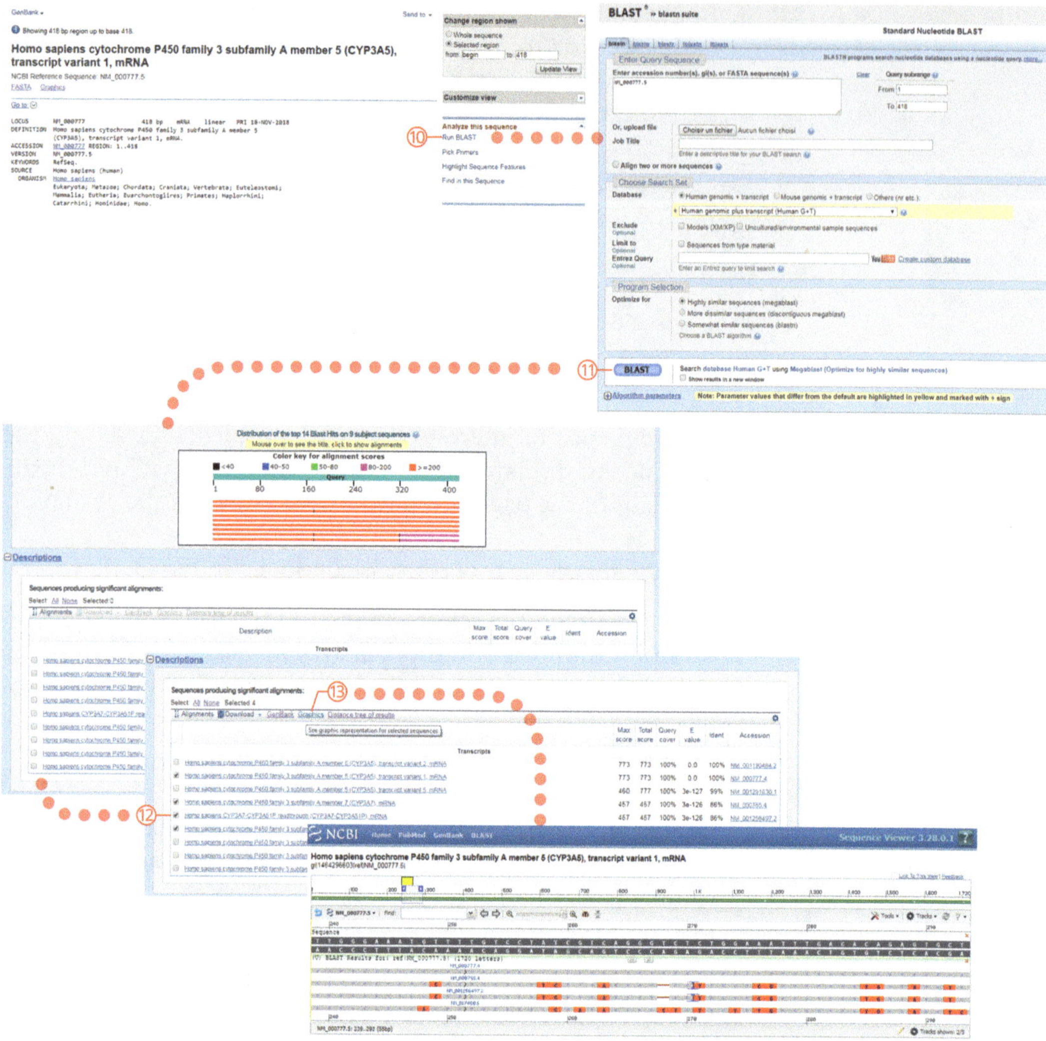

Fig. 3 Retrieval of RNA sequences from homologous genes by blasting the sequence of interest and making a multiple sequence alignment to define the specific nucleotide sequence regions containing the most numerous mismatches (**steps 10–13**)

The general recommended parameters for the design of hydrolysis probes (*see* **Note 3**) are already configured in Primer3Plus (*see* Fig. 5 for guided design). What is interesting about this tool is that it is completely possible to change the parameters in the options via the interface tabs, to adapt the parameters to very specific sequences (very short sequences, atypical GC content, etc.).

1. Access to https://primer3plus.com/ website and select the "Load server settings" option to "Probe." Click on Activate Settings in order to take into account hydrolysis probe assay design.
2. Paste the sequence that will be used for the design.

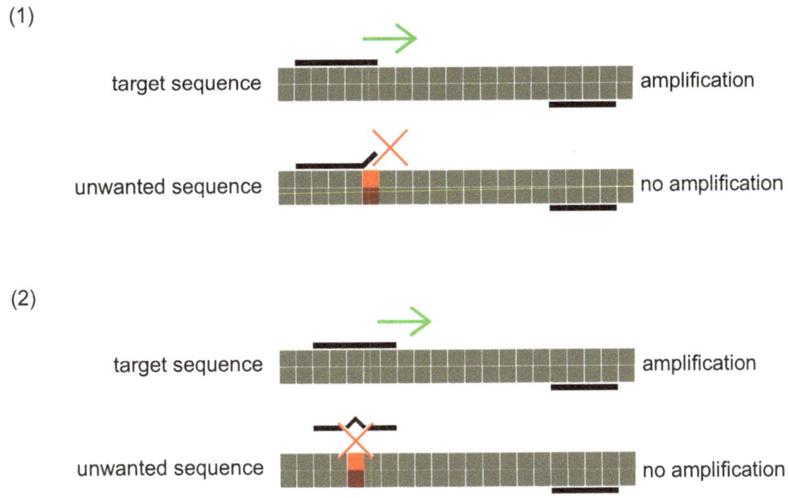

Fig. 4 Schematic representation of the strategies allowing to achieve specificity with annealing of primers with mismatch(es) on related sequences. Mismatches are either designed in 3′ (1) or in the middle of the primer (2)

Optionally, define the "Excluded Regions," "Targets," etc. parameters to force the assay to be designed or excluded from positions indicated with the start position number and the length of the region. For instance, tagging the last nucleotide of exon 2 at position 265 and the first nucleotide of exon 3 located at position 266 would result in indicating 265,2 in the corresponding text box.

3. Run the program by clicking on the Green "Pick Primers" button.

   The result window displays different assays ("Pair 1," "Pair 2," etc.) with the oligonucleotide sequences that have been selected by the software. Forward and reverse primers are indicated as left and right primers, respectively, and probe as "Internal Oligo." These oligonucleotides are highlighted in blue, yellow, and black colors on the provided sequence, respectively.

   The oligonucleotides can be further analyzed by selecting all the returned assays (or only those that have been chosen).

4. Clicking on "send to Primer3Manager" button will allow saving the selected assays in the Primer3Manager utility.

5. Clicking on the BLAST! Hyperlink will allow controlling their ability to amplify unspecific sequences by running a dedicated blast.

6. Configure Blast to search the Human genomic + transcript database. Note that Blast is already configured with Word size and expect threshold adjusted for blasting oligonucleotides (short sequences).

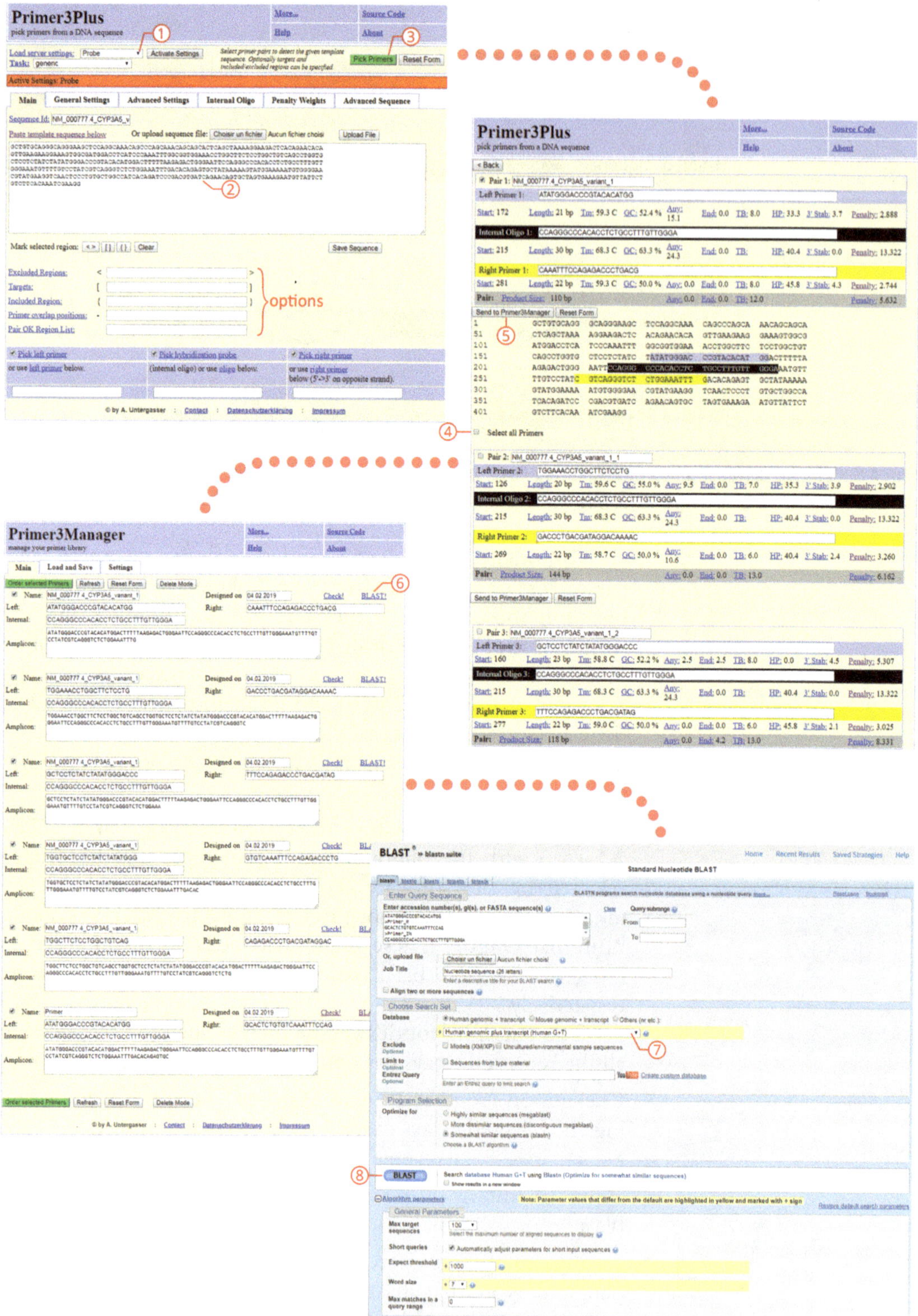

Fig. 5 Design of the oligonucleotide primers and probe for the assay (**steps 1–8**). The design is performed with Primer3Plus software. The result window allows saving the oligonucleotides into the Primer3Manager and further analyze with blast to check for primers and probe specificity

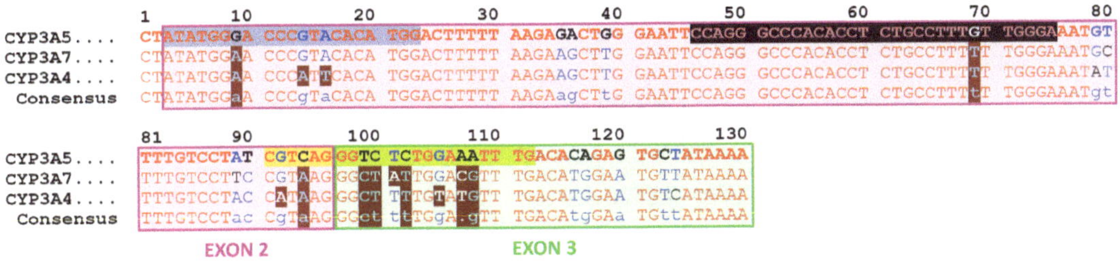

Fig. 6 Multiple sequence alignment of CYP3A5, CYP3A7, and CYP3A4 with representation of oligonucleotides used as primers and probe and the sequence differences that allow specific amplification. Exons 2 and 3 are also indicated

7. Clicking on the "BLAST" Button will open the result page, intended to make sure that at least one of the oligonucleotides (primers or probe) will allow the specificity of the amplification. Check which putative targets are identified and estimate how stable is the hybridization to the unspecific target: how many nucleotides are perfectly matched in the 3′ part of primer? If unsure, it is possible to compare the Tm of this 3′ sequence to the corresponding full-length primer using Primer3 to determine the Tm (might not be sufficiently stable to yield efficient priming under qPCR conditions. Are there any nucleotides floating in 3′ of the primer (priming not allowed), etc.
8. Figure 6 shows at which position the Pair 1 assay oligonucleotides are positioned with respect to the multiple sequence alignment of CYP3A5, CYP3A7, and CYP3A4. The multiple sequence alignment was performed with the Multalin software [31]. The figure shows that the cDNA specificity (with respect to genomic DNA contamination) is conferred by the right primer as it encompasses the exon 2-exon 3 junction. In addition, specificity to CYP3A5 (rather than other CYP3A isoforms) is mainly driven by the right primer as it displays the maximal level of mismatches with other CYP3A nucleotide sequences. However, the left primer also includes a mismatch with CYP3A7 and CYP3A4 in a rather centered and therefore destabilizing position, which should also favors CYP3A5 amplification. Lastly, there is also a mismatch within the probe sequence, and although it will not prevent the binding of the probe on homologous sequences, it might nevertheless slightly favor its hybridization on the CYP3A5 cDNA matrix. Even if the hybridization of left primer and probe might occur (although less efficiently) with CYP3A sequences other than CYP3A5, the right primer will ensure no exponential amplification of CYP3A4 and CYP3A7 will occur.
9. It is highly recommended to check if SNPs are present in the sequence of the selected primers and probe. One way to achieve this is to scroll to the alignment section within the blast result

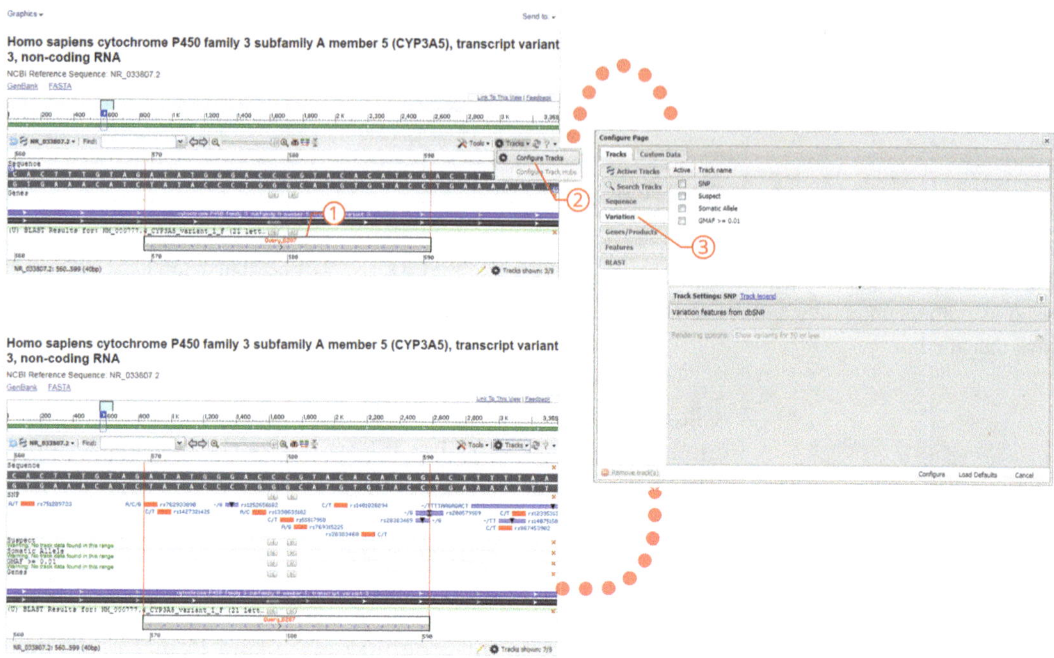

Fig. 7 Checking the sequence variations within oligonucleotide primers and probe (**steps 1–3**)

page and to consider the target gene alignment (top of the list, 100% match, here any CYP3A5 sequence will be suitable). On the top of the alignment, click on the "Graphics" hyperlink; this displays the page depicted in Fig. 7. Click on "Query" to visualize the sequence of the oligonucleotide. Then click on "Tracks" and "Configure tracks" and select "Variations" to display the SNP aligned with the oligonucleotide. Inspect each SNP one after another to check the frequency of the SNP and evaluate if the variation can be ignored or if the assay requires adjustment.

10. It is also recommended to check the secondary structure of the PCR product in order to avoid hard-to-amplify problematic sequences as target for the assay. This can be achieved using the prediction of the secondary structure of both strands of the PCR amplicon and avoid very stable secondary structure as predicted using unafold or RNAfold.

3.6 Ordering the Oligonucleotide Primers and Probe

It is essential to check the filters available on the device that will be used to carry out the qPCR in order to select a suitable fluorophore and a suitable quencher for the probe. Although the number and diversity of fluorescent labels have significantly increased, the following are some of the most common pairs of 5′ fluorophores and 3′ quencher labels:

1. 5′ FAM (Ex 494/495 nm, Em 516/518/520 nm) with 3′ BHQ-1, DABCYL, or TAMRA

2. 5′ HEX (Ex 530/535 nm, Em 553/556 nm) with 3′ BHQ-1, (BHQ-2) or TAMRA
3. 5′ JOE (Ex 520/522/529 nm, Em 548/550/555 nm) with 3′ BHQ-1, (BHQ-2), or TAMRA
4. 5′ ROX (Ex 575/580/586 nm, Em 602/605/607/610 nm) with 3′ (TAMRA), BHQ-2
5. 5′ TET (Ex 521 nm, Em 536/538 nm) with 3′ BHQ-1 or TAMRA.

Spectral properties and compatibilities between dyes are compiled from the following manufacturers/suppliers: abcam, atdbio, Bio-Rad, Eurofins, Eurogentec, Invitrogen/Molecular Probes, Qiagen, and Sigma-Aldrich.

One important parameter is to avoid G as probe 5′ nucleotide since it will display quenching properties with some fluorophores [38] (*see* **Note 3**, bullet point with information about the probe).

4 Notes

1. Hydrolysis probes are also marketed under several names, including double-dye probes and dual-labeled probe, and under the brand name TaqMan(R).
2. The search tag fields can be used for searches in all NCBI databases and are indicated in square brackets to the right of the search request (without spaces or characters as separators). In the case of the nucleotide database, a description of the available fields is documented in Entrez Sequences Help (NCBI help manual) available electronically in the NCBI bookshelf: https://www.ncbi.nlm.nih.gov/books/NBK49540/.
3. Primer3Plus has the advantage of being already preconfigured with parameters that are optimal for designing qPCR assays with hydrolysis probes. These parameters define constraints for designing optimal assays with maximal quality criteria to ensure the best chance to pick up a successful qPCR assay. Nevertheless, selecting the "Pick Anyway" option in "advanced settings" allows to select an assay even if one or more of the oligonucleotides used as primer or probe violates certain constraints. This situation may occur in cases where the sequence in which the design must be located is limited in size or its nucleotide composition and/or sequence is very specific. The following are some general rules:
   - Product:

     Product size should be between 75 and 150 bp, although it can be reduced to 50 bp.

- Primers:

  The primers must have a size of 18–23 bases (with an optimal target size of 20 nt). Their Tm must be in the range of 58–62 °C (with an optimal target of 60 °C Tm), and in addition the Tm difference must be less than 3 °C. The primer GC content should ideally be as balanced as possible and between 30 and 80%. In the default "Probe" settings, the Max_End_GC parameter is not restrictive for primers (set to 5, which means that the maximum number of Gs or Cs in the last five 3′ bases can reach 5); optionally, this parameter can be reduced to 2 to comply with the Applied Biosystems (Thermo Fisher Scientific) recommendations and thus minimize nonspecific priming. The Max_Poly-X parameter (maximum number of consecutive identical nucleotides allowed) can be maintained at the default value (i.e., 5). Among the most important parameters considered are TH_Max_Self_Complementarity (or Max_Self_Complementarity) and TH_Max_End_Self_-Complementarity (or Max_End_Self_Complementarity) since they ensure that oligonucleotides will not have the ability to form dimers or secondary structures that could compromise the PCR efficiency.
- Probe:

  The size of the probe should be between 18 and 30 nt with an optimal 25 nt size, most importantly with a Tm between 68 and 72 °C (optimal 70 °C) as this will allow the hybridization of the probe before annealing of the primers. This condition will allow the hydrolysis of the probe during the PCR elongation step and therefore release of the fluorophore after extension of the primers. To avoid quenching of the fluorescence [38], the 5′ nucleotide can be forced not to be a G using the Primer_Internal_Must_Match_Five_-Prime parameter set to hnnnn (the specified sequence must be 5 nt long, h means "not G," and n means "any nucleotide"; *see* letter code for nucleotide combination in Primer3 online help).
- Applied Biosystems (Thermo Fisher Scientific) also recommends other criteria for the probe, such as to also avoid G as the second nucleotide position in 5′ of the probe, which can be manually excluded from the result list or configured as a parameter as specified before, and to avoid four consecutive Gs in the probe.

Acknowledgments

This work was supported by University Paris Diderot and CNRS.

References

1. Southern EM (1975) Detection of specific sequences among DNA fragments separated by gel electrophoresis. J Mol Biol 98:503–517
2. Alwine JC, Kemp DJ, Stark GR (1977) Method for detection of specific RNAs in agarose gels by transfer to diazobenzyloxymethyl-paper and hybridization with DNA probes. Proc Natl Acad Sci U S A 74:5350–5354
3. Higuchi R, Dollinger G, Walsh PS, Griffith R (1992) Simultaneous amplification and detection of specific DNA sequences. Biotechnology (N Y) 10:413–417
4. Livak KJ, Flood SJ, Marmaro J et al (1995) Oligonucleotides with fluorescent dyes at opposite ends provide a quenched probe system useful for detecting PCR product and nucleic acid hybridization. PCR Methods Appl 4:357–362
5. Bassler HA, Flood SJ, Livak KJ et al (1995) Use of a fluorogenic probe in a PCR-based assay for the detection of Listeria monocytogenes. Appl Environ Microbiol 61:3724–3728
6. Heid CA, Stevens J, Livak KJ, Williams PM (1996) Real time quantitative PCR. Genome Res 6:986–994
7. Bustin SA, Benes V, Nolan T, Pfaffl MW (2005) Quantitative real-time RT-PCR—a perspective. J Mol Endocrinol 34:597–601. https://doi.org/10.1677/jme.1.01755
8. Kubista M, Andrade JM, Bengtsson M et al (2006) The real-time polymerase chain reaction. Mol Asp Med 27:95–125. https://doi.org/10.1016/j.mam.2005.12.007
9. Lyon E, Wittwer CT (2009) LightCycler technology in molecular diagnostics. J Mol Diagn 11:93–101. https://doi.org/10.2353/jmoldx.2009.080094
10. Pfaffl MW (2001) A new mathematical model for relative quantification in real-time RT-PCR. Nucleic Acids Res 29:e45
11. Navarro E, Serrano-Heras G, Castaño MJ, Solera J (2015) Real-time PCR detection chemistry. Clin Chim Acta 439:231–250. https://doi.org/10.1016/j.cca.2014.10.017
12. Busi F, Arluison V, Régnier P (2018) Absolute regulatory small noncoding RNA concentration and decay rates measurements in Escherichia coli. Methods Mol Biol (Clifton, NJ) 1737:231–248
13. Lie YS, Petropoulos CJ (1998) Advances in quantitative PCR technology: 5′ nuclease assays. Curr Opin Biotechnol 9:43–48
14. Huang Q, Li Q (2009) Characterization of the 5′ to 3′ nuclease activity of Thermus aquaticus DNA polymerase on fluorogenic double-stranded probes. Mol Cell Probes 23:188–194. https://doi.org/10.1016/j.mcp.2009.04.002
15. Emig M, Saussele S, Wittor H et al (1999) Accurate and rapid analysis of residual disease in patients with CML using specific fluorescent hybridization probes for real time quantitative RT-PCR. Leukemia 13:1825–1832
16. Didenko VV (2001) DNA probes using fluorescence resonance energy transfer (FRET): designs and applications. BioTechniques 31:1106–1121. https://doi.org/10.2144/01315rv02
17. Tyagi S, Kramer FR (1996) Molecular beacons: probes that fluoresce upon hybridization. Nat Biotechnol 14:303–308. https://doi.org/10.1038/nbt0396-303
18. Buh Gasparic M, Tengs T, La Paz JL et al (2010) Comparison of nine different real-time PCR chemistries for qualitative and quantitative applications in GMO detection. Anal Bioanal Chem 396:2023–2029. https://doi.org/10.1007/s00216-009-3418-0
19. Ryazantsev DY, Kvach MV, Tsybulsky DA et al (2014) Design of molecular beacons: 3′ couple quenchers improve fluorogenic properties of a probe in real-time PCR assay. Analyst 139:2867–2872. https://doi.org/10.1039/c4an00081a
20. Whitcombe D, Theaker J, Guy SP et al (1999) Detection of PCR products using self-probing amplicons and fluorescence. Nat Biotechnol 17:804–807. https://doi.org/10.1038/11751
21. Ugozzoli LA, Latorra D, Puckett R et al (2004) Real-time genotyping with oligonucleotide probes containing locked nucleic acids. Anal Biochem 324:143–152
22. Mouritzen P, Nielsen AT, Pfundheller HM et al (2003) Single nucleotide polymorphism genotyping using locked nucleic acid (LNA™). Expert Rev Mol Diagn 3:27–38. https://doi.org/10.1586/14737159.3.1.27
23. Schmidt A, Rott ME (2006) Real-time polymerase chain reaction (PCR) quantitative detection of *Brassica napus* using a locked nucleic acid TaqMan probe. J Agric Food Chem 54:1158–1165. https://doi.org/10.1021/jf052036m
24. Salvi S, D'Orso F, Morelli G (2008) Detection and quantification of genetically modified organisms using very short, locked nucleic acid TaqMan probes. J Agric Food Chem 56:4320–4327. https://doi.org/10.1021/jf800149j

25. Kutyavin IV, Afonina IA, Mills A et al (2000) 3′-minor groove binder-DNA probes increase sequence specificity at PCR extension temperatures. Nucleic Acids Res 28:655–661
26. Kennedy B, Arar K, Reja V, Henry RJ (2006) Locked nucleic acids for optimizing displacement probes for quantitative real-time PCR. Anal Biochem 348:294–299. https://doi.org/10.1016/j.ab.2005.10.037
27. Altschul SF, Gish W, Miller W et al (1990) Basic local alignment search tool. J Mol Biol 215:403–410. https://doi.org/10.1016/S0022-2836(05)80360-2
28. Untergasser A, Nijveen H, Rao X et al (2007) Primer3Plus, an enhanced web interface to Primer3. Nucleic Acids Res 35:W71–W74. https://doi.org/10.1093/nar/gkm306
29. Untergasser A, Cutcutache I, Koressaar T et al (2012) Primer3—new capabilities and interfaces. Nucleic Acids Res 40:e115–e115. https://doi.org/10.1093/nar/gks596
30. Koressaar T, Remm M (2007) Enhancements and modifications of primer design program Primer3. Bioinformatics 23:1289–1291. https://doi.org/10.1093/bioinformatics/btm091
31. Corpet F (1988) Multiple sequence alignment with hierarchical clustering. Nucleic Acids Res 16:10881–10890
32. Zuker M (2003) Mfold web server for nucleic acid folding and hybridization prediction. Nucleic Acids Res 31:3406–3415
33. Hofacker IL (2003) Vienna RNA secondary structure server. Nucleic Acids Res 31:3429–3431
34. Busi F, Cresteil T (2005) Phenotyping-genotyping of alternatively spliced genes in one step: study of CYP3A5*3 polymorphism. Pharmacogenet Genomics 15:433–439
35. Busi F, Cresteil T (2005) CYP3A5 mRNA degradation by nonsense-mediated mRNA decay. Mol Pharmacol 68:808–815. https://doi.org/10.1124/mol.105.014225
36. Bottema CD, Sommer SS (1993) PCR amplification of specific alleles: rapid detection of known mutations and polymorphisms. Mutat Res 288:93–102
37. Wu DY, Ugozzoli L, Pal BK, Wallace RB (1989) Allele-specific enzymatic amplification of beta-globin genomic DNA for diagnosis of sickle cell anemia. Proc Natl Acad Sci U S A 86:2757–2760
38. Seidel CAM, Schulz A, Sauer MHM (1996) Nucleobase-specific quenching of fluorescent dyes. 1. Nucleobase one-electron redox potentials and their correlation with static and dynamic quenching efficiencies. J Phys Chem 100:5541–5553. https://doi.org/10.1021/jp951507c

Chapter 19

Probing RNA–Protein Interactions and RNA Compaction by Sedimentation Velocity Analytical Ultracentrifugation

Somdeb Mitra and Borries Demeler

Abstract

Recent advances in multi-wavelength analytical ultracentrifugation (MWL-AUC) combine the power of an exquisitely sensitive hydrodynamic-based separation technique with the added dimension of spectral separation. This added dimension has opened up new doors to much improved characterization of multiple, interacting species in solution. When applied to structural investigations of RNA, MWL-AUC can precisely report on the hydrodynamic radius and the overall shape of an RNA molecule by enabling precise measurements of its sedimentation and diffusion coefficients and identify the stoichiometry of interacting components based on spectral decomposition. Information provided in this chapter will allow an investigator to design experiments for probing ion and/or protein-induced global conformational changes of an RNA molecule and exploit spectral differences between proteins and RNA to characterize their interactions in a physiological solution environment.

Key words RNA folding, Counterion, Multi-wavelength analytical ultracentrifugation, Sedimentation velocity, Hydrodynamic measurements

1 Introduction

RNA molecules, although synthesized as extended polymeric structures, more often than not, adopt compact conformations to execute their biological functions [1]. Such compactions are almost always induced by specific and nonspecific interactions with counterions and sometimes by proteins [2]. Even compact conformations, by interacting with small metabolites and cellular proteins, can undergo distinct global topological changes which are manifested as alterations in the shape and hydrodynamic properties of a given RNA [3, 4]. One of the most sensitive techniques for detecting RNA compaction and global conformational changes is analytical ultracentrifugation (AUC) [5–7]. Although AUC has been more widely used for analyzing protein conformations and protein–protein interactions, the easy adaptability of analytical detection methods, offered by AUC, to biological and nonbiological

Véronique Arluison and Frank Wien (eds.), *RNA Spectroscopy: Methods and Protocols*, Methods in Molecular Biology, vol. 2113, https://doi.org/10.1007/978-1-0716-0278-2_19,

samples has led to the adoption of this technique for structural characterization of RNA [5–11] and RNA–protein complexes [12–14], as well as investigations of RNA–protein interactions [15–18].

AUC is at an interesting crossroads where the development of custom-built open-source multi-wavelength (MWL) optics [19–21] has revolutionized the field by generating new information on macromolecular complexes in solution at unprecedented resolution [22]. Initially, the access to such technology was limited to a handful of laboratories that had the technical expertise to build the necessary hardware into existing ultracentrifuges. Excitingly, with the advent of the latest generation of Beckman AUC instruments, the Optima™ series, which has appeared in the market in 2017, more laboratories now potentially have access to a commercially available instrument with MWL detection and analysis capabilities in each experiment. The number of publications from studies using these new instruments remains low, mostly centered around characterization of protein–protein interactions [23] and none, till date, on RNA–protein interactions. Another parallel exciting advancement in the field has been the development of the open-source data-analysis software, Ultrascan III [24], which can be used for analyzing single-wavelength (SWL) and MWL data from both the older Proteome Lab™ and the latest Optima™ AUC instruments. A recent series of articles appearing in a special issue of the European Biophysics Journal [25] highlight several major technological breakthroughs in AUC over the last decade, ranging from instrumentation and sample detection to data-analysis software and structural modeling. In this chapter, we elaborate on the application of MWL-AUC to studies of RNA and RNA–protein interactions and highlight the importance of MWL analysis in relevant areas. While additional optical systems exist (the Aviv fluorescence detector and Rayleigh interference optics), this chapter focuses on the applications of MWL-AUC because of the favorable spectral characteristics of proteins and RNA, which make these optics particularly useful for studying their interactions.

The basic experimental premise for both SWL and MWL AUC experiments is the same. As shown in Fig. 1a, a typical AUC experiment involves subjecting a solution, contained within sector-shaped chambers in an ultracentrifuge cell, to centrifugal forces that lead to sedimentation of the solute particles, away from the rotor center and toward the bottom of the cell (Fig. 1b). Consequently, a solute-depleted region is created toward the top of the cell, near the meniscus, and a solvent–solution boundary is generated. This emerging boundary gives rise to a concentration gradient which triggers diffusional transport in addition to sedimentation transport. The Optima™ AUC's MWL optical system is designed to measure the intensity of monochromatic light between 190 and 800 nm passing through the windows of the ultracentrifuge cells, which allows real-time monitoring of the

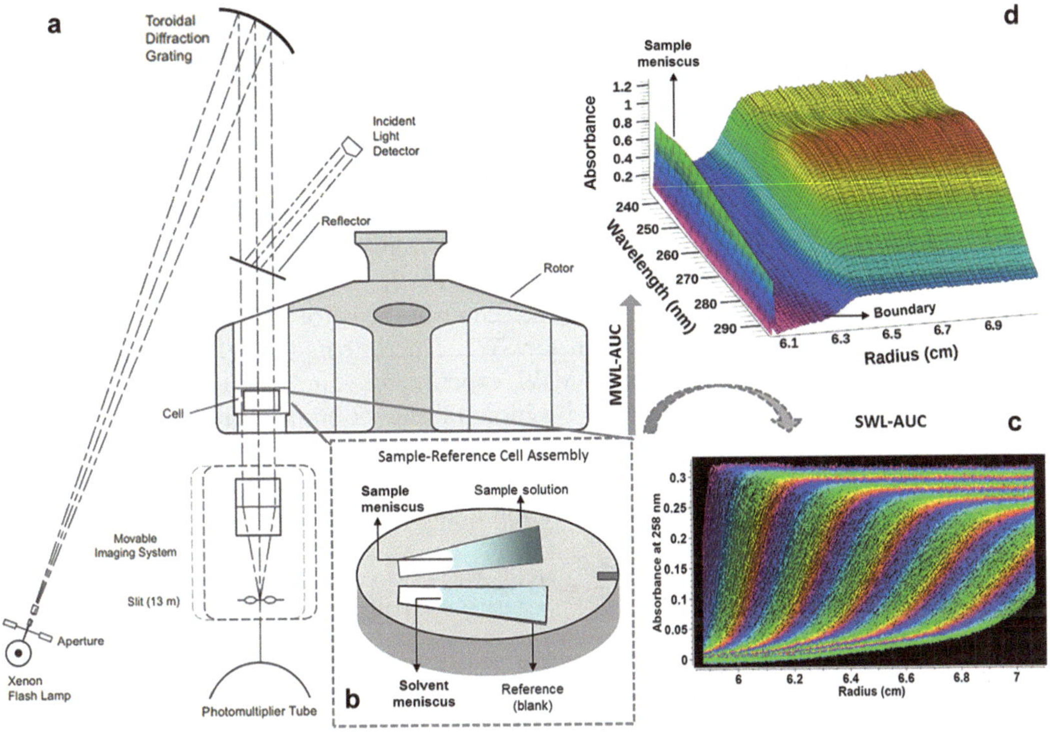

Fig. 1 Analytical ultracentrifugation: from the optical system to data output. (**a**) Overview of the optical system and rotor arrangement in the Beckman analytical ultracentrifuge (*courtesy of Beckman Coulter, Inc.* [Adapted from Greg Ralston; *Introduction to Analytical Ultracentrifugation*; http:/www.beckman.com/literature/Bioresearch/361847.pdf]). (**b**) One AUC cell is zoomed in, to show the schematic of a centerpiece inside an assembled cell. (**c**) Examples of SWL scans obtained from an AUC run, showing progressive movement of the boundary in a radial direction toward the bottom of the cell. Although the original scans are acquired in the intensity mode, the data, after processing, are presented as (pseudo)absorbance versus radius (cm) scans, as shown. (**d**) Data from MWL-AUC analysis of an RNA–protein interaction. A single time point of a four-dimensional data set as a function of time, radius, and wavelength is shown. The sample meniscus is at the left edge ([**a**, **b** adapted from Methods in Enzymology, Volume 469, *Using analytical ultracentrifugation (AUC) to measure global conformational changes accompanying equilibrium tertiary folding of RNA molecules*, Somdeb Mitra, Pages 209–236, Copyright 2009 [5], with permission from Elsevier; (**d**) adapted with permission from Analytical Chemistry 2017, 89 (1), *Spectral and Hydrodynamic Analysis of West Nile Virus RNA–Protein Interactions by Multiwavelength Sedimentation Velocity in the Analytical Ultracentrifuge*, Zhang, J. et al, Pages 862–870. Copyright (2017) American Chemical Society])

movement of the solvent–solution boundary in each cell by measuring the amount of absorbed light by the solution. If the experiment is allowed to proceed indefinitely, an equilibrium state will be reached where the material collecting at the outward facing wall of the cell reaches very high concentration at the bottom of the cell. The large concentration gradient formed at the bottom generates strong diffusional transport in the opposite direction of the sedimentation transport. Eventually, diffusion and sedimentation transport cancel each other, an equilibrium state is reached, and net flow of solute is no longer observed.

During the approach-to-equilibrium phase of the experiment, which is also called a sedimentation velocity (SV), the most useful data are obtained [26–28]. Molecules with more compact and spherical shapes, with a smaller hydrodynamic radius (Stokes radius or R_H), travel faster compared with molecules that are less compact and elongated, with a larger R_H [10]. Denser molecules with the same hydrodynamic radius experience less buoyancy and also travel faster. The experimentally recorded data include the time-dependent changes of the concentration profile as a function of the distance from the center of the rotor (x), either monitored at a fixed wavelength for SWL experiments (Fig. 1c) or at a series of wavelengths for MWL experiments (Fig. 1d). The time-dependent changes in the position of the boundary of the concentration gradient ($\partial c/\partial x$) are then fit to finite-element solutions of the Lamm equation, which describes the time-dependent evolution of the radial concentration gradient ($\partial c/\partial t$):

$$\frac{\partial c}{\partial t} = D\left[\frac{\partial^2 c}{\partial x^2} + \frac{1}{x}\frac{\partial c}{\partial x}\right] - s\omega^2\left[x\frac{\partial c}{\partial x} + 2c\right]$$

where ω is the angular velocity, c is the concentration of the solute, and s and D are two fundamental hydrodynamic properties of a molecule, its *sedimentation*, and *diffusion* coefficients, respectively [29–31].

The sedimentation coefficient, defined as the ratio of the linear velocity (u) to the angular acceleration ($\omega^2 x$) of the solute particle $\left(s = \frac{u}{\omega^2 x}\right)$, determines the speed of sedimentation under a given set of solution conditions. It is proportional to the molar mass (M) of the solute particle and inversely proportional to the frictional coefficient (f), which determines the viscous drag that a particle experiences in a direction opposite to its movement. Hence, the sedimentation coefficient, s, can be expressed as

$$s = \frac{m(1 - \bar{v}\rho)}{Nf} \tag{1}$$

where $\bar{v}$ is the partial specific volume of the solute particle, N is Avogadro's number, and ρ is the density of the solvent. According to Stokes law, the frictional coefficient of a particle (f) is proportional to the hydrated radius (R_H) of the particle and the viscosity (η) of the solvent:

$$f = 6\pi\eta R_H \tag{2}$$

The diffusion coefficient (D) of a spherical solute particle is a function of the universal gas constant (R), the absolute temperature (T), and the frictional coefficient (f), according to the following relationship:

$$D = \frac{RT}{Nf} \tag{3}$$

Dividing Eq. 1 by Eq. 3 yields the following expression for the mass of a solute, provided that the partial specific volume and solvent density are known:

$$M = \frac{sRT}{D(1 - \bar{\nu}\rho)} \tag{4}$$

Furthermore, we can define a hypothetical sphere with radius R_0 that has the same volume as the sedimenting particle:

$$R_0 = \left(\frac{3M\bar{\nu}}{4\pi N}\right)^{1/3} \tag{5}$$

Using Stokes law, a frictional coefficient, f_0, can be derived for this particle:

$$f_0 = 6\pi\eta R_0 \tag{6}$$

Dividing Eq. 2 by Eq. 6 gives rise to the frictional ratio or shape factor F (defined as the ratio, f/f_0), which denotes the anisotropy of the particle. $F = 1.0$ represents a perfect sphere, while higher anisotropies signify increasingly non-globular or asymmetric shapes [9, 32].

These relationships show how measurements of s and D, by SV-AUC, report on the mass, the anisotropy, and the hydrodynamic radius of the sedimenting particle [26, 33]. Readers who are interested in more detailed theoretical treatments of hydrodynamic variables governing the bulk transport of solutes under centrifugal force are referred to [26, 33–36].

The premise of detecting counterion or protein-induced RNA tertiary compaction or global conformational change by SV-AUC is to monitor the change in the overall size or anisotropy of the RNA molecule. A series of measurements of s and D at varying counterion or protein concentrations reports on the values of f and hence, R_H at each condition. If the investigator is monitoring global conformational changes of an already compact structure, such topological alterations can be detected as long as there is a change in the anisotropy [34].

The additional advantage offered by MWL AUC, in combination with new data analysis algorithms and high-performance computing, is the ability to spectrally resolve multiple interacting components in solution that differ in their molar absorptivity [19, 22, 23]. Therefore, SV-AUC experiments with MWL technology can directly report on molecular weights and stoichiometries of interacting partners [37]. In order to achieve clear spectral separation of the solutes, an investigator can either make use of the intrinsic ultraviolet–visible (UV–vis) absorptions of the macromolecules being studied [22, 23] or use macromolecules that have been tagged with a small fluorescent dye or fused to a fluorescent reporter protein [38].

Since the intrinsic UV absorption of RNA and proteins is significantly different, information on the sedimenting behavior of each of these components can be obtained from the same experiment by collecting the absorption signal over an entire accessible spectral range (depending on buffer composition, up to 215–300 nm) and then computationally deconvoluting the signals from each molecule. An excellent example of this approach has been described in detail in a recent publication [15]. To briefly summarize, this study investigated the interaction between a model RNA (a stem loop structure from the 3′ end of the negative-strand genomic RNA of the West Nile virus, WNV-RNA) and the human T-cell-restricted intracellular antigen-1-related protein (hTIAR). In each run of a MWL-AUC experiment, hTIAR was added at different molar ratios to a fixed molar concentration of WNV-RNA, which, for each titration point, yielded four-dimensional datasets in the form of intensity (converted to pseudo-absorbance) as a function of time, radius, and wavelength (Fig. 2a). Using the non-negatively constrained least squares (NNLS) algorithms [39], the 4D data were then decomposed into three sets of individual absorption components: RNA, protein, and buffer (if the buffer has any absorbing component in the chosen spectral range) (Fig. 2b). Treating the buffer absorption as time and radius invariant signal, the time-dependent changes of the separated molar concentrations of the WNV-RNA and the hTIAR were then plotted as a function of the radial distance from the center of the rotor and separately stored as two three-dimensional datasets (Fig. 2c) (pseudo-absorbance vs. radius and time). Then each dataset was initially analyzed independently, using two-dimensional spectrum analysis (2DSA) [40], implemented in UltraScan [24], reporting the *s*, *D*, and *F* values of the RNA and the protein separately. Finally, global fits of the RNA and protein datasets were performed, using genetic algorithm-Monte Carlo analysis [41, 42], to identify all distinct hydrodynamic species in solutions of RNA and proteins at varying molar ratios (Fig. 2d–g). Identical *s* values for hydrodynamic species observed in both the RNA and protein signals indicate the presence of complexes. Comparing the molar concentrations of these complex species, one obtains the absolute amounts of RNA and protein present in the RNA–protein complexes which uniquely identify their ratios [15, 37]. These ratios are then used to determine the contributions of protein and RNA to the partial specific volume of the complex, which in turn permits a more precise prediction of the molar mass from the sedimentation and diffusion coefficients. Both

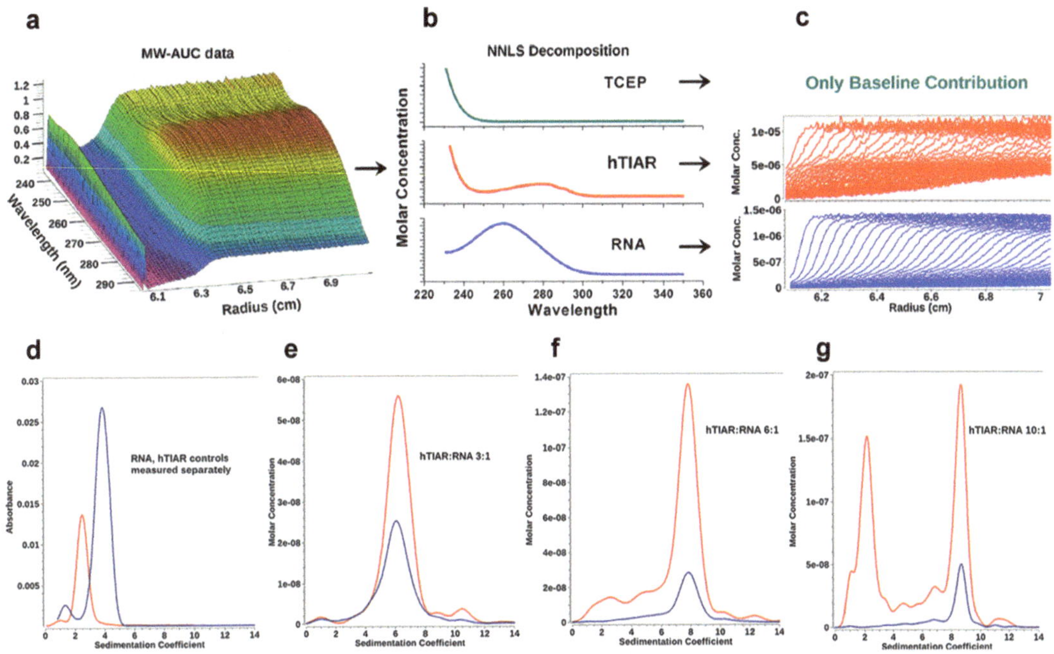

Fig. 2 MWL-AUC data and its analysis. Primary four-dimensional MWL-AUC data (**a**) (only one time point is shown; the meniscus is visible at the left edge of the radial range) are decomposed, by the nonnegatively constrained least square (NNLS) algorithms, into their spectral constituents (**b**), resulting in spectrally separated three-dimensional (3D) datasets (**c**). Each dataset only reflects the hydrodynamic contributions of one of the constituents, hTIAR and WNV-RNA, which have different chromophores in this case. The TCEP absorbance contribution must be considered in the spectral decomposition; however, since it does not sediment, it only contributes to the baseline, and a hydrodynamic analysis is not necessary. Panels **d–g** depict global 2DSA-Monte Carlo models obtained from decomposed RNA and hTIAR datasets, at different molar mixing ratios. Separate hTIAR and RNA controls (**d**), hTIAR:RNA 3:1 (**e**), hTIAR:RNA 6:1 (**f**), and hTIAR: RNA 10:1 (**g**). All plots show the protein data in red and the RNA data in blue. A clear shift to higher-molecular-weight complex formation is observed as the protein concentration is increased. At the highest ratio (10:1, **g**), the excess unbound protein appears as a peak at the same position as the free hTIAR control (**d**). Broad peaks indicate the presence of a reaction boundary, and absence of a free hTIAR peak, except when added in excess, indicates tight binding between hTIAR and WNV RNA. The control experiments and the 10:1 mixtures were performed at hTIAR and RNA loading concentrations of 11.6 μM and 1.16 μM, respectively. All plots are shown as *g*(*s*) distributions with *dC*/*ds* plotted on the *Y*-axis in the indicated units [Adapted with permission from Analytical Chemistry 2017, 89 (1), *Spectral and Hydrodynamic Analysis of West Nile Virus RNA–Protein Interactions by Multiwavelength Sedimentation Velocity in the Analytical Ultracentrifuge*, Zhang, J. et al, Pages 862–870. Copyright (2017) American Chemical Society]

the molar concentration contribution ratios and the molar mass, together, will then uniquely identify the complex stoichiometry.

In this chapter, we provide sufficient information for investigators to design, conduct, and analyze data from an SV-AUC experiment to measure counterion–protein-mediated tertiary compaction of RNA molecules, starting from an ensemble that possesses native secondary structure, as well as to detect global conformational changes in a folded RNA molecule upon ligand–protein binding. We will focus our discussions on the new

Optima™ instrument with its built-in MWL technology to highlight the additional insights into RNA–protein interactions that can be obtained with the new instrument. One of the authors (B.D.) offers recurring workshops on problem solving with AUC to help new users gain confidence in using this new technology effectively (for more information, please visit htp://ultrascan.aucsolutions.com).

2 Materials

To ensure the quality and integrity of the RNA samples, all buffers and solutions are prepared in ribonuclease (RNase)-free, deionized ultrapure water (resistivity 18.2 MΩ cm at 25 °C, TOC < 10 ppb) obtained from a Milli-Q® water purifier (Millipore). Molecular biology grade, nuclease-free buffer components are purchased from Ambion® (ThermoFisher Scientific®), Fisher Scientific®, or Sigma-Aldrich®, and the final buffer stock solutions are filtered through 0.2 μm filters (Pall Corporation®). Buffer exchanges, when needed, are performed using Amicon® Ultra-15/50 (MilliporeSigma®) Centrifugal Filter Units or EMD Millipore Ultra Centrifugal Filters, 3 kDa Ultracel, and 0.5 mL (Cole-Parmer Scientific Experts®), of the appropriate molecular cutoffs. All equipment are routinely cleaned with RNaseZap (Ambion®/ThermoFisher Scientific®) and only autoclaved glasswares are used for storing reagents.

2.1 RNA Transcription

1. A bacterial plasmid which is constructed by cloning the DNA sequence corresponding to the RNA of interest, downstream of a T7 promoter sequence. Changing the first nucleotide after the T7 promoter (+1) to G greatly improves transcription yield. Usually, the 3′ side of the RNA sequence is flanked by an HδV ribozyme sequence to generate homogeneous 3′ ends [43] (*see* **Note 1**).
2. Polymerase chain reaction (PCR) machine (thermal cycler) and forward and reverse primers for amplification of the linear transcription template from the plasmid. The reverse primers are often designed to carry 2′-O methylation modification at the first two positions (*see* **Note 2**).
3. MEGAscript® T7 Kit (Ambion®) for in vitro transcription of RNA.
4. Denaturing gel solution containing 5% urea-acrylamide/bisacrylamide (19:1) and 7 M urea in 1× Tris–borate–EDTA (TBE) buffer.
5. Ammonium persulfate (Sigma-Aldrich®) dissolved in water to make a 10% (wt/vol) solution.

6. *N,N,N′,N′*-tetramethylethylenediamine (TEMED >97%; Fisher Scientific).
7. Vertical gel running apparatus (Thermo Scientific, Owl Dual-Gel Vertical Electrophoresis Systems), glass plates (10 cm × 10 cm; one regular one notched), combs and spacers (1.5 mM thick), and a high-voltage programmable power supply (e.g., Thermo Scientific, Owl EC3000XL).
8. Hand-held UV torch and Fluor-coated TLC plate (Ambion®) for UV shadowing.
9. Fluor-coated TLC plate (10 × 10 cm; Invitrogen™ Ambion™).
10. Elution buffer: 10 mM sodium cacodylate, 0.1 mM EDTA, 500 mM KCl (or NaCl), pH 7.3.
11. 100 and 70% ethanol.
12. Refrigerated tabletop centrifuge (e.g., Eppendorf 5402 Refrigerated Centrifuge Benchtop).
13. Lyophilizer/SpeedVac (e.g., Savant DNA120 SpeedVac Concentrator DNA–RNA Sample Dryer) for drying RNA pellets.
14. RNA stock solutions of 10–50 μM concentration in CE buffer (*see* **Note 3**).

2.2 Protein Expression and Purification

1. Proteins used in these experiments are almost always expressed in *E. coli* BL-21 (DE3) RIPL Codon Plus electro-competent cells (Agilent technologies), codon-optimized for eukaryotic gene expression.
2. All the proteins used in our studies are tagged with either a standard N-terminal 6X-His tag or a C/N-terminal Intein–Chitin binding domain (CBD) tag (IMPACT™: Intein Mediated Purification with an Affinity Chitin-binding Tag; NEB®). For N-His-tagged proteins, the corresponding genes are generally cloned in a suitable pET vector (e.g., pET-45b(+); Novagen®) and for the C/N-Intein-CBD-tagged proteins, the corresponding genes are cloned in the pTYB1-2 (for C-tag) or pTYB11-12 (for N-tag) vectors (NEB®).
3. HisPur™ Ni-NTA Resin (Thermo Scientific™) or Chitin Resin (NEB®) for affinity purification of His-tagged and Intein-CBD-tagged proteins, respectively.
4. Luria-Bertani (LB) broth and agar powder from ThermoFisher Scientific®.
5. Isopropyl-β-D-thiogalactoside (IPTG) (Sigma-Aldrich®), stored as 10 mM stock at −20 °C.
6. Ampicillin (Sigma-Aldrich®), stored as 100 mg/mL stock at −20 °C.

7. Any standard refrigerated floor-centrifuge that can accommodate a JA-10 (or similar) and a SS-34 (or similar) fixed-angle rotors.
8. Any standard model of Sonicator (e.g., Fisherbrand™ Sonic Dismembrator) or French Press (e.g., French Press G-M™; GlenMills®) for bacterial cell lysis.
9. Amicon® Ultra-15 or Amicon® Ultra-50 Centrifugal Filters (MilliporeSigma®), for protein concentration and buffer-exchange reactions of volumes up to 15 or 50 mL, respectively. Choose appropriate molecular weight cutoff, 2–3 times smaller than the molecular weight of the protein being used in the experiments.
10. EMD Millipore UFC500324 Amicon Ultra Centrifugal Filters, 3 kDa Ultracel, 0.5 mL (Cole-Parmer Scientific Experts®) for buffer-exchange reactions of smaller volumes (up to 500 μL), immediately preceding RNA–protein MWL AUC experiments.

2.3 Buffers for RNA and Proteins

1. CE buffer: 10 mM sodium cacodylate trihydrate ($(CH_3)_2As$-$O_2Na.3H_2O$; Sigma-Aldrich®), 0.1 mM EDTA, pH 7.3. A 10× stock solution is made and stored in a 4 °C refrigerator. The CE buffer is the starting point for experiments in which RNA folding is induced by monovalent ions, like sodium or potassium [10, 44] (*see* **Note 3**). It is also the buffer in which the RNA samples are routinely dissolved and stored.
2. CEK buffer: 10 mM sodium cacodylate, 0.1 mM EDTA, 100 mM potassium chloride, pH 7.3. This is the buffer routinely used while conducting Mg^{2+} (or other divalent ion) induced RNA folding experiments [45–47] (*see* **Note 4**). A 10× stock solution is made and stored in a 4 °C refrigerator.
3. We purchase ribonuclease-free solutions of $MgCl_2$ (1 M), EDTA (0.5 M, pH 8.0), KCl (2 M), and NaCl (5 M) from Ambion®.
4. Standard protease inhibitors: 1 tablet Roche® Complete EDTA-free protease inhibitor cocktail tablet per 50 mL buffer, 1 mg/mL leupeptin, 1 mg/mL aprotinin, 1 mg/mL pepstatin, 1 mM benzamidine.
5. SUPERase• In™ RNase Inhibitor (20 U/μL; Invitrogen™).
6. Protein lysis buffer (common to His-tagged and Intein–CBD-tagged proteins): 20 mM Hepes·KOH or potassium salt of 4-(2-hydroxyethyl)-1-piperazineethanesulfonic acid ($C_8H_{17}KN_2O_4S$; Sigma-Aldrich®), pH 7.6, 0.5 M KCl, 0.1% Triton X100 (from 1% stock; Invitrogen™), 1 mM EDTA, standard protease inhibitors.

7. Ni-NTA equilibration buffer: 20 mM Hepes·KOH, pH 7.4, 0.3 M KCl, 10 mM imidazole (from freshly made 1 M stock; purchased in crystalline form from Sigma-Aldrich®).
8. Ni-NTA wash buffer: 20 mM Hepes·KOH, pH 7.4, 0.3 M KCl, 25 mM imidazole.
9. Ni-NTA elution buffer: 20 mM Hepes·KOH, pH 7.4, 0.3 M KCl, 250 mM imidazole.
10. Intein wash buffer: 20 mM Hepes·KOH, pH 7.4, 1 M KCl, 0.1% Triton X100, 1 mM EDTA.
11. Intein cleavage buffer (−DTT): 20 mM Hepes·KOH, pH 8.0, 0.5 M KCl, 1 mM EDTA (to initiate the Intein cleavage reaction, DTT is added later to a final concentration of 50 mM).
12. Protein storage buffer: 20 mM HEPES, pH 7.6, 200 mM KCl, 10 mM beta-mercaptoethanol (2-mercaptoethanol, ≥99.0% stock; Sigma-Aldrich®), and 10% glycerol (from ACS certified, ≥99.5% stock; Fisher Scientific®). A 10× stock solution is usually made without the BME and stored in a 4 °C refrigerator. BME is added freshly, to a final concentration of 10 mM, right before diluting the buffer to 1× concentration.
13. AUC phosphate buffer: 10 mM sodium phosphate, pH 7.0, 50 mM NaCl, and 1 mM TCEP·HCl (Tris(2-carboxyethyl) phosphine hydrochloride, BioUltra, ≥98%; Sigma-Aldrich®) (*see* **Note 5**). Stored at 4 °C as a 10× stock solution.

2.4 Analytical Ultracentrifugation

1. Beckman Optima™ AUC by Beckman Coulter (*see* **Note 6**).
2. A four-slot An-60 Ti rotor with max. angular velocity of 60,000 rpm (alternatively, an eight-slot An-50 Ti rotor can also be used with a max. angular velocity of 50,000 rpm).
3. AUC cells (Beckman Coulter): Epon charcoal centerpieces with sector-shaped channels of 1.2 cm optical path length (Fig. 1b), quartz or sapphire windows, window liners and gaskets, upper and lower window holders, cell housing, screw ring, red polyethylene gasket sheet for sealing sample loading ports, brass screws for closing loading ports. It is recommended to have a full complement of 4 or 8 cells, one for each rotor hole, on hand so a maximum number of conditions (8 or 16, depending on rotor type) can be examined under identical run conditions.
4. A counterbalance (Beckman Coulter) for radial calibration (*see* **Note 7**).
5. 1 mL Disposable syringe with 30-gauge hypodermic needles.
6. Analytical Torque Stand and Wrench and screwdrivers (Beckman Coulter).

7. RNaseZap (Ambion®) solution for cleaning the AUC cells.
8. Kimwipe® (Kimberly-Clark®) extra low-lint 2-ply paper wipers for delicate surfaces.

2.5 Software for AUC Run Setup, Data Acquisition, and Data Analysis

1. UltraScan-III (http://ultrascan.aucsolutions.com).
2. Origin® (Origin Labs®) or Prism (GraphPad®) or similar software for data plotting and curve fitting.

3 Methods

3.1 RNA Sample Preparation

1. Using a PCR-generated linear DNA template containing the RNA sequence downstream of a T7 promoter, transcribe the RNA in vitro by run-off transcription, using components of the MEGAscript® T7 Kit (Invitrogen™ Ambion®) or using a recombinant RNA polymerase, expressed and purified in our laboratory, with appropriate buffer conditions [43] at 37 °C for 2 h (*see* **Note 8**).
2. If a ribozyme is not used to obtain homogenous ends, the transcribed RNA can be directly purified by using spin-column-based methods, such as MEGAclear™ Transcription Clean-Up Kit (Invitrogen™ Ambion®).
3. When more than one RNA species is present in the transcription reaction, separate the reaction products on a denaturing 5% urea:polyacrylamide gel (19:1), and after the run is complete, transfer gel from the glass plates on to a Fluor-coated TLC plate (covered with plastic wrap), illuminate it from overhead with short wavelength UV, and cut out the gel piece containing the RNA band of interest with a clean razor blade (*see* **Note 9**).
4. Crush the gel piece by passing it through a 5 mL syringe, soak the pieces in the elution buffer in a 15 mL Falcon™ tube, and rock it overnight in a cold room, to elute the RNA.
5. Next morning (or after ~6–8 h), give the solution a quick spin to let the gel pieces settle to the bottom and pipette out the supernatant carefully. Precipitate the RNA from the supernatant solution by adding 100% ethanol (3× volume of the supernatant), incubating at −20 °C (or in dry ice) for 1 h, and spinning down the samples in a refrigerated tabletop centrifuge at 14,000 rpm for 1 h at 4 °C.
6. Rinse the pellet by spinning twice (14,000 rpm for 30 min at 4 °C) with 70% ethanol, dry it in a desiccator (SpeedVac), and dissolve it in 1× CE buffer to a concentration of 10–50 μM.
7. For RNA–protein studies, right before the AUC experiment, thaw an appropriate volume of the stored RNA sample and

buffer-exchange it five times into the 1× AUC Phosphate buffer, using the small (0.5 mL) EMD Millipore UFC500324 Amicon Ultra Centrifugal Filters.

3.2 Protein Sample Preparation

1. For genes cloned with a His-tag in pET vectors or with an Intein-CBD-tag in pTYB vectors, pick a single colony from a plate of transformed BL21-Codon Plus (RIPL) cells, inoculate 5 mL LB broth (containing ampicillin 100 μg/mL) to set up a primary culture, and grow by shaking at 37 °C overnight (or until $O.D._{600}$ is »1).
2. Inoculate a 1 L LB (+ampicillin) broth with ~100 μL of primary culture and grow cells by shaking at 37 °C until the cells are in the log phase of growth (OD_{600} ~ 0.5–0.8).
3. Induce cells for protein expression by adding IPTG to a final concentration of 0.5 mM and incubate with shaking either at 37 °C for ~8 h or at 15 °C overnight (incubation conditions, including optimum IPTG concentrations, temperature, and duration of expression, to obtain best soluble protein yields should be tested beforehand).
4. Spin down cells in a JA-10 rotor (spin 15 min at 8000 rpm), discard the supernatant, resuspended the pelleted cells in ice-cold lysis buffer, and lyse the cells either by sonication (typically 10–15 cycles of 15–30 s pulse and 30 s gap between pulses, on ice) or using a French Press (two rounds, each at 1800 psi, for efficient lysis).
5. Spin down the lysate in SS-34 type rotors at 13,000 rpm for 30 min at 4 °C, collect the supernatant (always keep it at 4 °C), and proceed to the appropriate column-purification.
6. For general purification strategies of His-tagged proteins, cloned in pET vectors, refer to the manufacturer's protocol for details (http://www.emdmillipore.com/US/en/product/pET-45b+-DNA-Novagen,EMD_BIO-71,327#documentation), and when using the HisPur™ Ni-NTA Resin, refer to https://assets.thermofisher.com/TFS-Assets/LSG/manuals/MAN0011700_HisPur_NiNTA_Resin_UG.pdf. Briefly, equilibrate the desired amount (typically 5–10 mL) of Ni-NTA resin in a gravity-flow column with the equilibration buffer (2× the bed volume), mixing the lysate with an equal volume of equilibration buffer, and incubate at 4 °C for 30 min, with constant rolling, and drain the flow-through. Wash the resin with the Ni-NTA wash buffer (3× bed volume) and elute the protein by adding the Ni-NTA elution buffer (1× bed volume). If using an FPLC system, briefly, charge a regenerated 5 mL HiTrap chelating column (Amersham®) with a freshly prepared 100 mM $NiSO_4$ (Sigma-Aldrich®), equilibrate with 5× column-volume (CV) of the Ni-NTA equilibration buffer,

and load the filtered lysate (filtered sequentially through 5 mm to 0.8 mm filters) onto the equilibrated nickel column at a flow rate of 3 mL/min. Wash the column with Ni-NTA wash buffer (5 CV) and elute with the Ni-NTA elution buffer (2 CV) at 5 mL/min.

7. For purifying Intein–CBD-tagged proteins, refer to the manufacturer's protocol for details (https://www.neb.com/products/e6901-impact-kit#Protocols,%20Manuals%20&%20Usage).

   Briefly equilibrate ~5 mL of the chitin resin in a 20 mL gravity flow column, with 1 column volume (CV) of the lysis buffer, apply the clarified supernatant to the resin, incubate with gentle shaking at 4 °C for 1 h, drain the flow-through by gravity flow, and wash the resin with 3 CV of the Intein wash buffer. Then wash the resin with 1 CV of the Intein cleavage buffer (without DTT) and elute the protein with ~3× chitin bed volume of the Intein cleavage buffer supplemented with 50 mM DTT.

8. Concentrate the eluted proteins using an Amicon® Ultra Centrifugal Filter of the appropriate molecular weight cutoff, carry out buffer exchange at least five times with the protein storage buffer, and further concentrate down to the desired concentration (typically 10–50 μM). Aliquot in small volumes, flash-freeze in liquid nitrogen, and store at −80 °C (*see* **Note 10**).

9. Right before the AUC experiment, thaw an appropriate volume of the stored protein sample and buffer-exchange it five times in to 1× AUC Phosphate buffer, using the small (0.5 mL) EMD Millipore UFC500324 Amicon Ultra Centrifugal Filters.

3.3 Setting Up Equilibrium Counterion/Protein-Mediated RNA Compaction Reactions

1. To study counterion (monovalent cations, divalent cations, polyamines, etc.) or protein-induced equilibrium RNA compaction, set up each titration point in a separate 1.5 mL microfuge tube, such that the total reaction volume is 500 μL, with a final RNA concentration of 0.1–1.0 OD (A_{260}). For monovalent and divalent cation titrations, use Tables 1 and 2, respectively, whereas for protein titrations, use Table 3, as a guideline to set up the reactions. We typically use 0.5–0.6 OD as the final RNA concentration, and the optimal loading volume in each sector-shaped compartment of an AUC cell is 460 μL (also *see* **Notes 11** and **12**).

2. Prepare sufficient volume of stock solutions of the counterions/proteins (e.g., 2 M KCl, 5 M NaCl, 1 M $MgCl_2$, 20 μM protein). Titrations should span an entire range over which the equilibrium compaction of the RNA is expected to occur. Generally, this range is 0 to 1.5 M for monovalent ions (M^+),

Table 1
Sample reaction setup for monovalent cation titrations (e.g., K^+; using a 2 M stock)

Final K^+ conc.	RNA (10–50 μM)	CE buffer (10×)	KCl (2 M)	H_2O	Total volume
X mM	Υ μL for (0.5 OD)	50 μL	0.25× μL	[500 – (Υ + 50 + 0.25X)] μL	500 μL

Table 2.
Sample reaction setup for divalent cation titrations (e.g., Mg^{2+}, using a 1 M stock)

Final Mg^{2+} conc.	RNA (10–50 μM)	CEK buffer (10×)	$MgCl_2$ (1 M)	H_2O	Total volume
X mM	Υ μL for (0.5 OD)	50 μL	0.5× μL	[500 – (Υ + 50 + 0.5X)] μL	500 μL

Table 3
Sample reaction setup for protein titrations (e.g., using a 20 μM stock protein)

Final protein conc.	RNA (10–50 μM)	AUC phosphate buffer (10×)	Protein (20 μM)	H_2O	Total volume
X μM	Υ μL for (0.5 OD)	50 μL	25× μL	[500 – (Υ + 50 + 25X)] μL	500 μL

0 to 100 mM for divalent ions (M^{2+}), and 0 to 1 to 10 μM for proteins (varies considerably depending on the protein being studied).

3. Using Tables 1, 2, and 3 as guidelines, prepare eight titration points (for An-60 Ti rotors), for each SV-AUC run (*see* **Note 7**). DO NOT add the RNA to the tubes at this stage. Therefore, at the end of this step, one should have a total of eight reaction tubes containing (500 – Υ) μL volume of each of the titration samples, with all the reagents added in except Υ μL of the RNA. We typically use the CE buffer for M^+ titrations, the CEK buffer for M^{2+}, and the AUC Phosphate buffer for protein titrations (*see* **Notes 3**, **4**, and **5**). The following tables show examples of reaction setups for M^+ (K^+), M^{2+} (Mg^{2+}) and protein titrations, using 2 M KCl, 1 M $MgCl_2$, and 10 μM protein stock solutions, respectively.
4. In a separate 1.5 mL tube, take slightly greater than 8Υ μL of the RNA stock solution (Υ is the amount to be added to each of the eight folding reaction points as described above). The slight excess volume accounts for evaporation during the following denaturation step.

5. Heat the RNA solution to 95 °C on a heating block for 1 min (to denature it) and cool it slowly to 50 °C (to allow equilibration to native secondary structure) (*see* **Note 13**). During this step, also incubate the reaction tubes containing all the reagents except the RNA, at 50 °C.
6. After briefly centrifuging the RNA solution in a tabletop centrifuge to spin down the condensates on the lid, put them back at 50 °C and pipette out *Y* μL of the RNA solution into each of the three reaction tubes containing the folding buffers at 50 °C. Mix the solutions gently by pipetting up and down and continue incubation at 50 °C for an additional 30 min (*see* **Note 14**).
7. Move the reaction tubes from 50 °C to another heat block that is set to temperature at which the equilibrium tertiary structure will be monitored (usually, 25 or 37 °C) and incubate for 1 h to allow the RNA molecules to equilibrate to their native tertiary structures at the given temperature (*see* **Note 15**).

step

3.4 Setting Up Reactions to Probe Equilibrium Conformational Change of an RNA Upon Ligand/Protein Binding

1. To probe whether binding to a small-molecule ligand or a protein alters the global conformation of an RNA, whose hydrodynamic parameters in the folded state have already been determined, set up eight reactions at a time, each with the fully folded RNA molecule (in CEK buffer +5 mM Mg^{2+}) and with eight different molar concentrations of the ligand, with respect to the RNA, to start with (e.g., 0.5:1, 1:1, 2:1) (*see* **Note 16**). Table 4, as presented below, may be used as a reference to set up the reactions:
2. For each reaction tube, first add and mix all the components except for the $MgCl_2$ (5 μL) and the ligand (5× μL), incubate the reactions on a heating block at 95 °C for 2 min, and slowly cool them down to ~ 50 °C. During this step, also incubate the $MgCl_2$ stock (1 M) at 50 °C and the ligand stock (100 μM) at the final reaction temperature (usually 25 or 37 °C).
3. After briefly centrifuging the reaction tubes in a tabletop centrifuge to spin down the condensates on the lid, put them back

Table 4
Sample reaction setup to monitor changes induced by a ligand (100 μM stock)

Final ligand conc.	RNA (10–50 μM)	CEK buffer (10×)	$MgCl_2$ (1 M)	Ligand (100 μM)	H_2O	Total volume
X μM	*Y* μL for (0.5 OD)	50 μL	5 μL	5× μL	[500 − (*Y* + 55 + 5*X*)] μL	500 μL

at 50 °C and pipette in 5 μL of the 1 M $MgCl_2$ solution into each of the three reaction tubes containing the RNA samples in CEK buffer at 50 °C. Continue incubation at 50 °C for an additional 15 min.

4. After briefly centrifuging the reaction tubes, transfer them to the heating block set at the final reaction temperature (at which the ligand is also being incubated), allow equilibration for ~ 5 min, and add 5× μL of the 100 μM ligand stock solution. Mix the solutions gently by pipetting up and down and continue incubation at the reaction temperature for an additional 1 h to allow the folded RNA to reach its new equilibrium conformation in the presence of the ligand. Treat the three corresponding blank solutions in exactly the same way.

3.5 AUC Cell Assembly and Loading Samples into the Cells

1. Before each experiment, soak all the components of the cells with a mild detergent (e.g., RNaseZap™; Invitrogen™ Ambion™), wash with ultrapure water, rinse with 70% ethanol, and dry carefully with a low-lint, nonabrasive tissue (e.g., low-lint Kimwipes®), to prevent scratching of the quartz windows and the epon-charcoal centerpieces.
2. Refer to the instruction manual accompanying the AUC instrument or to Backman Coulter's website under Product Documentation ("An-50 Ti and An-60 Ti Analytical Rotor, Cells, and Counterbalance"), for detailed instructions on AUC cell assemblies. Briefly place each quartz window in its corresponding holder, with the gasket and the liner, and then make a vertical assembly containing the centerpiece sandwiched between the two windows. Gently slide the metal cell housing over the vertical assembly (*see* **Note 17**).
3. Insert the O-ring and tighten the screw ring at the top (make sure the side with the marking "OUT" is facing outside), first by hand and then with the Analytical Torque Stand and Wrench (Beckman Coulter) to 120 inch-pounds, to prevent sample leakage during AUC runs.
4. With the screw ring facing the investigator, and the cell in a horizontal position with the fill holes at the top, load 460 μL of the reaction sample into each channel. Looking at the screw ring, and the fill holes at the top, the left channel will be marked "B" and the right channel will be marked "A" in the instrument (*see* **Note 18**). Pipette carefully to avoid air bubbles and be careful not to touch the channel surface with the hypodermic needle to avoid damaging the centerpieces.
5. After loading all four pairs of reaction samples, number the aluminum housing of each cell with a marker and note which cell and channel corresponds to each sample.
6. Cut small circular polyethylene gaskets and seal the loading ports of the centerpieces by first placing a gasket piece on

each port and then gently tightening the brass screws over them.

7. Clean the windows on either side using brief bursts of compressed air.

3.6 Loading AUC Cells into the Rotor

1. While the experimental samples are incubating at the folding temperature, it is advisable to equilibrate the rotor to the experimental temperature to minimize delays at the beginning of the experiment. For experiments to be performed at 4 °C, it is recommended to temperature-equilibrate the rotor in a cold room or refrigerator, or inside the Optima™ AUC set to the desired temperature and under vacuum.
2. Once the samples and blanks have been loaded, measure the weight of the sealed ultracentrifugation cells by balancing two cells against each other to minimize weight differences between opposing pairs of cells. Ideally, weight differences are between 0 and 0.2 g to maintain the rotor balance within tolerance, should one of the cells leak. If a leak occurs, the weight difference should not exceed 0.5 g and the experiment should be stopped immediately to prevent centerpiece damage.
3. Take out the rotor from the chamber, place it on the rotor stand (*see* **Note 19**), and insert the balanced cells across from each other, with at least one cell in hole "4" (for An-60 Ti rotors) or "8" (for An-50 Ti rotors). Place the remaining pair (s) of balanced cells at positions across from each other in the remaining slots. Make a note as to which samples (based on the numbering outside the ultracentrifuge cell) are placed into each rotor hole. The screw ring-sides should be facing upward, and the screw-capped fill-holes should point to the rotor center (*see* **Note 20**).
4. Very carefully tilt the entire loaded rotor sideways on a bench, and while looking from the bottom, use the supplier's alignment tool to perfectly align the markings on the lower edge of the metal cell-housings with the lines below each slot in the rotor. Note the rotor serial number which is needed for the data acquisition protocol.
5. Place the rotor into the centrifuge chamber, and swivel the UV-vis periscope into position such that the magnet lock is engaged. Place the lid on top of the rotor chamber, close the sliding door, and turn on the vacuum pump (*see* **Note 21**).
6. Allow the rotor to re-equilibrate to the run temperature until the desired temperature has been reached and is stable before accelerating the rotor.

3.7 Experimental Design

A critically important step in the experimental process is the development of a carefully crafted experimental design that best addresses the questions of the experimentalist. AUC experiments, and in particular MWL-AUC experiments, are not trivial to perform and design and require detailed planning to maximize information content of the resulting data. In addition to the preparation of the samples, a number of instrument variables affecting the experimental outcome need to be optimized, including temperature, rotor speed, experiment duration, wavelength(s) of observation, frequency of scanning, and solute concentration. We outline here some of the important factors that an investigator must consider prior to setting up an AUC run:

1. *Temperature*: For temperature settings, typical considerations include the stability of the sample (4 °C favors stability) or the need to maintain physiological conditions (37 °C). Although it was difficult to conduct experiments at temperatures above 25 °C with the older oil-diffusion pumps of the Proteome Lab™ instruments, the turbomolecular pumps used in the Optima™ AUC make it easier to conduct higher temperature experiments. If no other requirements exist, a 20 °C temperature avoids the need to convert water density and viscosity corrections to standard conditions (although this is handled automatically by UltraScan for experiments run at temperatures other than 20 °C).
2. *Rotor speed*: The rotor speed can be modulated to enhance certain characteristics of the experiment. For example, a higher rotor speed will favor better resolution of the sedimentation coefficient, useful for the characterization of heterogeneity and also for the separation of similar species. On the other hand, a slower rotor speed allows more time for the sample to diffuse before it is pelleted, improving the observed diffusion signal [48]. High-quality diffusion information is required for density, molar mass, or anisotropy determinations. If sufficient sample and instrument time are available, both a high-speed and a low-speed experiment can be performed and globally fitted with UltraScan to combine both optimized signals.
3. *Experimental duration*: The duration of the experiment should be chosen such that at the selected rotor speed the sample is sedimented long enough to approach the equilibrium state. It is important to realize that longer solution columns and hence, longer sedimentation distances also improve resolution. Hence, it is often better to run a more dilute sample with a higher loading volume than a smaller volume of a more concentrated solution. The length of the solution column, of course, also affects the length of time until equilibrium is approached. This varies greatly depending on the experimental needs and the sample characteristics. For highly anisotropic molecules, like nucleic acids, additional time is required to

reach an equilibrium state, whereas for small, globular proteins with larger diffusion coefficients, faster rotor speeds can be considered to obtain an acceptable resolution of the sedimentation coefficient as well as good diffusion signal. For multi-wavelength experiments, additional considerations come into play: The experiment needs to be collected over a sufficiently long time to support collection of a sufficient number of scans for each wavelength. In such a case, one may choose to extend the run-time and sacrifice sedimentation-resolution to increase the data quantity by collecting additional scans.

4. *Frequency of optical scans*: In most cases, the scan frequency should be set to collect as much data as possible. The intrinsic signal from the solutes in the experiment increases linearly with multiple observations, while the stochastic noise only increases with the square root of the number of observations, increasing the distance between the two metrics with additional observations. Hence, the signal-to-noise ratio (SNR) improves through the collection of more scans, although there is a limited return at some point when the SNR improves so little that it is not worth to extend the instrument time and expend the additional computational effort for fitting additional scans (larger datasets take more CPU time to fit).
5. *Concentration of the analyte*: Nucleic acids, which are polyanions that may carry unbalanced negative charges from backbone phosphates, depending on ionic strength, could exhibit considerable concentration-dependent nonideality arising from anisotropic shape and charge–charge repulsions between proximal molecules. If these charges are not shielded by cations, the repulsion leads to concentration-dependent nonideality which complicates analysis considerably and therefore should be avoided. Lowering concentration, or increasing ionic strength, or both, often remedies these problems.
6. *Instrument optical settings:* The factors to be considered for selecting scanning wavelengths include the light intensity produced by the Xenon flash lamp and the dynamic range of the detector (a photomultiplier system). The overall concentration of the analytes should be adjusted to match this range, with 0.5–0.6 OD typically producing the optimal signal-to-noise ratio (*see* **Notes 11** and **12**).
7. *Molar extinction coefficients*: For MWL experiments, the molar extinction coefficient of the analyte(s) as a function of wavelength should be carefully considered. Prior to mixing samples, pure spectra for each sample should be collected, ideally by diluting the analyte with a non-absorbing buffer like sodium phosphate, which allows experiments to extend to 215 nm without significant background absorbance (Fig. 3). Such

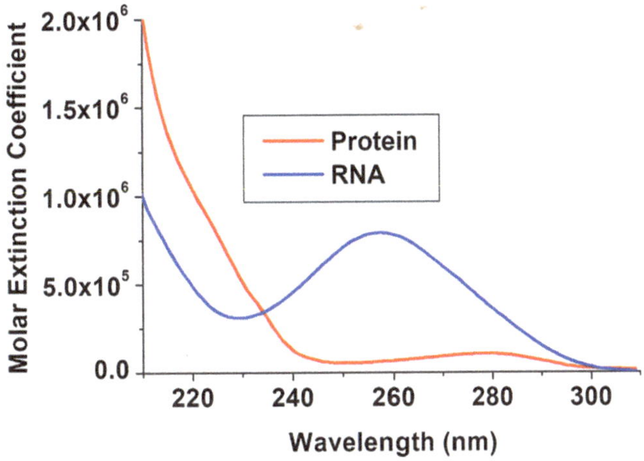

Fig. 3 Estimating RNA and protein absorbance for wavelength selection. Typical RNA (blue) and RNA-binding protein (red) extinction profiles, showing good complementarity that aids spectral separation between RNA (260 nm) and proteins (280 nm). Due to smaller molar extinction coefficient of proteins at 280 nm, compared with nucleic acids at 260 nm, the absorbance signal at 280 nm is much smaller compared with that at 260 nm

pure spectra can later be used to deconvolute composite spectral patterns using UltraScan. For multi-analyte solutions, consider that the molar extinction coefficients at each wavelength are mostly additive, assuming that hypochromic effects are minimal (*see* **Note 5**).

8. *Wavelength selection for RNA–protein interactions*: For cases where protein and RNA interactions are to be studied, it is often difficult to match absorbances in the 260–280 nm range for proteins and RNA due to the strong discrepancy in their molar extinction coefficients. Especially if stoichiometric measurements are performed, the size of each molecule plays a significant role. Generally, due to the much higher molar extinction of nucleic acids around 260 nm compared with proteins at 280 nm, protein absorbance can be negligible in comparison with the nucleic acid and signals cannot be properly resolved. If a phosphate buffer is used, it is often possible to obtain sufficient protein absorbance by simply shifting the wavelength range to exploit the backbone peptide bond extinction. Each peptide bond adds significantly to the absorbance between 215 and 230 nm producing an absorbance that is proportional to the protein size (*see* Fig. 3 for a relative comparison between RNA and protein extinction spectral characteristics). Whenever samples are mixed at different ratios, it is important to consider the relative extinction coefficients for different molecules participating in the reaction as they typically add to potentially move the total extinction outside of the

dynamic range of the detector. In the experimental design, it should be attempted to maintain total absorbance always below 1 OD at any wavelength to be scanned.

3.8 Setting Up a Sedimentation Velocity Run and Data Acquisition

Start the UltraScan data acquisition software (US-DA) and make sure it can communicate with the Optima™ AUC and the UltraScan LIMS database. Refer to the UltraScan user manual to configure the experimental protocol and submit it to the Optima instrument. Several important design choices should be considered as follows:

1. Make sure the rotor temperature is well equilibrated with the experimental run temperature before starting the experiment. The instrument is programmed not to start the rotor until the rotor temperature is within 0.4 °C of the target temperature. This is not sufficient for complete temperature equilibration. An additional 30–60 min of equilibration at the target temperature should be included in the protocol. This delay is set in the US-DA software.
2. Use the UltraScan adaptive space-time finite element method simulator (ASTFEM) [49, 50] to predict the appropriate run duration for your experiment. The US-DA software will automatically calculate the total number of scans available for the determined duration based on the chosen rotor speed. The number of scans that can be collected over a given duration varies significantly as a function of rotor speed since the flash lamp's flash rate needs to be synchronous with the rotor hole positioning to achieve the maximum scan rate. Refer to the UltraScan ASTFEM simulator manual for proper operation of this software.
3. Select investigator, instrument, rotor, rotor calibration, cell type and window type, and solutions (containing preselected buffer and analyte entries) from the relevant database menus and populate the optical system and range settings for radial scan limits (typically 5.75–7.25 cm), and program the selected wavelength(s), depending on SWL or MWL operation. For MWL, design the experiment such that at least 30–40 scans are collected per wavelength to obtain a good statistical representation of the experimental data (*see* **Note 22**). It should be kept in mind that both channels in one cell can contain different samples, so even if only one cell is scanned, one can still compare two different conditions in a single multi-wavelength experiment.
4. Submit the experimental protocol to the Optima AUC and select the protocol from the Optima's touch screen. Start the protocol and return to the computer to monitor the experiment remotely from the US-DA software terminal. In particular, monitor the meniscus position from each channel during the first 2–4 scans to detect any leaks in any channel. The

telltale sign of a leak is that for a given channel, the meniscus will appear to move to the right with successive scans. If this is observed, immediately abort the experiment, since unequal menisci on either side of the centerpiece-septum can generate significant pressure differentials that could damage the centerpiece.

5. At the end of the experiment, the software will automatically retrieve the data from the instrument and convert it to the Open AUC data format used in the UltraScan software [21]. At this point, data analysis may commence.
6. After the SV run is complete and the rotor stops spinning, turn off the vacuum pump and let the pressure return to ambient conditions. Open the door and remove the lid and retrieve the rotor. Remove the cells from the rotor, disassemble the parts, and clean them thoroughly with RNaseZap® and ultrapure water before assembling them for loading the next set of samples (*see* **Note 17**).

3.9 Analyzing SV Data to Obtain Hydrodynamic Properties

Several software are available for analyzing SV data. These software include DCDT+ [51], SEDPHAT [52, 53], LAMM [54], SVEDBERG [55], SEDFIT [56, 57], SEDANAL [58], and UltraScan [24]. We focus on the use of UltraScan because it is the only software that can take advantage of high-performance computing to handle the much higher data density generated by the Optima™ AUC instrument for MWL-AUC experiments and fits intensity data directly without the requirement for scan-pair subtractions that amplify the stochastic noise by a factor of ~1.4 (*see* **Note 23**). Please note that investigators who are using the older Proteome Lab™ XL instruments can also use UltraScan for data analysis as the software is compatible with both the new and the older data file formats.

3.9.1 Steps Common to MWL and SWL Data Analysis

1. After uploading of data by the US-DA software is complete, edit the data to define an initial estimate of the meniscus position and the outer limit of the data at the bottom of the cell. Very steep boundary regions should be excluded either by deleting such scans (toward the beginning of the experiment) or by limiting the data range near the bottom to avoid refractive artifacts occurring in these regions (Fig. 4a).
2. Process the edited MWL (and SWL) data in batch mode by submitting all wavelengths simultaneously to the UltraScan Science Gateway [59, 60] for analysis according to the two-dimensional spectrum analysis (2DSA) workflow [40] (Fig. 4a–d). This process is identical for SWL and MWL experiments and further described in the analysis flowchart at http://

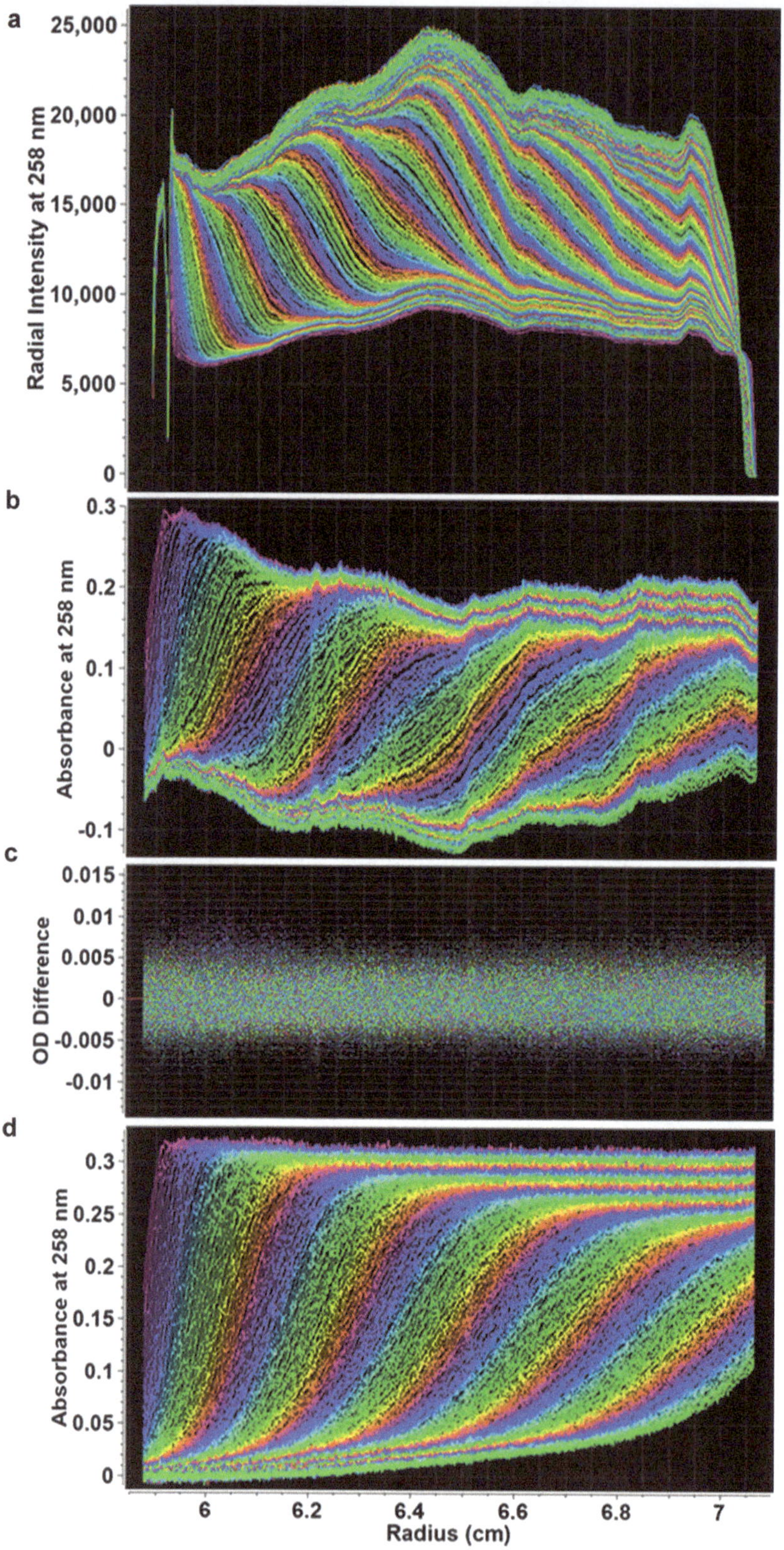

Fig. 4 Steps in the processing of raw data at a single wavelength. (**a**) RNA data acquired in intensity mode (258 nm) on the Optima AUC with UltraScan. (**b**) Same data converted to pseudo-absorbance mode and edited to exclude unnecessary scans and meniscus defined, plus top and bottom data range. (**c**) Residuals from ASTFEM [49, 50] fit with UltraScan. (**d**) Data after fitting with ASTFEM/2DSA solutions with UltraScan

www.ultrascan3.aucsolutions.com/sed-veloc-flowchart.php. (For more information, you can also see http://www.ultrascan3.aucsolutions.com/documentation.php.)

3. Check the RMSD values of the processed data in the *Finite Element Model Viewer* implemented in UltraScan, to confirm the presence of random residuals fits to all the wavelength scans.

3.9.2 Steps Specific to MWL Analysis (RNA–Protein Interactions)

For MWL, some additional analysis steps are required. The Optima AUC acquires all wavelength scans sequentially, which interferes with spectral deconvolution, since wavelength scans are not on a synchronous time grid. To solve this problem, the obtained 2DSA iterative fitting models are mapped on to an identical time grid for all wavelengths with the *Optima MWL.Fit Simulation* module.

1. Replicate actual run conditions for the simulation (i.e., rotor speed, duration, meniscus position) to create a three-dimensional surface for the wavelength decomposition. This is performed with the intrinsic molar extinction profiles of all absorbing species in the solution using the *MWL Species Fit* module of UltraScan.
2. Generate a separate traditional dataset with dimensions of time, radius, and molar concentration for each species present in the previous step.
3. Analyze these datasets with the 2DSA-Monte Carlo analysis [42] to generate 2D hydrodynamic profiles for each species. If complexes were formed between the species, identical hydrodynamic species will be present in those profiles. To facilitate interpretation of the results, the resulting dC/ds distributions can be overlaid with the function "Combine Discrete Distributions" from the "Utility" menu in UltraScan.
4. Should additional refinement be required, perform a global genetic algorithm analysis by simultaneously fitting all species to a discrete model [15, 41].

3.9.3 Steps for SWL Analysis (CounterIon or Ligand-Induced RNA Compaction)

1. For each titration point, repeat **steps 1–3** of *Steps common to MWL and SWL data analysis*, to obtain the equilibrium $s_{20,w}$ and the $D_{20,w}$ of the RNA at each counterion/protein/ligand concentration used for detecting tertiary folding or for measuring ligand/protein binding.
2. Plot the $s_{20,w}$ values as a function of cation concentration to demonstrate that the RNA molecules assume an increasingly compact shape (higher $s_{20,w}$) with increasing cation concentration (Fig. 5a). Also, plot the $s_{20,w}/D_{20,w}$ value, at each cation concentration, to check for counterion-dependent RNA aggregation (Fig. 5e) (*see* **Note 24**).

3. Both R_H and F report on tertiary compaction of an RNA induced by a counterion or a global conformational change induced by a ligand/protein; the midpoint of the transition and the cooperativity of the folding reaction can be obtained by phenomenological fits (*see* **Note 25**) of the plots of R_H (or F) vs. counterion concentration to the Hill equation: $\Upsilon = \frac{K_d{}^{n_H}[M]^{n_H}}{1+K_d{}^{n_H}[M]^{n_H}}$, where Υ denotes the signal quantitated (R_H or F), [M] is the counterion concentration, K_d is the midpoint of the equilibrium folding transition, and n_H is the Hill coefficient that indicates the cooperativity of the folding transition. Fit the data using the "nonlinear least squares fitting" routine implemented in standard software like Origin® and report the best-fit values at 65% confidence intervals (*see* **Note 26**).

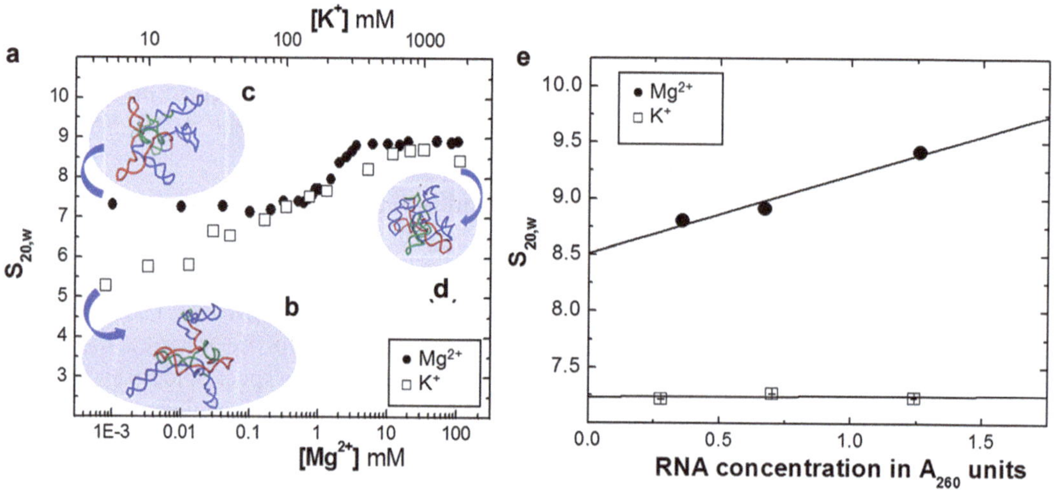

Fig. 5 Counterion-induced folding of the *Twort* group I intron ribozyme by SV-AUC. (**a**) SV-AUC-derived $s_{20,w}$ values for Mg^{2+}-mediated (filled black circles) and K^+-mediated (open squares) equilibrium folding of the *Twort* ribozyme in CEK and CE buffer, respectively. The lower and upper abscissa represent Mg^{2+} or K^+ concentrations, respectively, while the ordinate represents $s_{20,w}$ in Svedberg units ($S = 10^{-13}$ s); increasing $s_{20,w}$ depicts that the molecules assume a more compact, faster sedimenting conformation with increasing ionic strength. The insets in panel **a** depict representative models of the *Twort* ribozyme at three different ionic strengths, consistent with the measured hydrodynamic properties: (**b**) CE buffer, extended structures ($s_{20,w} = 4.47$ S, $R_H = 72.5$ Å); (**c**) CEK buffer, partly compact structures ($s_{20,w} = 7.26$ S, $R_H = 44.8$ Å); (**d**) 80 mM Mg^{2+} or 2 M K^+, tertiary folded structure ($s_{20,w} = 8.91$S, $R_H = 36.5$ Å). The blue region around each model schematically depicts its global shape; ellipsoid structures become progressively spherical ($f/f_s = 2.8$, 1.76, 1.43 for **b**, **c,** and **d**, respectively). The data presented in **a** have been replotted from [45]. The structural models and the hydrodynamic measurements have been extracted from [75]. (**e**) Test for concentration-dependent aggregation of the *Twort* ribozyme, either in the folded state (CEK with 10 mM Mg^{2+}; filled black circles) or in the unfolded state (CEK; open squares). The upward slope of the $s_{20,w}$ values of the folded RNA, with increasing RNA concentration (measured in 260 nm absorbance units), indicates formation of higher molecular weight species, suggestive of aggregation. The constant $s_{20,w}$ values of the unfolded ensemble, with increasing RNA concentration, suggests that the aggregation is specific to the folded state of the ribozyme [Reprinted by permission from Springer; RNA Folding: Methods and Protocols, Methods in Molecular Biology, Volume 1086, Pages 265–288, *Detecting RNA Tertiary Folding by Sedimentation Velocity Analytical Ultracentrifugation*, Somdeb Mitra, Pages 862–870. Copyright (2014)]

4 Notes

1. To obtain RNA molecules with precise and homogenous 5′ and 3′ ends, often the DNA sequence corresponding to the RNA of interest is cloned immediately downstream of a DNA sequence encoding a Hammerhead (HH) ribozyme and immediately upstream of a DNA sequence encoding a hepatitis δ virus (HδV) ribozyme. The HH ribozyme sequence is immediately preceded by the promoter sequence for the RNA polymerase used for the in vitro transcription reaction, such that transcription from this template generates the RNA of interest with a 5′ HH ribozyme and a 3′ HδV ribozyme. The ribozyme sequences usually cleave themselves out co-transcriptionally, thereby leaving a homogenous population of RNA molecules with well-defined 5′ and 3′ nucleotides [61]. However, often, after the transcription reaction is over, an additional 20 mM $MgCl_2$ is added, and the reaction is incubated at 60 °C for 15 min to ensure maximum self-cleavage of the both the ribozymes.
2. Heterogeneity at the 3′ end, which is a more pressing issue, especially when using T7 RNA polymerase, can be significantly reduced by using modified reverse primers during PCR amplification of the linear DNA template for transcription. Such modified primers generate templates in which the last two nucleotides are methylated at 2′-O position, which prevents addition of extra nucleotides at the 3′ end by the RNA polymerase [43]. Finally, for RNA sequences <100 nucleotides, high transcription yields are often obtained by using a single stranded antisense template DNA, containing a sequence at its 3′ end (immediately preceding the complementary sequence of the RNA of interest), complementary to the T7 promoter sequence. When hybridized to an oligonucleotide that encodes the sense-strand sequence of the T7 promoter, the double-stranded promoter sequence is restored, leading to recognition and transcription initiation by T7 RNA polymerase [43].
3. If the goal of the experiment is to measure the midpoint and cooperativity of an RNA tertiary folding transition induced by monovalent ions, the starting buffer of choice is CE buffer, which has approximately 8 mM of free Na^+ ions. Although the amount of free Na^+ ions from 10 mM Na-cacodylate is negligible (~8 mM) compared with what is required for either RNA backbone charge neutralization or tertiary folding, to be technically accurate while conducting monovalent ion titrations with other counterions (e.g., potassium), one might choose

to use the corresponding salt of cacodylic acid for making the experimental buffer.

4. If Mg^{2+} (or other multivalent cation)-induced tertiary folding is being studied, it is recommended to use buffers like CEK, which have at least 100 mM monovalent ions. The 100 mM K^+ ions in the CEK buffer neutralizes the negatively charged phosphodiester backbone of RNA molecules, thereby reducing intermolecular electrostatic repulsion that can generate artifacts during measurements of hydrodynamic properties. Furthermore, it also screens intramolecular repulsive forces between the backbones of the helices, thereby allowing for greater conformational flexibility (Fig. 5b vs. Fig. 5c) [62]. Finally, it helps avoid the idiosyncratic Mg^{2+}-mediated folding behavior observed for some large structured RNA molecules, like the self-splicing group I intron ribozymes (*Tetrahymena* [63] and *Azoarcus* [64]), under conditions of very low ionic strength, most likely due to due to non-native docking of secondary structure modules in a rigid ensemble.
5. The choice of the AUC phosphate buffer RNA–protein MWL experiments minimizes the background absorption from buffer components and extends the optical detection range to 215 nm (suitable for the detection of the peptide bond absorbance of the protein backbone). Moreover, it is recommended to eliminate reductants in the buffer to avoid background absorbance in the 250–285 nm absorbance range. If a reductant is required, it is recommended to replace thiol-containing reducing reagents present in protein-storage buffers, like DTT and BME, with the non-thiol reducing reagent TCEP (tris(2-carboxyethyl)phosphine)), in the AUC phosphate buffer. However, TCEP does absorb below 250 nm so it should be added only if needed and preferably only when conducting experiments at wavelengths greater than 250 nm. Always check the absorbance of the buffer to make sure it is not contributing to the absorbance in the wavelength range to be examined.
6. For all RNA and most RNA–protein interaction studies, since sample detection is based on UV absorption, investigators can use the standard Optima™ AUC instrument. For detection of unlabeled molecules that do not absorb significantly in the UV–visible range, investigators can also use the Rayleigh interference optics of the Optima™ AUC (or the older Proteome Lab™ XL-I instruments). Based on detecting differences in refractive indices between the sample and the reference solution, the interference method can also provide higher signal–noise ratio and a greater coverage of sample concentration range (beyond the 0.1–1 OD range). Finally, the ability of both the Optima™ AUC and the older Proteome Lab™ instruments to detect absorbance in the visible region

(400–700 nm) extends the scope of analytical investigations to molecules that are either naturally absorbed in the visible region of the spectrum or labeled with fluorescent dyes/proteins that do so [38, 65].

7. The direct integration of data acquisition software within UltraScan-III for the Optima™ AUC instrument permits the use of all 4 positions in the An-60 Ti rotor (or 8 positions in the An-50 Ti) for the collection of 8 or 16 samples, respectively, in UV–vis data collection mode only. A counterbalance is only used for an initial, one-time radial calibration of the UV–vis detection system (the Rayleigh interference optics still require a counterbalance in each experiment). Once calibrated, the information is stored in the Optima™ AUC, and the instrument can be used without a counterbalance in positions 4 or 8. Occasional recalibration is advised.

8. When using a recombinant T7 RNA polymerase, overexpressed and purified in our laboratory, in vitro transcription can be carried out in standard reaction buffers like 1× transcription buffer (50 mM Tris HCl (pH 8.1), 30 mM $MgCl_2$, 2 mM spermidine, 0.1% Triton-X-100, 0.05 mg/ml BSA, usually made and stored as a 10× stock at 4 °C) [43]. A standard transcription reaction is set up in the desired preparative volume (1–10 mL) by adding an appropriate amount of RNA polymerase (optimized using smaller analytical scale reactions prior to setting up preparative scale reactions) and transcription buffer (to a final 1× concentration) to the following components: 5 mM DTT, 2.5 mM of each NTP, and 1 μM linear DNA template. The reactions are incubated at 37 °C for 3–5 h. Usually, a white precipitate of magnesium phosphate is formed as a side product, which should be spun down before purifying the RNA by spin columns or denaturing PAGE.

9. The standard gel-loading buffers (e.g., gel-loading buffer II; Invitrogen™/Ambion™) contain loading dyes like xylene cyanol and bromophenol blue, which appear as dark bands in UV-shadowed gels. To distinguish them from actual RNA bands, it is always recommended to run only the loading buffer in one of the lanes, which can be used as reference during UV-shadowing and band excision. In case of RNA transcripts containing ribozyme sequences, it is also advised to run either an RNA ladder or an RNA of known molecular weight in a separate lane, which can be used as a reference to distinguish the RNA of interest from the cleaved HH and/or HδV ribozyme bands in the gel.

10. Since the proteins are used in combination with RNA and it is not uncommon for protein preparations to be contaminated with RNases, an aliquot from each batch of protein preparation should be tested for the presence of RNases. Generally, this

involves incubating the RNA with the highest concentration of protein used in the experiments, in the experimental buffer, at 50 °C, and at the experimental temperature (37 °C/25 °C), for 2–3 h, and running the samples in a denaturing Urea-PAGE. Presence of smeared bands or multiple bands in the RNA + protein lane, compared with an RNA only lane, is indicative of significant RNase contamination in the protein sample.

11. The optimal absorbance range for operating the Optima AUC is an absorbance range between 0.2 and 0.8 OD units. Samples should be scanned on a high-quality spectrophotometer with a 1 cm quartz cuvette (and not with a Nanodrop-like device which is too imprecise) prior to performing the AUC experiment. The RNA stock solution should be sufficiently concentrated to give an O.D. >0.1 upon dilution into the reaction buffer. The optical pathlength of the sector-shaped channels in the centerpieces of the AUC cells is 1.2 cm, as opposed to the typical 1 cm pathlength of generic cuvettes. This pathlength difference should be considered while calculating the amount of RNA required to achieve the desired OD (~0.5). For any RNA, the 0.1–1.0 A_{260} range in a 1.2 cm optical path length spans a concentration range of approximately 4 to 48 μg/mL (using the conversion factor 1 A_{260} is equivalent to 40 μg/mL of RNA). Therefore, for a 100-nucleotide-long RNA molecule, with a 50% G-C content and a molecular weight (MW) of approximately 32.66 KDa, this spans a molar concentration range of 0.12–1.47 μM.

12. Before beginning an actual titration experiment, one should verify by preliminary SV-AUC runs that there is no concentration-dependent RNA aggregation in the 0.1–1 OD range, both with and without the counterion/protein. In the absence of aggregation, the hydrodynamic properties of the RNA (*s* and/or *D* values) should remain invariant as a function of increasing RNA concentration. An upward slope of the *s* vs. RNA concentration plot, for the counterion containing sample, is typically indicative of the RNA aggregating into larger molecular weight species (Fig. 5e). It is also typical to observe a downward slope of the *s* vs. RNA concentration plot for the samples lacking counterions (especially those that are just dissolved in the CE buffer that lacks even moderate concentrations of monovalent cations), most likely due electrostatic repulsion between polyanionic molecules, resulting in deviation from ideal solute behavior [32].

13. RNA molecules can often form highly stable alternate (non-native) secondary structures [66]. Therefore, in the absence of divalent cations (or insufficient monovalent cations), the starting point for RNA tertiary compaction experiments can contain a heterogenous ensemble of RNA

molecules with the native and non-native secondary structures. The heat denaturation, followed by slow cooling, helps in generating a relatively homogenous ensemble, as a starting point for the tertiary compaction reaction. The slow cooling from 95 to 50 °C is critical because it allows equilibration to the native secondary structure. It is always advisable to conduct chemical or enzymatic structure-probing experiments (e.g., SHAPE [67] DMS footprinting [68], RNase mapping [69]), either prior to or in parallel with SV-AUC experiments) to verify formation of native secondary structure under the experimental conditions of SV-AUC.

14. Addition of the counterion that mediates tertiary compaction at the higher temperature of 50 °C, followed by the short incubation at that temperature, helps RNA tertiary-folding intermediates overcome kinetic traps by using the higher thermal energy [70]. However, when working with divalent cations (e.g., Mg^{2+}) as the folding counterion, the incubation time should be kept as short as possible (no more than 10–15 min) to avoid divalent cation-mediated RNA hydrolysis at higher temperatures.

15. Since RNA tertiary compaction happens on second to minute timescales [47], the hour long incubation time at the folding temperature is generally sufficient for RNA molecules to reach their native tertiary structures. However, in a few instances, very long-lived misfolded intermediates that can persist for several hours have been detected [71]. Tertiary structure-probing experiments (e.g., hydroxyl radical footprinting) [46] and/or functional assays (e.g., ribozyme activity) [72] often help in determining the appropriate time and solution conditions for equilibrium compaction into the native tertiary structure.

16. Magnesium ions heavily influence the dynamics of RNA structures [73, 74]. The end points of Mg^{2+} titrations, intended to maximize the population of fully folded molecules, have >10 mM Mg^{2+}. Such high Mg^{2+} concentrations bias the equilibrium tertiary structure heavily toward the folded state, thereby compromising the structural dynamics of the RNA molecule. In the experiments designed to probe equilibrium conformational changes of RNA, mediated by small-molecule ligands or proteins, the Mg^{2+} concentration should be kept around the midpoints of typical Mg^{2+}-mediated RNA tertiary folding reactions (1–5 mM Mg^{2+}, when conducting the experiments in CEK buffer) to render the RNA sufficiently flexible for conformational switching.

17. Sometimes it might be hard to push the assembly into the cell housing. It is recommended to not apply too much force which can damage the windows and the centerpiece; instead, all the cell components, including the housing, are placed inside a

refrigerator at 4 °C for about 15 min, taken out, reassembled, and then gently pushed into the housing; it should go in easily. Similarly, it is often hard to push the assembly out of the housing after an AUC run, in which case, the cells are again placed at 4 °C for about 15 min to allow the assembly to be gently pushed out.

18. Each centerpiece has two sector-shaped channels with 1.2 cm optical path length, both of which can be used for samples. In the older Proteome Lab™ AUC instruments, the channel names are reversed ("A" is the left, "B" is the right channel when facing the screw-ring with the fill holes at the top). Also, for the Proteome Lab™ instruments, it is important to never exceed a total OD of 0.4 in the "A" channel to avoid resetting the photomultiplier tube voltage in the middle of a run.
19. A photosensitive analytical overspeed detection disk, with embedded magnets, is attached to the bottom of the rotor. Along with a sensor in the chamber, this device is used by the instrument to time the flash of the Xenon lamps and to check the rotor speed and make sure it does not cross the upper speed limit. Investigators are urged to exercise extreme caution to place the rotor only on the rotor stand and not on any other surface, to avoid damage to the analytical disk.
20. If you encounter difficulties inserting the cells, confirm that the fill-hole screws do not extend beyond the cell's surface and that the fill-hole screws and the screw ring were not tightened excessively, which will bulge the entire cell body and prevent a smooth fit in the rotor.
21. Special care should be taken while attaching the Proteome Lab™ monochromator to its socket. The users should keep in mind that it is inserted into the socket at an acute angle. After securing the device in its socket, the base ring should be carefully hand-tightened to avoid damaging the thread. Before closing the chamber door, the monochromator should be thoroughly checked to ensure it is not wobbling. Serious damages to the instrument can occur if the monochromator gets dislodged during an AUC run. These considerations, however, do not apply to the Optima™ AUC which uses a periscope that can be swiveled into place and is locked into position by a magnetic lock.
22. The number of wavelengths selected for a given experiment also determines the number of scans that can be recorded for each wavelength. For example, if 3000 scans are available for the entire experiment, and 60 wavelengths are scanned, only 50 scans can be collected at each wavelength for a single cell. If two cells are scanned, this number drops to 25 scans per cell.

Therefore, it is often advantageous to only scan one cell in a multi-wavelength experiment.

23. Subtraction of pairwise scans as done in SedAnal and DCDT+ to eliminate time invariant noise introduces amplification of stochastic noise by a factor equal to the square root of two which is undesirable. The same problem exists when absorbance data are used instead of intensity data. Absorbance data incur higher stochastic noise due to the subtraction of the reference scan. The UltraScan workflow automatically eliminates time- and radially invariant noise and therefore effectively deals with the time-invariant noise typically observed from window imperfections or photomultiplier tube response variations across its surface.

24. UltraScan will automatically transform s and D measured under a given experimental condition to standard conditions (20 °C, water) and report them as $s_{20,w}$ and $D_{20,w}$ values. This standardization is especially important in counterion-mediated RNA folding studies as it reports on the changes in the intrinsic properties of an RNA molecule and therefore, facilitates comparison between folding behaviors studied in different solution conditions. Appearance of more than one peak in the s-value distribution computed by UltraScan could indicate the existence of distinct conformational states with different hydrodynamic properties or self-association of the RNA molecules, or both. Users are advised to carefully check the distribution to rule out the presence of multiple peaks. Since, s/D is proportional to the molecular weight, the plot of $s_{20,w}/D_{20,w}$ versus counterion concentration should produce a straight line parallel to the abscissa, provided that the molecules do not aggregate at high ionic strengths. It is not uncommon to observe deviation from this behavior in samples containing very high concentrations of Mg^{2+}.

25. Cooperative ligand binding to a molecule is quantitatively described by the Hill equation, $\Upsilon = \frac{Kd^{n_H}[M]^{n_H}}{1+Kd^{n_H}[M]^{n_H}}$, in which K_d represents the equilibrium dissociation constant, [M] is the ligand concentration, and n_H is the Hill coefficient that depicts the cooperativity of the process. The fitting of equilibrium RNA folding data points to the Hill equation is purely phenomenological, and the fitted curves do not represent counterion binding to RNA; instead, the fitted curve at a given temperature, also called the folding isotherm, simply represents a transition between two states, mediated by the counterion. For the purpose of analyzing counterion-mediated RNA folding isotherms, K_d and n_H represent, respectively, the midpoint and the cooperativity of the folding transition, whereas [M] denotes the concentration of the counterion.

26. In some cases where the RNA folding reaction is not a simple two-state transition, the folding isotherms may not be

adequately fit by a single-site binding model. In such cases, the data points should be fit to modified Hill equations that represent multiple-site cooperative ligand binding. For example, an isotherm that is best described by two transitions can be phenomenologically fit to a two-site binding Hill equation:

$$\Upsilon = f\left\{\frac{Kd1^{n_{H1}}[M]^{n_{H1}}}{1 + Kd1^{n_{H1}}[M]^{n_{H1}}}\right\} + (1-f)\left\{\frac{Kd2^{n_{H2}}[M]^{n_{H2}}}{1 + Kd2^{n_{H2}}[M]^{n_{H2}}}\right\}$$

where Υ denotes the signal quantitated; f represents the fraction of the first transition; K_{d1}, K_{d2} and n_{H1}, n_{H2} are the midpoints and cooperativities of the two transitions, respectively; and [M] is the counterion concentration.

Acknowledgments

S.M. is grateful to the Chemistry Department of New York University to host him as a faculty during the preparation of this manuscript. The *Twort* intron work described here was originally funded by 1RO1-GM085130 from the National Institute of General Medical Sciences of the National Institutes of Health to Prof. Michael D. Brenowitz at the Albert Einstein College of Medicine. B.D. wishes to credit NIH-NIGMS grant RO1-120600 and the Canada Research Chairs program for financial support of this work.

References

1. Gesteland RF, Cech TR, Atkins JF (2006) The RNA world, vol 43. Cold Spring Harbor Press, New York
2. Fresco JR (1998) RNA structure and function, vol 35. Cold Spring Harbor Press, New York
3. Dethoff EA, Chugh J, Mustoe AM, Al-Hashimi HM (2012) Functional complexity and regulation through RNA dynamics. Nature 482(7385):322–330
4. Mustoe AM, Brooks CL, Al-Hashimi HM (2014) Hierarchy of RNA functional dynamics. Annu Rev Biochem 83:441–466
5. Mitra S (2009) Using analytical ultracentrifugation (AUC) to measure global conformational changes accompanying equilibrium tertiary folding of RNA molecules. Methods Enzymol 469:209–236
6. Brautigam CA, Wakeman CA, Winkler WC (2009) Methods for analysis of ligand-induced RNA conformational changes. Methods Mol Biol 540:77–95
7. Mitra S (2014) Detecting RNA tertiary folding by sedimentation velocity analytical ultracentrifugation. Methods Mol Biol 1086:265–288
8. Chaires JB, Dean WL, Le HT, Trent JO (2015) Hydrodynamic models of G-Quadruplex structures. Methods Enzymol 562:287–304
9. Kieft JS, Costantino DA, Filbin ME, Hammond J, Pfingsten JS (2007) Structural methods for studying IRES function. Methods Enzymol 430:333–371
10. Takamoto K, He Q, Morris S, Chance MR, Brenowitz M (2002) Monovalent cations mediate formation of native tertiary structure of the Tetrahymena thermophila ribozyme. Nat Struct Biol 9(12):928–933
11. Chillon I, Marcia M, Legiewicz M, Liu F, Somarowthu S, Pyle AM (2015) Native purification and analysis of long RNAs. Methods Enzymol 558:3–37
12. Wang X, Xi W, Toomey S, Chiang YC, Hasek J, Laue TM, Denis CL (2016) Stoichiometry and change of the mRNA closed-loop factors as translating ribosomes transit from initiation to elongation. PLoS One 11(3):e0150616
13. Luque D, Mata CP, Gonzalez-Camacho F, Gonzalez JM, Gomez-Blanco J, Alfonso C, Rivas G, Havens WM, Kanematsu S,

Suzuki N, Ghabrial SA, Trus BL, Caston JR (2016) Heterodimers as the structural unit of the T=1 capsid of the fungal double-stranded RNA Rosellinia necatrix quadrivirus 1. J Virol 90(24):11220–11230

14. Patel TR, Chojnowski G, Astha KA, McKenna SA, Bujnicki JM (2017) Structural studies of RNA-protein complexes: a hybrid approach involving hydrodynamics, scattering, and computational methods. Methods 118-119:146–162
15. Zhang J, Pearson JZ, Gorbet GE, Colfen H, Germann MW, Brinton MA, Demeler B (2017) Spectral and hydrodynamic analysis of West Nile virus RNA-protein interactions by multiwavelength sedimentation velocity in the analytical ultracentrifuge. Anal Chem 89 (1):862–870
16. Wong CJ, Launer-Felty K, Cole JL (2011) Analysis of PKR-RNA interactions by sedimentation velocity. Methods Enzymol 488:59–79. https://doi.org/10.1016/B978-0-12-381268-1.00003-3
17. Berke IC, Modis Y (2012) MDA5 cooperatively forms dimers and ATP-sensitive filaments upon binding double-stranded RNA. EMBO J 31(7):1714–1726
18. Mayo CB, Wong CJ, Lopez PE, Lary JW, Cole JL (2016) Activation of PKR by short stem-loop RNAs containing single-stranded arms. RNA 22(7):1065–1075
19. Pearson JZ, Krause F, Haffke D, Demeler B, Schilling K, Colfen H (2015) Next-generation AUC adds a spectral dimension: development of multiwavelength detectors for the analytical ultracentrifuge. Methods Enzymol 562:1–26
20. Pearson J, Walter J, Peukert W, Colfen H (2018) Advanced multiwavelength detection in analytical ultracentrifugation. Anal Chem 90(2):1280–1291
21. Colfen H, Laue TM, Wohlleben W, Schilling K, Karabudak E, Langhorst BW, Brookes E, Dubbs B, Zollars D, Rocco M, Demeler B (2010) The open AUC project. Eur Biophys J 39(3):347–359
22. Gorbet GE, Pearson JZ, Demeler AK, Colfen H, Demeler B (2015) Next-generation AUC: analysis of multiwavelength analytical ultracentrifugation data. Methods Enzymol 562:27–47
23. Johnson CN, Gorbet GE, Ramsower H, Urquidi J, Brancaleon L, Demeler B (2018) Multi-wavelength analytical ultracentrifugation of human serum albumin complexed with porphyrin. Eur Biophys J 47(7):789–797
24. Demeler B, Gorbet GE (2016) Analytical ultracentrifugation data analysis with UltraScan-III. In: Uchiyama S, Arisaka F, Stafford W, Laue T (eds) Analytical ultracentrifugation. Springer, Cham, pp 119–143
25. Byron O, Nischang I, Patel TR (2018) European biophysics journal. In: Byron O, Nischang I, Patel TR (eds) Special issue: 23rd international analytical ultracentrifugation workshop and symposium, AUC 2017, vol 693. Springer International Publishing, Cham
26. Fujita H (1975) Foundations of ultracentrifugal analysis. Wiley, New York
27. Philo JS (2000) A method for directly fitting the time derivative of sedimentation velocity data and an alternative algorithm for calculating sedimentation coefficient distribution functions. Anal Biochem 279(2):151–163
28. Stafford WF 3rd (1994) Boundary analysis in sedimentation velocity experiments. Methods Enzymol 240:478–501
29. Laue TM, Stafford WF 3rd (1999) Modern applications of analytical ultracentrifugation. Annu Rev Biophys Biomol Struct 28:75–100
30. Stafford WF 3rd (1992) Boundary analysis in sedimentation transport experiments: a procedure for obtaining sedimentation coefficient distributions using the time derivative of the concentration profile. Anal Biochem 203 (2):295–301
31. Correia JJ, Stafford WF (2015) Sedimentation velocity: a classical perspective. Methods Enzymol 562:49–80
32. Costantino D, Kieft JS (2005) A preformed compact ribosome-binding domain in the cricket paralysis-like virus IRES RNAs. RNA 11(3):332–343
33. Tanford C (1961) Physical chemistry of macromolecules. Wiley, New York
34. Cantor CR, Schimmel PR (1980) Ultracentrifugation. In: Bartlett AC (ed) Biophysical chemistry, Part II: techniques for the study of biological structure and function, vol II. W.-H. Freeman and Company, San Francisco
35. Scott DJ, Schuck P (2005) A brief introduction to the analytical ultracentrifugation of proteins for beginners. Analytical ultracentrifugation: techniques and methods. Royal Society of Chemistry, Cambridge, UK
36. Uchiyama SA (2016) Important and essential theoretical aspects of AUC. Analytical ultracentrifugation. Springer, Tokyo
37. Demeler B, Brookes E, Wang R, Schirf V, Kim CA (2010) Characterization of reversible associations by sedimentation velocity with UltraScan. Macromol Biosci 10(7):775–782
38. MacGregor IK, Anderson AL, Laue TM (2004) Fluorescence detection for the XLI

analytical ultracentrifuge. Biophys Chem 108 (1–3):165–185

39. Lawson CLH, Hanson RJ (1974) Solving least squares problems. Automatic computation. Prentice-Hall, Englewood Cliffs
40. Brookes E, Cao W, Demeler B (2010) A two-dimensional spectrum analysis for sedimentation velocity experiments of mixtures with heterogeneity in molecular weight and shape. Eur Biophys J 39(3):405–414
41. Brookes E, Demeler B (2007) Parsimonious regularization using genetic algorithms applied to the analysis of analytical ultracentrifugation experiments. In: GECCO ACM proceedings of the 9th annual conference on genetic and evolutionary computation, pp 361–368
42. Demeler B, Brookes E (2008) Monte Carlo analysis of sedimentation experiments. Colloid Polym Sci 268(2):129–137
43. Beckert B, Masquida B (2011) Synthesis of RNA by in vitro transcription. Methods Mol Biol 703:29–41
44. Shcherbakova I, Gupta S, Chance M, Brenowitz M (2004) Monovalent ion-mediated folding of the Tetrahymena thermophila ribozyme. J Mol Biol 342(5):1431–1442
45. Mitra S, Laederach A, Golden BL, Altman RB, Brenowitz M (2011) RNA molecules with conserved catalytic cores but variable peripheries fold along unique energetically optimized pathways. RNA 17(8):1589–1603
46. Shcherbakova I, Mitra S (2009) Hydroxyl-radical footprinting to probe equilibrium changes in RNA tertiary structure. Methods Enzymol 468:31–46
47. Kwok L, Shcherbakova I, Lamb J, Park H, Andresen K, Smith H, Brenowitz M, Pollack L (2006) Concordant exploration of the kinetics of RNA folding from global and local perspectives. J Mol Biol 355(2):282–293
48. Williams TL, Gorbet GE, Demeler B (2018) Multi-speed sedimentation velocity simulations with UltraScan-III. Eur Biophys J 47 (7):815–823
49. Cao W, Demeler B (2008) Modeling analytical ultracentrifugation experiments with an adaptive space-time finite element solution for multicomponent reacting systems. Biophys J 95 (1):54–65
50. Cao W, Demeler B (2005) Modeling analytical ultracentrifugation experiments with an adaptive space-time finite element solution of the Lamm equation. Biophys J 89(3):1589–1602
51. Philo JS (2006) Improved methods for fitting sedimentation coefficient distributions derived by time-derivative techniques. Anal Biochem 354(2):238–246
52. Schuck P (2003) On the analysis of protein self-association by sedimentation velocity analytical ultracentrifugation. Anal Biochem 320 (1):104–124
53. Brautigam CA (2011) Using Lamm-equation modeling of sedimentation velocity data to determine the kinetic and thermodynamic properties of macromolecular interactions. Methods 54(1):4–15
54. Behlke J, Ristau O (1997) Molecular mass determination by sedimentation velocity experiments and direct fitting of the concentration profiles. Biophys J 72(1):428–434
55. Philo JS (1997) An improved function for fitting sedimentation velocity data for low-molecular-weight solutes. Biophys J 72 (1):435–444
56. Schuck P, MacPhee CE, Howlett GJ (1998) Determination of sedimentation coefficients for small peptides. Biophys J 74(1):466–474
57. Brown PH, Schuck P (2008) A new adaptive grid-size algorithm for the simulation of sedimentation velocity profiles in analytical ultracentrifugation. Comput Phys Commun 178 (2):105–120
58. Sherwood PJ Stafford WF (2016) SEDANAL: model-dependent and model-independent analysis of sedimentation data. In: Uchiyama S, Arisaka F, Stafford W, Laue T (eds) Analytical ultracentrifugation. Springer, Tokyo, pp 81–102
59. Memon S, Riedel M, Janetzko F, Demeler B, Gorbet G, Marru S, Grimshaw A, Gunathilake L, Singh R, Attig N, Lippert T (2014) Advancements of the UltraScan scientific gateway for open standards-based cyberinfrastructures. Concurr Comput Pract Exp 26 (13):2280–2291
60. Pierce M, Marru S, Demeler B, Singh R, Gorbet G (2014) The apache airavata application programming interface: overview and evaluation with the UltraScan science gateway. In: 9th gateway computing environments workshop (GCE 2014), New Orleans, LA, USA, 2014. IEEE Press, Piscataway
61. Kieft JS, Batey RT (2004) A general method for rapid and nondenaturing purification of RNAs. RNA 10(6):988–995
62. Takamoto K, Das R, He Q, Doniach S, Brenowitz M, Herschlag D, Chance M (2004) Principles of RNA compaction: insights from the equilibrium folding pathway of the P4-P6 RNA domain in monovalent cations. J Mol Biol 343(5):1195–1206
63. Sclavi B, Sullivan M, Chance MR, Brenowitz M, Woodson SA (1998) RNA folding at millisecond intervals by synchrotron

hydroxyl radical footprinting. Science 279 (5358):1940–1943

64. Chauhan S, Woodson SA (2008) Tertiary interactions determine the accuracy of RNA folding. J Am Chem Soc 130(4):1296–1303
65. Nelson TG, Ramsay GD, Perugini MA (2016) Fluorescence detection system. In: Uchiyama S, Arisaka F, Stafford W, Laue T (eds) Analytical ultracentrifugation. Springer, Tokyo, pp 39–61
66. Pan J, Woodson SA (1998) Folding intermediates of a self-splicing RNA: mispairing of the catalytic core. J Mol Biol 280(4):597–609
67. Wilkinson KA, Merino EJ, Weeks KM (2006) Selective 2′-hydroxyl acylation analyzed by primer extension (SHAPE): quantitative RNA structure analysis at single nucleotide resolution. Nat Protoc 1(3):1610–1616
68. Grohman J, Del Campo M, Bhaskaran H, Tijerina P, Lambowitz A, Russell R (2007) Probing the mechanisms of DEAD-box proteins as general RNA chaperones: the C-terminal domain of CYT-19 mediates general recognition of RNA. Biochemistry 46 (11):3013–3022
69. Carey MF, Peterson CL, Smale ST (2013) The RNase protection assay. Cold Spring Harb Protoc 2013(3):pdb.prot071910
70. Herschlag D, Cech TR (1990) Catalysis of RNA cleavage by the Tetrahymena thermophila ribozyme. 1. Kinetic description of the reaction of an RNA substrate complementary to the active site. Biochemistry 29 (44):10159–10171
71. Russell R, Das R, Suh H, Travers KJ, Laederach A, Engelhardt MA, Herschlag D (2006) The paradoxical behavior of a highly structured misfolded intermediate in RNA folding. J Mol Biol 363(2):531–544
72. Wan Y, Mitchell D 3rd, Russell R (2009) Catalytic activity as a probe of native RNA folding. Methods Enzymol 468:195–218
73. Mitra S, Brenowitz M (2008) Metal ions and RNA folding kinetics. In: Hud NV (ed) Nucleic-acid metal ion interactions. Royal Society of Chemistry, Cambridge, pp 221–265
74. Hud NV (2008) Nucleic acid-metal ion interactions. Royal Society of Chemistry, Cambridge
75. Chen C, Mitra S, Jonikas M, Martin J, Brenowitz M, Laederach A (2013) Understanding the role of three-dimensional topology in determining the folding intermediates of group I introns. Biophys J 104 (6):1326–1337

Chapter 20

RNA Nanostructure Molecular Imaging

Olivier Piétrement, Véronique Arluison, and Christophe Lavelle

Abstract

Atomic force and transmission electron microscopies (AFM/TEM) are powerful tools to analyze RNA-based nanostructures. While cryo-TEM analysis allows the determination of near-atomic resolution structures of large RNA complexes, this chapter intends to present how RNA nanostructures can be analyzed at room temperature on surfaces. Indeed, TEM and AFM analyses permit the conformation of a large population of individual molecular structures to be observed, providing a statistical basis for the variability of these nanostructures within the population. Nevertheless, if double-stranded DNA molecular imaging has been described extensively, only a few investigations of single-stranded DNA and RNA filaments have been conducted so far. Indeed, technique for spreading and adsorption of ss-molecules on AFM surfaces or TEM grids is a crucial step to avoid disturbing RNA conformation on the surface. In this chapter, we present a specific method to analyze RNA assemblies and RNA–protein complexes for molecular microscopies.

Key words RNA nanostructure, Molecular microscopy, RNA–protein complex, RNA self-assembly

1 Introduction

RNAs play major roles in gene expression: information transfer (messenger mRNA), cellular components (transfer tRNA or ribosomal rRNA), or catalytic function (for instance, in ribosomes, to form peptide bond during translation) [1]. More recently, regulatory noncoding RNA ncRNAs have also emerged as an important class of RNAs that regulate the expression of genes in all living organisms, from bacteria to eukaryotes [2, 3]. Although a variety of RNA molecules exist, their basic structures are similar and always begin as a polymer of nucleotides. Similarly to DNA, nitrogenous bases are covalently bound to a sugar–phosphate backbone. RNA mainly contains four nitrogenous bases: adenine A, guanine G (purines), uracil U, and cytosine C (pyrimidines), with U structurally similar to T found in DNA. Occasionally, RNA may also contain chemically modified bases that may influence their folding and global structure [4]. Like T, U can pair with A, while C pairs with G. Indeed, although RNA is produced as a single-stranded

Véronique Arluison and Frank Wien (eds.), *RNA Spectroscopy: Methods and Protocols*, Methods in Molecular Biology, vol. 2113, https://doi.org/10.1007/978-1-0716-0278-2_20,

molecule, it can form double-stranded structures, resulting in the formation of intramolecular base pairs (called secondary structure), such as hairpins or of intermolecular base pairs with other RNAs (same or different [5]). Such intermolecular base pairing is critical not only for important RNA functions, for instance, for the tRNA to bind mRNA during translation, but also to form large structures from multiple RNAs in the cell [6–8].

In vitro, DNA- and RNA-based nucleic acid nanostructures have been extensively explored by a wide range of techniques. Among them, molecular microscopy (i.e., atomic force microscopy (AFM) and transmission electron microscopy (TEM)) represents a powerful technique to image these molecular assemblies with a nanometer-scale resolution [9–11]. Whereas 3D-structure analysis using cryo-TEM, a cryogenic technique allowing the determination of near-atomic resolution structures of very large RNA complexes, is presented in Chap. 21, the current chapter intended to present how RNA nanostructures can be analyzed at room temperature. Indeed, TEM and AFM imaging analyses allow observing a population of individual molecular structures, permitting a statistical treatment of the variability of these structures within the population. Nevertheless, if double-stranded (ds) DNA molecular imaging has been described extensively [12, 13], less investigations of single-stranded (ss) DNA and RNA filaments have been conducted so far [7, 10, 14]. In this chapter, we present a method to analyze RNA and RNA–protein complexes using molecular microscopy.

2 Materials

2.1 Buffers and Solutions for In Vitro Transcription

All solutions have to be RNAse-free.

1. Sterile H_2O (double distilled endotoxin-free).
2. TAE buffer: 40 mM Tris–acetate, 1 mM EDTA pH 8.0.
3. *E. coli* RNA polymerase (RNAP) and transcription buffer containing 20 mM Tris–HCl pH 8, 20 mM KCl, 10 mM $MgCl_2$, 10 mM DTT, 300 μM NTPs, and 50 μg/mL BSA.
4. Denaturing gel-loading buffer (2×): 60% formamide, 12 mM EDTA, 0.03% bromophenol blue, 0.03% xylene cyanol.
5. Phenol–chloroform–isoamyl alcohol 25:24:1 (v/v/v) mixture (SIGMA #77617); ethanol 70% and 100%.
6. SYBR Green II (10,000×, Life Technologies ref. S7568).
7. RNeasy column for RNAs purification (Qiagen).
8. Phusion® High-Fidelity DNA Polymerase (ThermoScientific) or iProof DNA polymerase (Biorad) to produce DNA template for transcription.

9. RNAse-free DNAse (RQ1-DNAse, Promega).
10. Elution buffer: 10 mM Tris–HCl pH 7.5 containing 0.5 M ammonium acetate, 1 mM EDTA and 0.1% SDS.

2.2 TEM Imaging

1. Cryo-transmission electron microscope able to work in dark-field imaging mode [15] and equipped with a slow-scan CCD camera.
2. Copper grids 600 mesh (G600HH-C3 from Glider Grids, UK) for positive staining and copper grids (G200-Cu from Glider Grids, UK) for negative staining.
3. Uranyl acetate solution (2%).
4. Vacuum Evaporator Balzers MED20.
5. Amylamine (171409, Sigma-Aldrich).
6. TEM analysis software, such as iTEM software (Olympus Soft Imaging).

2.3 AFM Imaging

1. AFM microscope.
2. Mica sheets (Muscovite Mica V-1 quality (EMS, USA)).
3. Spermidine (85558 from Sigma-Aldrich, USA).
4. Ashless filter papers (204009 from Macherey-Nagel, Germany).
5. AFM cantilevers (Olympus AC200TS for air imaging, or SNL from Bruker for liquid imaging).
6. AFM free image analysis software such as WsXM (http://www.wsxmsolutions.com/) and Gwyddion (http://gwyddion.net/).

3 Methods

3.1 In Vitro Transcription for AFM and TEM Analysis

RNA can be prepared by in vitro transcription. To rule out the possibility of a read through during transcription (resulting in longer and abnormal RNA), PCR fragments terminating just at the end of desired RNA sequence need to be produced. Furthermore, T7 RNAP has been shown to turn around at the end of the template and read back on the other strand of DNA and therefore should be avoided. Instead, the transcription should be performed using *E. coli* RNAP that is not processive enough to switch DNA template strands. DNA template can be constructed by PCR as follows:

3.1.1 Design of the Primers

Forward primer including T7A1 promoter (the *E. coli* RNA polymerase is able to recognize and bind specifically to the A1 promoter of T7 phage to initiate transcription):

5′ATATAAAGCTTAAAAGATTAATTTAAAATTTATCAAAAAG AGTATTGACTTAAAGTCTAACCTATAGGATACTTACAG CC-*sRNAspecific*3′

Reverse primer, which is RNA-specific, corresponds to the 3′ end of the RNA.

The PCR results in a blunt end matrix encoding the RNA under the control of T7A1 promoter.

3.1.2 In Vitro Transcription

E. coli RNAP can be purchased from USB Corporation. Transcription is performed in 20 μL of transcription buffer (20 mM Tris–HCl pH 8, 20 mM KCl, 10 mM $MgCl_2$, 10 mM DTT) containing 300 μM NTPs, 50 μg/mL BSA with 1 pmol of DNA matrix, and one unit of RNA polymerase. The reaction is incubated for 6 h at 37 °C.

3.1.3 DNA Template Digestion

After transcription, the PCR DNA template used for the production of RNA in vitro has to be digested by an RNase-free DNase (RQ1-DNAse, Promega). The digestion is performed with 1 unit of enzyme per μg of DNA for 15 min at 37 °C, followed by phenol–chloroform–isoamyl alcohol extraction and ethanol precipitation. The transcript is resuspended in few μl of sterile water. It is usually better to perform the phenol–chloroform extraction to remove proteins used for in vitro transcription, as they can be observed by molecular imaging later. RNA concentrations are measured by UV spectrophotometry at 260 nm (*see* **Note 1**).

3.1.4 RNA Purification

In order to check the efficiency of transcription, the RNA transcript is loaded on a denaturing PAGE. This goal can be achieved with a 15% acrylamide/bis (19:1) gel containing 8 M urea in Tris–acetate–EDTA buffer (TAE). The sample is preheated at 80 °C in denaturing gel-loading buffer. The electrophoresis is run for 1 h with TAE buffer at ~50 °C. After migration, the RNA is stained with SYBR Green II. The visualization is achieved with a transilluminator. Two results can be observed: if there are no abortive products, the RNA can be purified directly using the RNeasy Mini Kit (Qiagen) according to the manufacturer's protocol. Alternatively, if there are some abortive products, the transcript needs to be purified from the PAGE. For this goal, the band corresponding to the RNA has to be cut from the gel, ground and eluted overnight at 37 °C under agitation with two volumes of elution buffer (10 mM Tris–HCl pH 7.5 containing 0.5 M ammonium acetate, 1 mM EDTA, and 0.1% SDS). The supernatant is recovered, and SDS is removed by phenol–chloroform/chloroform extraction followed by ethanol precipitation [16]. The transcript is resuspended in water.

3.1.5 RNA Folding

The transcripts are usually analyzed after relaxation: the RNA is heated and slowly cooled down at 20 °C [17].

3.2 RNA Chemical Synthesis

When short RNAs are analyzed, they can be chemically synthetized (RNA oligonucleotides). Similarly to transcript, the oligonucleotides are usually used after relaxation (*see* Subheading 3.1.5). Note that short RNAs can only be analyzed by molecular microscopy techniques TEM/AFM when they form polymers.

3.3 TEM Positive Staining

Single-stranded RNA or DNA is usually hard to visualize by these techniques, unless coated with proteins [14]. However, for large enough (>200 nt) RNA with clear folding motifs, TEM and AFM can be used conveniently to provide low-resolution images (Fig. 1, *see* **Note 2**). In the case of small RNAs, only their propensity for spontaneously forming large structures enables their visualization (Fig. 1).

A quick way to check the presence and structural characteristics of RNA nanostructures (length, diameter, interweaving), as well as the homogeneity of the sample, is positive staining and observation by transmission electron microscopy in dark-field imaging mode [15]. The protocol for such an analysis is described below:

3.3.1 Preparation of Carbon-Coated Grids

1. To make carbon films of a few nanometers thick, evaporate a carbon thread onto mica in the vacuum evaporator at distance of about 100 mm from the carbon source at 10^{-4} Pa. The thickness of the carbon layer should be as thin as possible. It varies according to the heating of the carbon thread and the square of the carbon source–mica distance and could be checked with a quartz balance.
2. Plunge the carbon-coated mica in water trough containing the copper grids to float off the carbon deposit. Allow dry carbon-coated grids to dry in the air under a lamp.
3. Draw up slowly the water with a vacuum pump to cover with the carbon film the grids previously set at the bottom of the trough, with the carbon film.

3.3.2 Activation of the Carbon-Coated Grids by Glow Discharge

1. Grids are deposited on a glass slide coated with Parafilm and placed on the lower steel plate. At 1 kPa vacuum, a small amount of amylamine (a few μL) is introduced up to about 2 kPa.
2. Ionization is performed over 30 s by switching on the voltage (400 V). A faint violet glow discharge light should be visible around the plates. Grids prepared in this manner retain their DNA-binding efficiency for about 1 day.

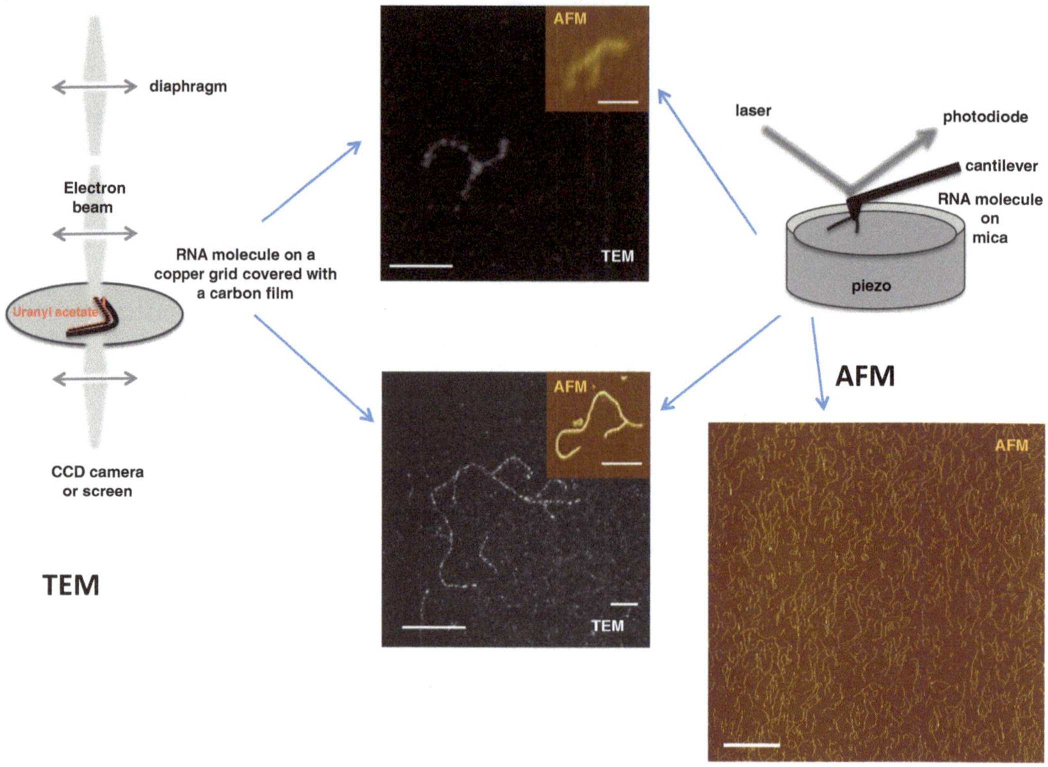

Fig. 1 Visualization of sRNA by molecular microscopy. Transmission electron microscopes (TEM) and atomic force microscopes (AFM) are efficient tools to investigate nucleic acids structures and interactions. In a TEM experiment (left, black and white images), a few nM of sRNA in TE or water is deposited on a copper grid (covered with carbon film and positively charged), stained with uranyl acetate (that will make electron-dense regions along DNA path), and the grid is subjected to an electron beam under vacuum in the microscope. In an AFM experiment (right, brown-yellow images), the same sample is deposited on freshly cleaved mica, and gently dried in air; the surface of the mica is then scanned with a tip. Top: an image obtained with 7SK, a human 331 nt ncRNA, is shown as an example of non-polymerizing sRNA. One can clearly recognize its characteristic folding in three loop domains important for its function as a negative regulator of RNA pol II transcription [20]. Scale bar: 10 nm. Middle: When an *E. coli* small noncoding RNA is spread in the same conditions, monomers self-assemble to form long polymorphic chains of several hundreds of nm [7]. Note that even if some free monomers might exist, they are too small to be resolved. Scale bar: 50 nm. Left: A large AFM scan of the noncoding RNA spread at a higher concentration (25 nM, compare to 5 nM in bottom right). Scale bar: 500 nm. sRNA pictures are adapted from [7]

3.3.3 Spreading and Staining

1. Deposit 5 μL of RNA or RNA–protein complexes on the activated carbon-coated grid for about 1 min. Starting from stock solutions, RNA molecules are diluted to about 5–20 nM in water, TE, or Tris–HCl (pH 7.5) with various amounts of monovalent (50–200 mM) and divalent (2–5 mM) ions (depending on the stability of the structures to be observed).
2. Grids are washed with aqueous 2% (w/v) uranyl acetate, dried, and observed in dark-field (annular or spectroscopic) imaging mode with transmission electron microscope.
3. Images are captured with a MegaView III camera CCD controlled by iTEM software (Olympus Soft Imaging).

3.4 AFM Imaging

AFM has become a common tool for the analysis of protein and nucleic acid fibrils [18]. It provides height information that complements structural information from two-dimensional projection images obtained using electron microscopy (*see* **Note 3**). An AFM setup consists of a microtip attached to a cantilever held on a piezo scanner that moves along the surface of the sample with high precision over a defined distance within the micrometer range. The deflection of the cantilever follows the surface topography and is monitored following the position of a laser spot reflecting from its surface on a photodiode. The motion of the cantilever can also be used to interrogate the mechanical properties of the sample. AFM imaging thus provides information about the dimension of biological objects at a nanometer resolution [19]. As nucleic acid fibrils are usually in the range of μm in length with lateral dimension between around few nm, AFM is thus perfectly suited to probe RNA-based assembly structure.

To image RNA in air, the protocol below can be used:

1. Prepare the solution of the RNA sample in the appropriate reaction buffer (must be optimized for each RNA, e.g., salts and divalent cations). The RNA concentration should be between 0.2 and 1 μg/mL.
2. Prepare a spermidine solution at a concentration that will depend on the ionic strength: for low ionic strength (NaCl concentration lower than 50 mM), a spermidine concentration of 50 μM is sufficient; for higher ionic strength, a spermidine concentration up to 300 μM would be required [12]. Please be aware that using a too-highly concentrated spermidine solution leads to RNA condensation on the mica surface, especially with in the presence of magnesium ions in buffer solution.
3. Cut a mica sheet to a desired size (1 cm diameter for the MultiMode AFM instrument) and cleave it with a Scotch tape several times to remove weakly bound layers.
4. Deposit 5 μL of the spermidine solution in the middle of the substrate for 1 min. This step is very sensitive to the quality of the mica cleavage. If cleavage is good, the shape of the drop stays symmetric on the mica surface. If the drop shape is not symmetric or if the drop slides to a corner of the mica surface, this means the cleavage is not good. Dry the mica and cleave it again.
5. Dry the mica surface with Ashless filter paper or clean nitrogen gas.
6. Deposit 5 μL of the reaction solution for 2 min.
7. Rinse softly the sample first with 30 μL of 0.02% uranyl acetate solution to stabilize the bound molecules on the mica surface in their 3D conformation, and then rinse again with 1 mL of buffer solution to remove free molecules.

8. Dry the sample with a filter paper or clean nitrogen gas. Stick it with an epoxy resin on a metal disk. Then, the sample can be mounted on AFM.
9. Imaging is performed in TappingMode with a Multimode AFM (Veeco, Santa Barbara, CA) operating with a Nanoscope IIIa controller with Olympus (Hamburg, Germany) silicon cantilevers AC160TS with nominal spring constants between 36 and 75 N/m.

   As shown in Fig. 1, monomers of a bacterial sRNA assembled into filamentous structures may be observed [7].

4 Notes

1. In average, 1 OD unit at 260 nm for RNA molecules = 40 ng/μL of RNA [16].
2. TEM and AFM are efficient and complementary tools to investigate macromolecular structures formed by nucleic acids. While TEM has the advantage of allowing quick imaging (therefore the possibility to test for various conditions and take many pictures in a short time), AFM is a more time-consuming technique (typically a few minutes are needed to complete a scan), but enables the visualization of the same samples in a much more biologically relevant environment (no staining, no vacuum; the molecules are directly observed in air). The combination of these two techniques, by enabling spreading in various conditions and visualization of large area with hundreds of molecules, provides a statistics for the efficiency of the self-assembly process and for the polymorphism–dispersity of the structures obtained.
3. Another advantage of AFM is the possibility to visualize a large area (up to several μm, Fig. 1).

Acknowledgments

This work was supported by CNRS, CEA, Université Paris Diderot, and Université de Bourgogne. We acknowledge Dr. E. Le Cam for TEM imaging and R. Lease for critical reading of the chapter.

References

1. Rodnina MV (2013) The ribosome as a versatile catalyst: reactions at the peptidyl transferase center. Curr Opin Struct Biol 23:595–602
2. Gottesman S (2011) Dicing defence in bacteria. Nature 471:588–589
3. Patil VS, Zhou R, Rana TM (2014) Gene regulation by non-coding RNAs. Crit Rev Biochem Mol Biol 49:16–32
4. Arluison V, Buckle M, Grosjean H (1999) Pseudouridine synthetase pus1 of Saccharomyces cerevisiae: kinetic characterisation, tRNA structural requirement and real-time analysis of its complex with tRNA. J Mol Biol 289:491–502
5. Lease RA, Arluison V, Lavelle C (2012) Twins, quadruplexes, and more: functional aspects of native and engineered RNA self-assembly in vivo. Front Life Sci 6:19–32
6. Lavelle C, Busi F, Arluison V (2015) Multiple approaches for the investigation of bacterial small regulatory RNAs self-assembly. Methods Mol Biol 1297:21–42
7. Cayrol B, Geinguenaud F, Lacoste J, Busi F, Le Dérout J, Piétrement O, Le Cam E, Régnier P, Lavelle C, Arluison V (2009) Auto-assembly of E. coli DsrA small noncoding RNA: molecular characteristics and functional consequences. RNA Biol 6:434–445
8. Busi F, Cayrol B, Lavelle C, LeDerout J, Piétrement O, Le Cam E, Geinguenaud F, Lacoste J, Régnier P, Arluison V (2009) Auto-assembly as a new regulatory mechanism of noncoding RNA. Cell Cycle 8:952–954
9. Dupaigne P, Lavelle C, Justome A, Lafosse S, Mirambeau G, Lipinski M, Piétrement O, Le Cam E (2008) Rad51 polymerization reveals a new chromatin remodeling mechanism. PLoS One 3:e3643
10. Hamon L, Pastré D, Dupaigne P, Le Breton C, Le Cam E, Piétrement O (2007) High-resolution AFM imaging of single-stranded DNA-binding (SSB) protein—DNA complexes. Nucleic Acids Res 35:e58–e58
11. Sorel I, Piétrement O, Hamon L, Baconnais S, Le Cam E, Pastré D (2006) The EcoRI-DNA complex as a model for investigating protein-DNA interactions by atomic force microscopy. Biochemistry 45:14675–14682
12. Pastré D, Hamon L, Landousy F, Sorel I, David M-O, Zozime A, Le Cam E, Piétrement O (2006) Anionic polyelectrolyte adsorption on Mica mediated by multivalent Cations: a solution to DNA imaging by atomic force microscopy under high ionic strengths. Langmuir 22:6651–6660
13. Lyubchenko YL, Gall AA, Shlyakhtenko LS (2014) Visualization of DNA and protein-DNA complexes with atomic force microscopy. Methods Mol Biol 1117:367–384
14. Veaute X, Jeusset J, Soustelle C, Kowalczykowski SC, Le Cam E, Fabre F (2003) The Srs2 helicase prevents recombination by disrupting Rad51 nucleoprotein filaments. Nature 423:309–312
15. Dupaigne P, Tavares EM, Piétrement O, Le Cam E (2018) Recombinases and related proteins in the context of homologous recombination analyzed by molecular microscopy. Methods Mol Biol 1805:251–270
16. Maniatis T, Stambrook J, Fritsch EF (1982) Molecular cloning: a laboratory manual; manual for genetic engineering. Cold Spring Harbor Laboratory, Cold Spring Harbor
17. Sun X, Li JM, Wartell RM (2007) Conversion of stable RNA hairpin to a metastable dimer in frozen solution. RNA 13:2277–2286
18. Creasey RG, Gibson CT, Voelcker NH (2012) Characterization of fiber-forming peptides and proteins by means of atomic force microscopy. Curr Protein Pept Sci 13:232–257
19. Dufrêne YF, Ando T, Garcia R, Alsteens D, Martinez-Martin D, Engel A, Gerber C, Müller DJ (2017) Imaging modes of atomic force microscopy for application in molecular and cell biology. Nat Nanotechnol 12:295–307
20. Eilebrecht S, Brysbaert G, Wegert T, Urlaub H, Benecke B-J, Benecke A (2011) 7SK small nuclear RNA directly affects HMGA1 function in transcription regulation. Nucleic Acids Res 39:2057–2072

Chapter 21

Grad-cryo-EM: Tool to Isolate Translation Initiation Complexes from Rabbit Reticulocyte Lysate Suitable for Structural Studies

Javier Rol-Moreno, Lauriane Kuhn, Stefano Marzi, and Angelita Simonetti

Abstract

Since its development, single-particle cryogenic electron microscopy (cryo-EM) has played a central role in the study at medium resolution of both bacterial and eukaryotic ribosomal complexes. With the advent of the direct electron detectors and new processing software which allow obtaining structures at atomic resolution, formerly obtained only by X-ray crystallography, cryo-EM has become the method of choice for the structural analysis of the translation machinery. In most of the cases, the ribosomal complexes at different stages of the translation process are assembled in vitro from purified components, which limit the analysis to previously well-characterized complexes with known factors composition. The initiation phase of the protein synthesis is a very dynamic process during which several proteins interact with the translation apparatus leading to the formation of a chronological series of initiation complexes (ICs). Here we describe a method to isolate ICs assembled on natural in vitro transcribed mRNA directly from rabbit reticulocyte lysate (RRL) by sucrose density gradient centrifugation. The Grad-cryo-EM approach allows investigating structures and composition of intermediate ribosomal complexes prepared in near-native condition by cryo-EM and mass spectrometry analyses. This is a powerful approach, which could be used to study translation initiation of any mRNAs, including IRES containing ones, and which could be adapted to different cell extracts.

Key words Translation initiation, Native ribosome complexes, Cryo-electron microscopy, Mass spectrometry

1 Introduction

The ribosome (80S particle in eukaryotes) is a dynamic molecular machine at the heart of the translation process, which is composed of a large and a small subunit (40S and 60S) made of ribosomal rRNA and proteins. Accompanied by several conformational changes and guided by a multitude of protein factors, the ribosome incorporates amino acids into proteins selecting aminoacylated tRNAs as specified by the mRNA codons. The active protein synthesis phase (translation elongation) is fast and accurate, and its mechanism is conserved though all organisms. In this phase, the

Véronique Arluison and Frank Wien (eds.), *RNA Spectroscopy: Methods and Protocols*, Methods in Molecular Biology, vol. 2113, https://doi.org/10.1007/978-1-0716-0278-2_21,

ribosome incorporates up to 20 amino acids per second with a very low error rate [1–5]. The initiation phase is a much more sophisticated and slow process; the ribosome assembles on the mRNA in the order of seconds [6, 7]. This long time is the result of a highly regulated process to carefully search the start site, usually an AUG codon, along the mRNA molecule. For bulk protein synthesis, the mRNA start codon is found by a base-by-base inspection starting from the 5′ end (usually an m^7G "cap" nucleotide), which positions the proximal AUG in the 40S subunit, a process that has been termed scanning. At least 13 initiation factors (eIFs) are required to ensure accurate AUG localization [8–11]. The first ribosomal initiation complex 43SIC, which is competent for mRNA 5′ → 3′ scanning, is recruited on the cap and is composed of 40S ribosomal subunit and several "core" factors: eIF2/Met-tRNAi^{Met}, eIF1, eIF1A, and eIF3. The scanning ends when the codon–anticodon interaction stabilizes a second ribosomal initiation complex, the 48SIC on the start codon. The third complex is the 80SIC, formed upon subunit joining and release of the initiation factors.

Several molecular aspects of the translation initiation process have been clarified by structural studies on in vitro-reconstituted complexes with a limited set of purified factors and short nonbiologically relevant mRNA mimicking fragments. Nevertheless, several "accessory" factors might be required depending on the structural and sequence features of the mRNAs [12], and their identity, roles, and mechanisms still need to be discovered. Moreover, alternative mechanisms relying on specific mRNA structural features can escape this classical model. Two examples are provided by the translation initiation of IRES containing mRNAs and the tethering mechanism of histone H4 mRNA. Internal ribosomal entry sites (IRES) characterize a heterogeneous class of mechanisms used by some viral and cellular mRNAs to improve translation efficiency and escape global control. mRNA structures located at their 5′ untranslated regions (5′ UTRs) directly guide the ribosome on the start codon without scanning and with the involvement of only a partial set of eIFs, but sometimes requiring IRES trans-acting factors (ITAFs) [9, 13–15]. Similarly to the IRES, also in the "ribosome tethering", the AUG is found directly without scanning, thanks to two structural RNA elements which are located in the coding region of H4 mRNA [16, 17].

These examples indicate how the eukaryotic translation initiation process is often mRNA-specific, depending on the presence of different signals. Furthermore, new initiation factors are continuously discovered which are involved in specific assembly processes. Therefore, the development of strategies allowing the structural analysis of complexes formed on natural mRNAs and isolated directly from cell extracts, without any a priori prejudice on the eIF composition, would represent a remarkable advance to understand their specific translation initiation mechanisms.

The investigation of labile multifactor translation initiation complexes represents a major challenge for a structural biologist. Single-particle cryo-electron microscopy (cryo-EM) technique had played a fundamental role in the structural study of the ribosome and of the process of translation initiation in both bacteria and eukaryotes. On the other hand, as ribosomes are bulky, electron dense, and distributing nicely on cryo-EM grids, they are producing perfect images and have assisted the development of several image processing methods. Despite this happy *connubium* since the dawn of cryo-EM in the early 1980s [18], the first reconstructions of eukaryotic translation initiation complexes appeared only a little more than 10 years ago [19, 20]. More recently, several studies have clarified the action of factors involved in ribosome scanning and tRNA accommodation [21–23] and the localization of the huge factor platform eIF3 important for mRNA binding [24, 25]. These studies, focused mainly on the canonical cap-dependent mechanism, represent partial or pre-initiation in vitro-assembled complexes. To obviate the need of studying native complexes by cryo-EM, we have recently set up two specific protocols to isolate from rabbit reticulocyte lysates (RRLs) ribosomal particles at different stages during initiation.

The first method [26, 27] allows purification of stable complexes, as the 80SICs, where no factors are bound. This method is based on mRNA ligation to a biotinylated DNA oligo that is immobilized on streptavidin-coated magnetic beads and involves extensive salt washes. In this way, using a combination of cycloheximide–hygromycin to prevent the first translocation step, 80SIC formed on H4 mRNA was purified and its structure solved at 9.3 Å [16], revealing how proper start codon positioning is achieved by a direct mRNA–rRNA interaction.

In this chapter, we provide a detailed protocol of the second method (Grad-cryo-EM) developed to isolate stalled 48SICs assembled on natural in vitro-transcribed mRNAs by sucrose den sity gradient centrifugation. Sucrose gradients are commonly used to profile ribosomes distribution on mRNA (polysomes), and the presence of specific inhibitors results in the accumulation of translation initiation intermediates, which sediment at the corresponding sucrose densities. Figure 1 shows the different radioactivity profiles on a 7–47% sucrose gradient obtained by incubating radiolabeled β-globin mRNA with nuclease-treated RRL in the absence of any inhibitor (blue line) or in the presence of cycloheximide (green line) or GMP-PNP (red line). While polysomes are detectable without any inhibitor, the addition of cycloheximide, inhibiting the eEF2-mediated translocation, blocks the progression of the translation to the elongation phase with the consequent accumulation of 80S monomer ribosomes [28, 29]. To block translation initiation before 60S association, non-hydrolysable nucleotides can be used. When nuclease-treated RRLs are supplemented with a non-hydrolysable analog of GTP

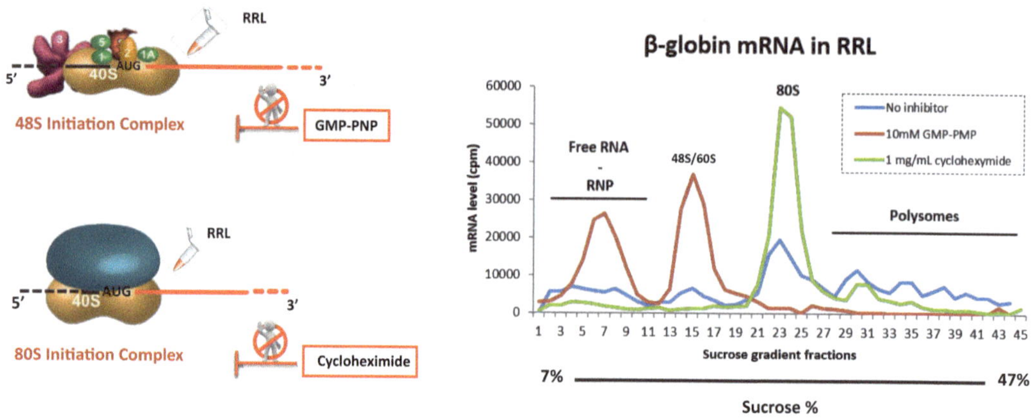

Fig. 1 Grad-cryo-EM strategy to isolate 48S and 80S initiation complexes. The initiation complexes are stalled on the start codon of β-globin mRNA from rabbit reticulocyte lysate (RRL), in the presence of different translation initiation inhibitors, GMP-PNP for the 48SIC, and cycloheximide for the 80SIC. 7–47% sucrose gradient analysis of the initiation complexes is used to monitor the formation of initiation complexes using P^{32} radiolabeled-capped β-globin mRNA. In the absence of any inhibitor (blue line), polysomes are detectable, while in the presence of cycloheximide (green line) or GMP-PNP (red line) could be observed a consequent accumulation of 80S monomer ribosomes and 48SIC, respectively

(GMP-PNP), the assembling of the 48SIC on the mRNA of interest arrests after start codon recognition, prior to eIF2 dissociation and large ribosomal subunit joining [30, 31]. These complexes can be easily isolated by centrifugation on appropriate sucrose gradients, which are both resolutive (even a 48SIC could be separated from 60S large ribosomal subunit) and gentle, preserving factor association. Composition of the isolated complexes will be then determined by mass spectrometry analysis, while their structure is analyzed by cryo-EM using a state-of-the-art framework, allowing to sort out the structures of different complexes present in a single sample. This strategy is adaptable to study translation initiation on any mRNA of interest, including IRES containing ones, with any cell extract that can be made from different cell lines, tissues, plants, insects, etc.

2 Materials

2.1 Cell Extracts, Buffers, Solutions, and Reagents

1. Rabbit reticulocyte lysate system, nuclease treated (L4960—Promega).
2. TMSDT buffer: 40 mM Tris–HCl pH 8.1, 22 mM $MgCl_2$, 5 mM DTT (dithiothreitol), 0.01% Triton-X100, 1 mM spermidine. This buffer is used for in vitro transcription.
3. RNA elution buffer: 0.5 M NH_4OAc pH 6.5, 0,1% sodium dodecyl sulfate (SDS), 1 mM ethylenediaminetetraacetic acid (EDTA).
4. Buffer A: 25 mM HEPES-KOH pH 7.6, 79 mM KOAc, 0.5 mM $Mg(OAc)_2$, 2 mM DTT. This buffer is prepared as

5× stock solution and used to obtain sucrose solutions at different concentrations to make the gradients.

5. 60% sucrose stock solution: Dissolve 60 g of D(+)-saccharose (VWR International) in a total volume of 100 mL of ultrapure water. A 60% sucrose stock solution is used to prepare in Buffer A 1×: (1) 7–47% linear sucrose gradient for analytical polysome profiling and (2) 5–25% linear sucrose gradient for the IC isolation.
6. Buffer B: 10 mM HEPES-KOH pH 7.6, 50 mM K OAc, 10 mM NH_4Cl, 5 mM $Mg(OAc)_2$, 2 mM DTT. This buffer is used to resuspend pelleted isolated complexes for cryo-EM studies.
7. [α-^{32}P]-GTP (6000 Ci/mmol).

Prepare all the solutions using ultrapure RNase-free water (Milli-Q deionized water attaining a sensitivity of 18 MΩ cm at 25 °C). Filter them with 0.22 μm membrane sterile syringe filter (SARSTED) and a 10 mL syringe (*see* **Note 1**).

2.2 Antibiotics and Translation Inhibitors

1. Cycloheximide (Sigma-Aldrich) powder is dissolved in water. Stock solution of 10 mg/mL can be stored at −80 °C for 1 month.
2. GMP-PNP (Sigma-Aldrich) powder is dissolved in water. Stock solution of 50 mM is stored at −80 °C.

2.3 Instrument and Accessories

1. Gradient Master™ device (BioComp Instruments, Fredericton, Canada) is used to generate linear sucrose gradients.
2. Piston Gradient Fractionator™ implemented with Triax™ Flow Cell is used for precise fractionation of sucrose gradients from the top down.
3. Beckman SW41 Ti Swinging-Bucket Rotor and Optima XE-90 ultracentrifuge.
4. Open-Top Polyclear™ centrifuge tubes (Seton Scientific).
5. Sorvall-Hitachi S140AT fixed angle rotor and M150 SE ultracentrifuge.
6. NanoDrop ND-1000 Spectrophotometer (Thermo Fisher Scientific, Waltham, MA, USA) or equivalent.
7. Micro Bio-Spin™ 6 Columns (Bio-Rad).
8. Refrigerated centrifuge.
9. High-precision balance.

3 Methods

3.1 mRNA Transcription and Capping

The template for human β-globin mRNA transcription (494 nt; accession number BC007075) was generated by PCR amplification from a plasmid containing the gene, essentially as described in [27].

The PCR product was phenol-extracted and precipitated with ethanol. The purified PCR product (5–8 μg) was directly used as template for the in vitro transcription of β-globin mRNA using 0.2 mg/mL of T7 RNA polymerase in a 100 μL volume of TMSDT buffer containing 5 mM of each NTP and 40 units of human placenta RNase inhibitor (RNasin, Promega). The reaction was incubated at 37 °C for 2 h followed by a further 1 h in the presence of RNase-free DNase I (4 units, New England Biolabs) to degrade the PCR template. Transcription products were purified to homogeneity from abortive transcription or degradation products using denaturing polyacrylamide-urea gel electrophoresis (urea-PAGE) on 8% polyacrylamide/8 M urea slab gels. After migration, the gel is removed from glass plates, RNA is visualized by UV shadowing on UV fluorescent TLC plate, and the band is cut from the acrylamide gel using a clean scalpel. The RNA is passively eluted at 4 °C overnight by gently mixing in RNA elution buffer and then extracted by phenol–chloroform treatment, precipitated in ethanol, washed with 85% ethanol, and vacuum-dried. The pure transcript was resuspended in ultrapure water, capped using the ScriptCap m^7G Capping System (Epicenter), and stocked at −20 °C. Radiolabeled transcripts were obtained by substituting the GTP from the capping kit with [α^{32}P]GTP. Micro Bio-Spin™ 6 Column is used to remove unincorporated GTP or [α^{32}P]GTP in case of a radiolabeled mRNA.

3.2 Linear Sucrose Gradient Preparation

About 5–25% linear sucrose density gradient centrifugation is used to isolate pure β-globin/48SIC directly from RRL incubated with 10 mM (final concentration) GMP-PNP. To obtain a linear sucrose density gradient, the Gradient Master™ device (BioComp) is used with the “SW41 Short Gradient” program and the corresponding range of sucrose percentage following the manufacturer’s manual. The use of the Gradient Master™ device offers the great advantage of preparing six reproducible and identical linear gradients of 12.5 mL (final volume) in one run. Tubes are cupped (short cups) and placed in a magnetic tube holder that adheres to the rotary steel plate on the instrument (*see* **Notes 2–4**). The percentage of sucrose used for linear gradients can be adjusted to optimize the separation of polysomal fractions according to the biological sample or experiment.

3.3 β-Globin/48SIC Formation and Isolation from Sucrose Gradient

In order to stall functional intermediate initiation complexes, compounds that inhibits specific steps of the translation machinery assembly need to be used. As mentioned above, in the presence of GMP-PNP, the GTPase eIF2 remains bound to the ribosome impairing subunits joining and thus blocking the ribosome assembly at the 48S late stage. To identify on the gradient the peak corresponding to the 48SIC, radiolabeled β-globin mRNA (250,000 cpm per gradient) is incubated with nuclease-treated

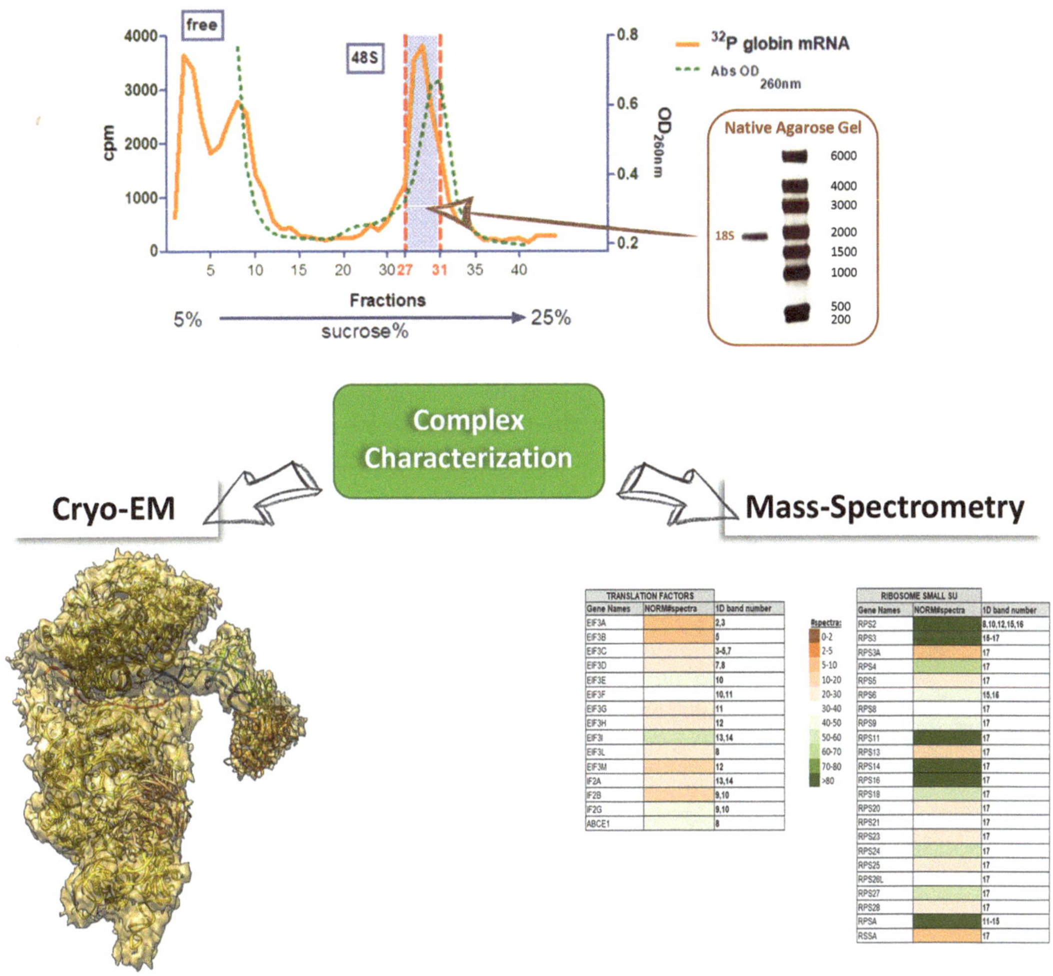

Fig. 2 48SIC isolation, factor composition, and structure determination. β-Globin mRNA is incubated with RRL in the presence of GMP-PNP to accumulate 48SIC. The RRL is fractionated on a 5–25% sucrose gradient to isolate stalled 48SIC, monitored using radioactive-capped β-globin mRNA (orange line). Ribosome localization is confirmed by optical density at 260 nm (green dashed line). After isolation, native agarose gel electrophoresis analysis is performed to assess the purity of the complex. The isolated β-globin-48SIC is then analyzed by nano-LC-MS/MS to get insights into its composition and by cryo-EM to determine its structure [31].

RRL in the presence of 10 mM GMP-PNP as described in [31]. The reaction is fractionated on 5–25% linear sucrose density gradient by centrifugation using SW41 Ti rotor, yielding optimal separation of the 48SIC late-stage particles from the 60S ribosomal subunits, as detected on native agarose gel electrophoresis (Fig. 2). Sample buffer components such as glycerol, sucrose, and detergent can reduce specimen contrast in cryo-EM. To completely remove sucrose, β-globin/48SIC containing fractions are pooled and the ribosomal complexes pelleted by centrifugation. The pellet is resuspended in Buffer B, which is suitable for both cryo-EM and mass spectrometry analysis (Fig. 2). The composition of the isolated β-globin/48SIC is determined by nano-liquid chromatography (LC) tandem mass spectrometry (MS/MS), helping the interpretation of the 3D density map obtained using cryo-EM.

Grad-cryo-EM approach can be adapted to also isolate the 80SICs by incubating RRL with the β-globin mRNA in the presence of cycloheximide, instead of GMP-PNP.

3.3.1 The Experiment

1. 5–25% linear sucrose gradient should be prepared and stored at 4 °C at least 2 h before loading of the sample.
2. Prior to stall late-stage 48SIC, ribosomes need to be reactivated. Thus, nuclease-treated RRL (Promega) should be incubated for 5 min at 30 °C in presence of a mixture of amino acids and then chilled on ice for 10 min. Amino acid mixture is provided with the Rabbit Reticulocyte Lysate System Kit to optimize translation.
3. Cycloheximide and hygromycin are thawed slowly on ice during ribosome activation step.
4. 60 μL of nuclease-treated RRL is incubated at 30 °C in the presence of 10 mM of GMP-PNP. After 2 min of incubation, 13 μg of β-globin mRNA is added, and the reaction mix is furthered incubated 5 min at 30 °C and chilled on ice for 10 min.
5. The reaction mix is layered (157 μL volume) onto 12.5 mL of 5–25% linear sucrose gradient and centrifugated at 37,000 rpm for 4 h and 37 min in a Beckman SW41 Ti Swinging-Bucket Rotor (average RCF 200,000 *g*; *see* **Note 4**).
6. The gradient is fractionated from the top down using BioComp Piston Gradient Fractionator™ devise implemented with Triax™ Flow Cell. This system allows monitoring the A_{260} absorbance throughout the fractionation process (*see* **Notes 5** and **6**).
7. An identical experiment is performed using 5′-labeled capped β-globin mRNA to identify fractions containing β-globin/48SIC.
8. Once identified, the fractions are centrifugated at 108,000 rpm for 2 h at 4 °C in a Sorvall-Hitachi S140AT fixed angle rotor (average RCF 600,000 *g*) . Ribosome pellet is dissolved in Buffer B, which is compatible for both mass spectrometry and cryo-EM analysis (*see* **Note** 7).
9. Composition of the obtained cryo-EM 3D reconstruction of the purified β-globin/48SIC is confirmed by nanoLC MS/MS analysis.

Grad-cryo-EM method yielded 2106 pmol of pure β-globin/48SIC from the incubation of 60 μL of RRL with 13 μg (108 pmol) of β-globin mRNA in the presence of 10 mM of GMP-PNP. 4 μL of β-globin/48SIC concentrated 70 nM was applied on 400 mesh holey carbon Quantifoil 2/2 grids (Quantifoil Micro Tools) to perform cryo-EM analysis. The cryo-EM structure of the β-globin/48SIC at 5.86 Å resolution [31] has shown the molecular details of the first decoding event on the start codon occurring at the end of the scanning process. eIF2/Met-tRNAi^{Met} is present, while no density for eIF1 could be observed, thus suggesting its dissociation from the complex upon start codon recognition, as confirmed by mass spectrometry analysis. Surprisingly, the 3D reconstruction has also shown an extra density located on the 40S intersubunit face. Both shape of the density and mass spectrometry data allowed assigning this density to ABCE1, an ATP-binding protein implicated in subunits disjoining after termination of translation [32]. This unexpected result demonstrates the importance of adopting methods for native complexes purification, but at the same time also indicates its limits. In fact, to unambiguously identify new factors when the obtained resolution of the structure is not enough to distinguish amino acid side chains, a deep characterization of the complex should be obtained by combining other techniques. A successful strategy has been attained by using immunolabeled cryo-EM in combination with Grad-cryo-EM [33].

4 Notes

1. All buffers used here are filtered using a 0.22 μm membrane sterile syringe filter (SARSTED) to ensure buffer sterilization and stored at 4 °C. The buffer solutions are made fresh every week.
2. BioComb Gradient Master™ device is provided with two types of tube cup. For our propose, we have used a short tube cup which leaves a 4 mm gap above the finished gradient.
3. Once prepared, the gradients should be handled with care to avoid disturbance and stored at 4 °C for at least 2 h before the layering of the sample.
4. Pay attention not to form air bubbles when the tube cap is removed and during sample layering on the sucrose gradient, since it causes gradient perturbation throughout the centrifugation step.
5. Fractionate the gradient top down using Piston Gradient Fractionator™ connected to a fraction collector. Fast flow rates can reduce resolution and make it more difficult to collect precise fractions. For the isolation of β-globin/48SIC, approximately 250 μL per Eppendorf tube were collected using piston speed of 0.12 mm/s.

6. In case several gradients should be fractionated, a washing step of the Piston Gradient Fractionator system is strongly suggested to avoid contamination of 60S ribosomal subunit.
7. During the ultracentrifugation step, the ribosome complex precipitates, forming an unstable pellet that should be handled carefully. Thus, the supernatant containing sucrose should be quickly and completely removed, avoiding the perturbation of the pellet.

References

1. Sorensen MA, Kurland CG, Pedersen S (1989) Codon usage determines translation rate in Escherichia coli. J Mol Biol 207:365–377
2. Wintermeyer W, Peske F, Beringer M, Gromadski KB, Savelsbergh A, Rodnina MV (2004) Mechanisms of elongation on the ribosome: dynamics of a macromolecular machine. Biochem Soc Trans 32:733–737
3. Lovmar M, Ehrenberg M (2006) Rate, accuracy and cost of ribosomes in bacterial cells. Biochimie 88:951–961
4. Rodnina MV, Beringer M, Wintermeyer W (2007) How ribosomes make peptide bonds. Trends Biochem Sci 32:20–26
5. Zaher HS, Green R (2009) Fidelity at the molecular level: lessons from protein synthesis. Cell 136:746–762
6. Kennell D, Riezman H (1977) Transcription and translation initiation frequencies of the Escherichia coli lac operon. J Mol Biol 114:1–21
7. Gold L (1988) Posttranscriptional regulatory mechanisms in Escherichia coli. Annu Rev Biochem 57:199–233
8. Hinnebusch AG (2011) Molecular mechanism of scanning and start codon selection in eukaryotes. Microbiol Mol Biol Rev 75:434–467, first page of table of contents
9. Jackson RJ, Hellen CU, Pestova TV (2010) The mechanism of eukaryotic translation initiation and principles of its regulation. Nat Rev Mol Cell Biol 11:113–127
10. Parsyan A, Svitkin Y, Shahbazian D, Gkogkas C, Lasko P, Merrick WC, Sonenberg N (2011) mRNA helicases: the tacticians of translational control. Nat Rev Mol Cell Biol 12:235–245
11. Sonenberg N, Hinnebusch AG (2009) Regulation of translation initiation in eukaryotes: mechanisms and biological targets. Cell 136:731–745
12. Aylett CH, Ban N (2017) Eukaryotic aspects of translation initiation brought into focus. Philos Trans R Soc Lond Ser B Biol Sci 372:20160186
13. Spahn CM, Kieft JS, Grassucci RA, Penczek PA, Zhou K, Doudna JA, Frank J (2001) Hepatitis C virus IRES RNA-induced changes in the conformation of the 40s ribosomal subunit. Science 291:1959–1962
14. Spahn CM, Jan E, Mulder A, Grassucci RA, Sarnow P, Frank J (2004) Cryo-EM visualization of a viral internal ribosome entry site bound to human ribosomes: the IRES functions as an RNA-based translation factor. Cell 118:465–475
15. Hashem Y, des Georges A, Dhote V, Langlois R, Liao HY, Grassucci RA, Pestova TV, Hellen CU, Frank J (2013) Hepatitis-C-virus-like internal ribosome entry sites displace eIF3 to gain access to the 40S subunit. Nature 503:539–543
16. Martin F, Menetret JF, Simonetti A, Myasnikov AG, Vicens Q, Prongidi-Fix L, Natchiar SK, Klaholz BP, Eriani G (2016) Ribosomal 18S rRNA base pairs with mRNA during eukaryotic translation initiation. Nat Commun 7:12622
17. Martin F, Barends S, Jaeger S, Schaeffer L, Prongidi-Fix L, Eriani G (2011) Cap-assisted internal initiation of translation of histone H4. Mol Cell 41:197–209
18. Dubochet J, Lepault J, Freeman R, Berriman JA, Homo J-C (1982) Electron microscopy of frozen water and aqueous solutions. J Microsc 128:219–237
19. Gilbert RJ, Gordiyenko Y, von der Haar T, Sonnen AF, Hofmann G, Nardelli M, Stuart DI, McCarthy JE (2007) Reconfiguration of yeast 40S ribosomal subunit domains by the translation initiation multifactor complex. Proc Natl Acad Sci U S A 104:5788–5793
20. Passmore LA, Schmeing TM, Maag D, Applefield DJ, Acker MG, Algire MA, Lorsch JR, Ramakrishnan V (2007) The eukaryotic translation initiation factors eIF1 and eIF1A induce

an open conformation of the 40S ribosome. Mol Cell 26:41–50
21. Llacer JL, Hussain T, Marler L, Aitken CE, Thakur A, Lorsch JR, Hinnebusch AG, Ramakrishnan V (2015) Conformational differences between open and closed states of the eukaryotic translation initiation complex. Mol Cell 59:399–412
22. Hashem Y, Georges A d, Dhote V, Langlois R, Liao HY, Grassucci RA, Hellen CU, Pestova TV, Frank J (2013) Structure of the mammalian ribosomal 43S preinitiation complex bound to the scanning factor DHX29. Cell 153:1108–1119
23. Hussain T, Llacer JL, Fernandez IS, Munoz A, Martin-Marcos P, Savva CG, Lorsch JR, Hinnebusch AG, Ramakrishnan V (2014) Structural changes enable start codon recognition by the eukaryotic translation initiation complex. Cell 159:597–607
24. des Georges A, Dhote V, Kuhn L, Hellen CU, Pestova TV, Frank J, Hashem Y (2015) Structure of mammalian eIF3 in the context of the 43S preinitiation complex. Nature 525:491–495
25. Aylett CH, Boehringer D, Erzberger JP, Schaefer T, Ban N (2015) Structure of a yeast 40S-eIF1-eIF1A-eIF3-eIF3j initiation complex. Nat Struct Mol Biol 22:269–271
26. Prongidi-Fix L, Schaeffer L, Simonetti A, Barends S, Menetret JF, Klaholz BP, Eriani G, Martin F (2013) Rapid purification of ribosomal particles assembled on histone H4 mRNA: a new method based on mRNA-DNA chimaeras. Biochem J 449:719–728
27. Chicher J, Simonetti A, Kuhn L, Schaeffer L, Hammann P, Eriani G, Martin F (2015) Purification of mRNA-programmed translation initiation complexes suitable for mass spectrometry analysis. Proteomics 15:2417–2425
28. Garreau de Loubresse N, Prokhorova I, Holtkamp W, Rodnina MV, Yusupova G, Yusupov M (2014) Structural basis for the inhibition of the eukaryotic ribosome. Nature 513:517–522
29. Schneider-Poetsch T, Ju J, Eyler DE, Dang Y, Bhat S, Merrick WC, Green R, Shen B, Liu JO (2010) Inhibition of eukaryotic translation elongation by cycloheximide and lactimidomycin. Nat Chem Biol 6:209–217
30. Anthony DD, Merrick WC (1992) Analysis of 40 S and 80 S complexes with mRNA as measured by sucrose density gradients and primer extension inhibition. J Biol Chem 267:1554–1562
31. Simonetti A, Brito Querido J, Myasnikov AG, Mancera-Martinez E, Renaud A, Kuhn L, Hashem Y (2016) eIF3 peripheral subunits rearrangement after mRNA binding and start-codon recognition. Mol Cell 63:206–217
32. Mancera-Martinez E, Brito Querido J, Valasek LS, Simonetti A, Hashem Y (2017) ABCE1: a special factor that orchestrates translation at the crossroad between recycling and initiation. RNA Biol 14:1279–1285
33. Brito Querido J, Mancera-Martinez E, Vicens Q, Bochler A, Chicher J, Simonetti A, Hashem Y (2017) The cryo-EM structure of a novel 40S kinetoplastid-specific ribosomal protein. Structure 25:1785–1794 e3

Chapter 22

Application of NMR Spectroscopy to Determine Small RNA Structure

Pengzhi Wu, Xiaodan Liu, Qingguo Gong, Yunyu Shi, and Jihui Wu

Abstract

Nuclear magnetic resonance (NMR) spectroscopy is a key experimental method to investigate the structure and dynamics of RNA. RNA often has only partially ordered structures responsible for its function, which makes it difficult to crystallize. In this chapter, we present the methodologies for RNA structure determination by liquid-state NMR, including the preparation of isotopically labeled RNA by in vitro transcription, NMR resonance assignment strategy, and structure calculation. Selected examples of NMR spectra are given for the first stem-loop of DsrA RNA (23 nt).

Key words RNA structure, NMR, Resonance assignment, Structure calculation

Abbreviations

NMR	Nuclear magnetic resonance
nt	Nucleotide
PAGE	Polyacrylamide gel electrophoresis
SL	Stem-loop

1 Introduction

RNA is not only functional as a messenger between DNA and protein but also involved in the regulation of genome organization and gene expression [1]. Knowledge of the RNA three-dimensional structure is indispensable for describing and understanding its regulation mechanism [2]. Many regulatory RNAs undergo conformational changes that play essential roles in their biological functions [3–5]. Conformational plasticity of the RNA restricts application of X-ray crystallography or cryo-electron microscopy (cryo-EM) [2]. NMR spectroscopy is uniquely suited to answer important questions in this area by looking at dynamic ensembles of

Véronique Arluison and Frank Wien (eds.), *RNA Spectroscopy: Methods and Protocols*, Methods in Molecular Biology, vol. 2113, https://doi.org/10.1007/978-1-0716-0278-2_22,

RNA structures [6, 7]. NMR is also a powerful tool for studying the interactions between RNAs and different ligands such as proteins, other nucleic acids, small molecules, and ions [8].

The production of large amounts of isotopically labeled RNA remains critical to the success of NMR-based structure studies [9]. Large quantities of RNA for NMR spectroscopy can be routinely prepared from either DNA template-directed in vitro transcription using T7 RNA polymerase (as well as T3 or SP6) or phosphoramidite-based chemical synthesis [10]. This chapter focuses on the synthesized small RNAs of defined length and sequence using T7 RNA polymerase and templates of synthetic DNA which contain the T7 promoter [11]. The desired RNAs can be purified by denaturing polyacrylamide gel electrophoresis (PAGE) or denaturing HPLC [12, 13].

After the production of isotope-labeled RNA, we subsequently collect NMR spectra and assign every NMR resonances for all the atoms (^{1}H, ^{13}C, ^{15}N, and ^{31}P) in the RNA molecule [14]. There is a limitation on the size of RNA for NMR studies, due to the increase of the number of signals with increased molecular weight. Further, RNA is a biopolymer containing only four different nucleotides, which makes its chemical-shift dispersion significantly reduced in comparison with protein [15]. Therefore, the resonance assignment of RNA is more difficult than protein. To overcome the limited chemical-shift dispersion, a set of multidimensional heteronuclear NMR experiments are applied for appropriately sized A-form helical elements [16].

The aim of this chapter is to describe how to use NMR spectroscopy to determine small RNA structure. The 23 nt first stem-loop of DsrA (DsrA-SL1) is used as a model system with high spectral resolution. This chapter is divided into three different parts, including (a) the transcription and purification of unlabeled or isotopically labeled RNA in milligram quantities, (b) the assignment strategies of RNA resonances, and (c) RNA structure calculation.

2 Materials

2.1 Enzymes, Buffers, and Chemicals

All solutions have to be RNase-free.

1. Template DNA: Complementary DNA oligonucleotide template for RNA transcription, 60 μM in water, store at −20 °C (*see* Fig. 1 for sequence parameters).
2. T7 promoter DNA: T7 RNA polymerase promoter DNA oligonucleotide, 60 μM in water, store at −20 °C (*see* **Note 1**).
3. DEPC-treated ddH_2O.

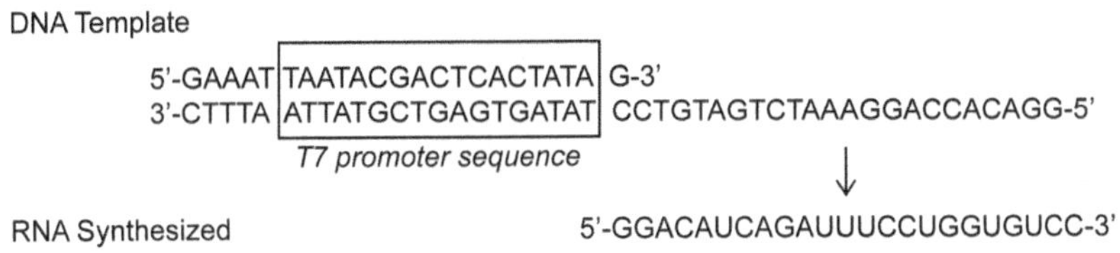

Fig. 1 Single-stranded DNA template sequence for the in vitro transcription of the DsrA-SL1 RNA

4. T7 RNA polymerase (T7 RNAP): 3 mg/mL in 25 mM sodium phosphate pH 7.0, 150 mM NaCl, 50% (v/v) glycerol, 10 mM DTT; store at −80 °C.
5. 10× T7 RNAP transcription buffer: 400 mM Tris–HCl pH 8.1, 10 mM Spermidine, 0.1% (v/v) Triton X-100.
6. 50 mM ATP, CTP, GTP, and UTP: Dissolve nucleotides in water and adjust to pH 7.0 using 1 M NaOH; store in 1 mL aliquots at −20 °C.
7. 500 mM $MgCl_2$.
8. 100 mM DTT.
9. 0.5 M EDTA stock solution: adjust to pH 8.0 using solid NaOH pellets.
10. Denaturing acrylamide gel solution: 16% acrylamide–bisacrylamide (19:1), 1× TBE, and 7 M urea.
11. Denaturing loading buffer: 95% (v/v) formamide, 2% (v/v) 0.5 M EDTA (pH 8.0), 0.02% (w/v) bromophenol blue.
12. 10× TBE running buffer: For 1 L of 10× TBE running buffer, dissolve 108 g of Tris base, 55 g of boric acid, and 7.5 g of disodium EDTA salt.
13. 10% Ammonium persulfate.
14. TEMED.
15. FastRuler Ultra Low Range DNA Ladder (Thermo Scientific) (*see* **Note 2**).
16. GelRed (10,000× Biotium).
17. 100% Ethanol: store at −20 °C.
18. 70% Ethanol: store at room temperature.
19. Elutrap Electroelution System (GE Healthcare Life Sciences).
20. NMR buffer: 10 mM sodium phosphate, pH 6.5, 50 mM NaCl.

2.2 NMR Spectroscopy

1. NMR spectrometer equipped with HCN cryoprobe.
2. Standard and Shigemi NMR tubes.
3. 99.99% D_2O.

4. nmrPipe: software to process, display, and analyze NMR spectra (https://www.ibbr.umd.edu/nmrpipe/nus.html).
5. Sparky: software for graphical NMR assignment and integration (https://www.cgl.ucsf.edu/home/sparky/).
6. CYANA: software for automated structure calculation (http://www.cyana.org/wiki/index.php/Main_Page).
7. AMBER: software for NMR structure refinement (http://ambermd.org/index.php).

3 Methods

3.1 In Vitro Transcription for NMR Studies

1. Prepare the in vitro transcription mixture (10 mL) which contains 40 mM Tris–HCl (pH 8.1), 1 mM spermidine, 10 mM dithiothreitol, 0.01% (v/v) Triton X-100, 20–50 mM $MgCl_2$, 1 mg T7 RNA polymerase, 5 mM each of unlabeled or uniformly $^{13}C,^{15}N$-labeled NTPs, 0.3 μM template DNA, and 0.3 μM T7 promoter DNA. Optimal concentration of $MgCl_2$ is determined in small-scale transcription reactions (50 μL) for every new DNA template (*see* **Note 3**).
2. Incubate the reaction solution for 4 h in a water bath at 37 °C, and then quench by the addition of 0.1 volume of 0.5 M EDTA (pH 8.0) (*see* **Note 4**).
3. Add 0.1 volume of 5 M NaCl, swirl to mix, and then add 3 volumes of cold 100% ethanol to the transcription solution.
4. Place the tube containing the RNA product at −20 °C overnight to precipitate.
5. Centrifuge the reaction at 21,600 × *g* for 30 min at 4 °C. Decant the supernatant, and very gently wash the pellet with 10 mL 70% ethanol.
6. Resuspend the pellet with 1.5 mL DEPC water, and add 1.5 mL RNA denaturing loading buffer.
7. Heat the RNA sample at ~95 °C for 5 min, and then incubate it on ice for 15 min.
8. Prepare a 16% acrylamide/bis 19:1 denaturing gel containing 7 M urea. Pre-run the gel in 0.5× TBE running buffer at 120 W for at least 30 min prior to loading.
9. Load the RNA sample on the gel, and run the gel at 120 W. The gel is monitored based on the location of the dye and previous observation from test transcription analytical gels.
10. Visualize the target RNA band by UV_{256} shadowing, and excise it from the gel with a clean razor blade.
11. Extract the RNA from the gel by running an Elutrap Electro-elution System in 0.5× TBE buffer at 4 °C overnight.

12. Combine the RNA-containing fractions, and wash it with 2 M NaCl three times. Then wash the RNA with DEPC-treated water for another three times.
13. Concentrate the RNA sample to ~ 1 mL. Transfer it to 50 mL tube with the addition of 30 mL NMR buffer.
14. Heat the RNA to form the native structure by incubation at ~95 °C for 5 min, and then immediately put it on ice for 20 min (*see* **Note 5**).
15. Concentrate the RNA down to the desired final volume (e.g., 500 μL for 5 mm NMR tube). The RNA is ready for NMR studies and should be stored at −80 °C.

3.2 RNA Chemical Synthesis

In some cases, short RNA sequences (<15 nt) are needed, which are usually difficult to obtain in good yield and purity by in vitro transcription. Further, nonstandard nucleotide derivatives may have to be incorporated site-specifically into the sequence. Such modifications are, for example, m^6A-modified RNA. Then these RNAs can be chemically synthetized and purchased from different biotechnology companies, such as TAKARA. Major limitations of the RNA chemical synthesis are the length of the sequence, the amount of RNA which can be produced, and the price [17].

3.3 NMR Resonance Assignment of RNA

The main strategy of NMR resonance assignments of DsrA-SL1 RNA is based on NOE contact and through-bond correlation (Fig. 2).

1. Imino protons of Watson–Crick base pairs tend to be found in the region of 12–15 ppm, whereas imino protons of noncanonical base pairs often can be found in the upfield region (Fig. 3a, b). Stable G:U wobble base pairs can be identified by a strong NOE contact between the guanine imino and uracil imino protons (*see* **Note 6**). The first step for NMR resonance assignment of RNA is to sequentially assign the imino protons of the guanine and uracil between 10 and 15 ppm in a 2D ^1H-^1H NOESY experiment (Fig. 3c). In the regions of A-form helical RNA, neighboring imino protons are ~4 Å apart and give rise to a moderate NOE cross peak in 2D NOESY. Counting the number of imino proton signals present should yield the approximate number of base pairs in the RNA. The chemical shift of imino protons can be confirmed by 2D ^1H-^{15}N HNN-COSY (Fig. 3d) [18].
2. Assign the amino protons of cytosine or H2 proton of adenine through the NOEs with the imino protons of the base-pairing cytosine or uracil, respectively (Fig. 4). G:C base pair has two strong NOE cross peaks between the two amino protons of cytosine and the guanine H1 imino proton (*see* **Note 7**). In A:

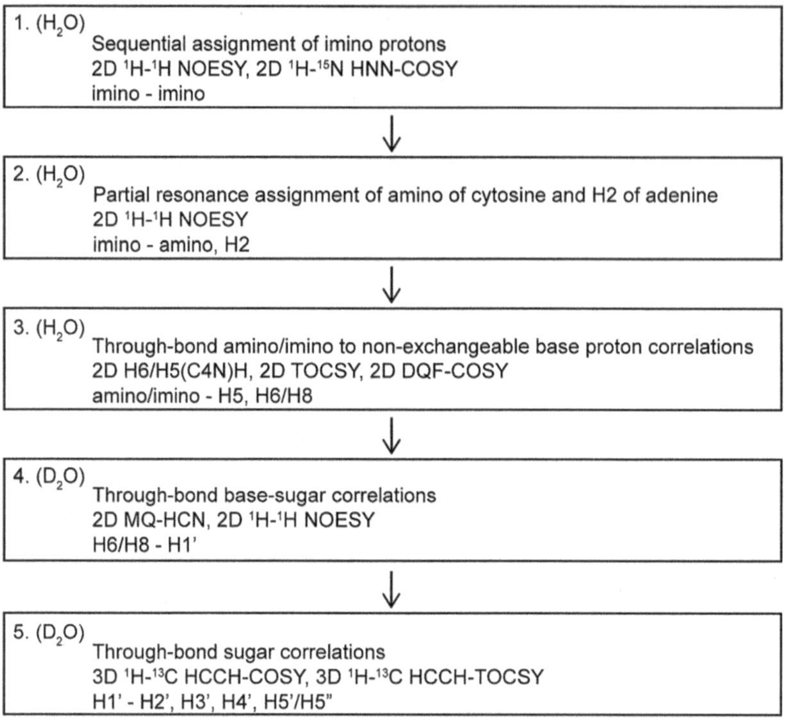

Fig. 2 Flowchart for resonance assignments of DsrA-SL1 based on NOE contact and through-bond coherence transfer

U base pair, the H2 proton of adenine has a strong cross peak with the uracil H3 imino proton.

3. Assign the aromatic protons of RNA. Triple-resonance 2D H6/H5(C4N)H experiment can provide through-bond H6/H5 protons of pyrimidines and H8 proton of guanine to imino/amino correlations in RNA bases (Fig. 5) [19]. Both the homonuclear 2D TOCSY and DQF-COSY experiments yield cross peaks for each pyrimidine spin system (H5 and H6 protons) (*see* **Note 8**).

4. Assign H1' protons through their correlation with H6/H8 in 2D MQ-HCN experiment (Fig. 6a) [20]. The HCN experiment correlates the ribose H1′ and H6/H8 protons via scalar rather than dipolar interactions, and is very useful for unambiguous assignment of intranucleotide connectivities between aromatic and sugar protons across the glycosidic bond. In the helical A-form RNA, all H6/H8 protons show NOE cross peaks to their own ribose and to the preceding nucleotide in the 5′ direction. Therefore, interresidual sequential assignment can also be obtained by observation of NOE contacts between H6/H8 and H1' protons in a 2D ^{1}H-^{1}H NOESY experiment (Fig. 6b).

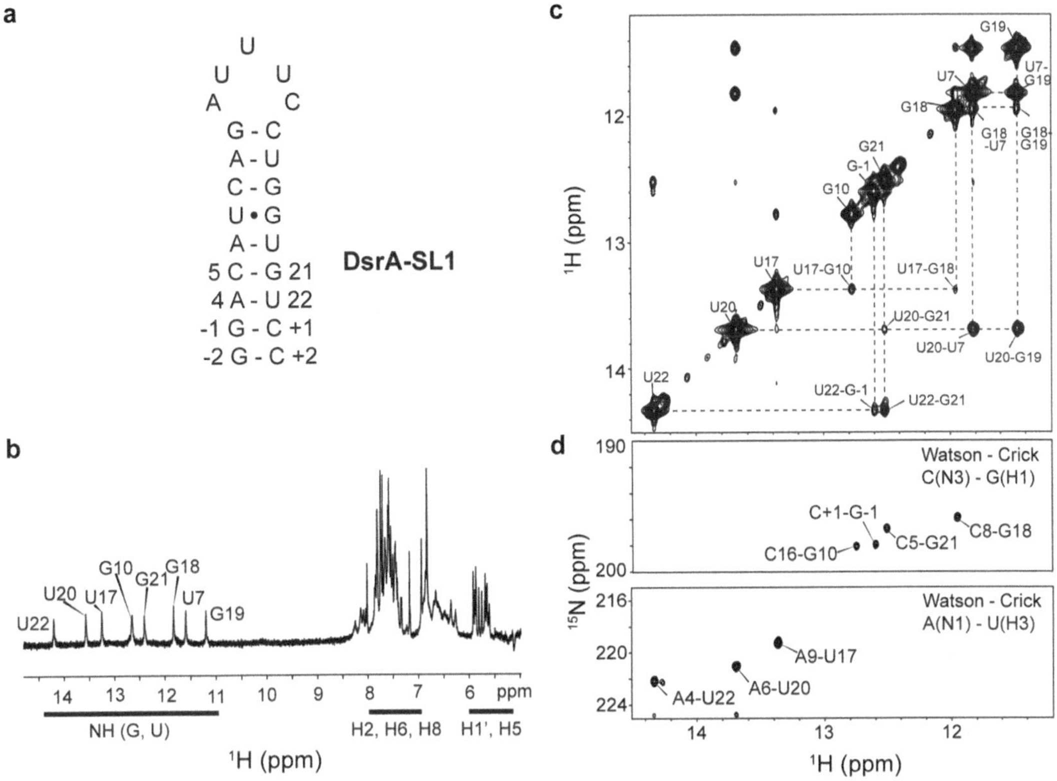

Fig. 3 Assignment of the imino proton resonances of the guanine and uracil. (**a**) Secondary structure of DsrA-SL1. (**b**) 1D proton spectrum of DsrA-SL1. Imino protons region of 2D ^{1}H-^{1}H NOESY (**c**) and ^{1}H-^{15}N HNN-COSY (**d**) experiments

5. Assign all the remaining atoms in the sugar moieties (H2′, H3′, H4′, and H5′/H5″) using 3D ^{1}H-^{13}C HCCH-COSY and HCCH-TOCSY experiments [21, 22]. The HCCH-TOCSY experiment on an RNA duplex identifies all protons in the same ribose spin system, allowing for straightforward assignment of ^{13}C and ^{1}H resonances based on the chemical shifts of the C1′/H1′ resonance (Fig. 7). The C2′/H2′ and C3′/H3′ have almost the same chemical shifts; therefore, HCCH-TOCSY cannot unambiguously assign them. The HCCH-COSY experiment, which only transfers magnetization through a single carbon–carbon bond, helps to assign the H2′, H3′, H4′, and H5′/H5″.

3.4 Structure Calculation

1. Calculate the initial structures of the RNA by adding the manually assigned NOEs. Semiquantitatively determine NOE-derived distance restraints. The NOE cross peaks are qualitatively categorized as strong, medium, weak, and very weak and used to assign upper distance limits of 2.7, 3.3, 5.0, and 6.0 Å, respectively.

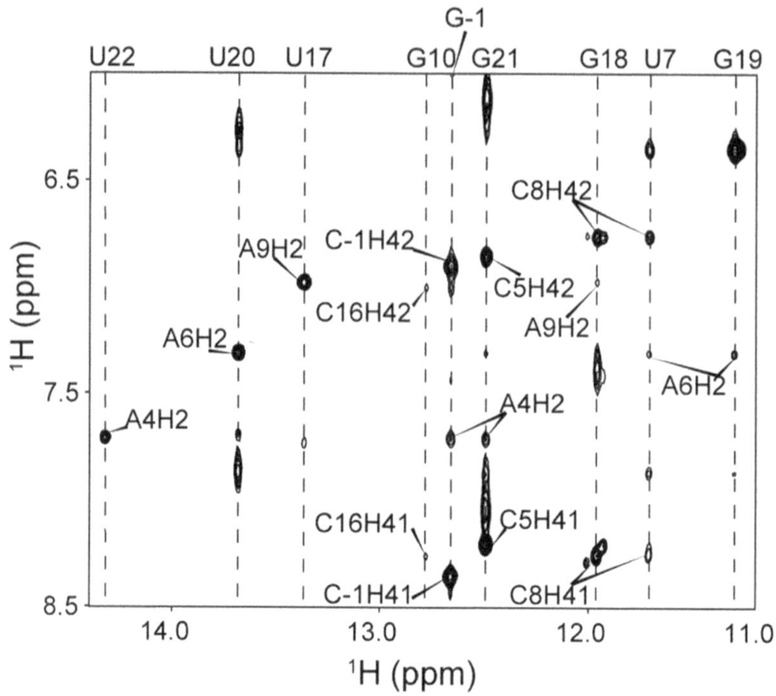

Fig. 4 Portion of the 2D NOESY spectrum obtained for DsrA-SL1 showing NOEs from the imino protons to amino protons of cytosine and H2 proton of adenine

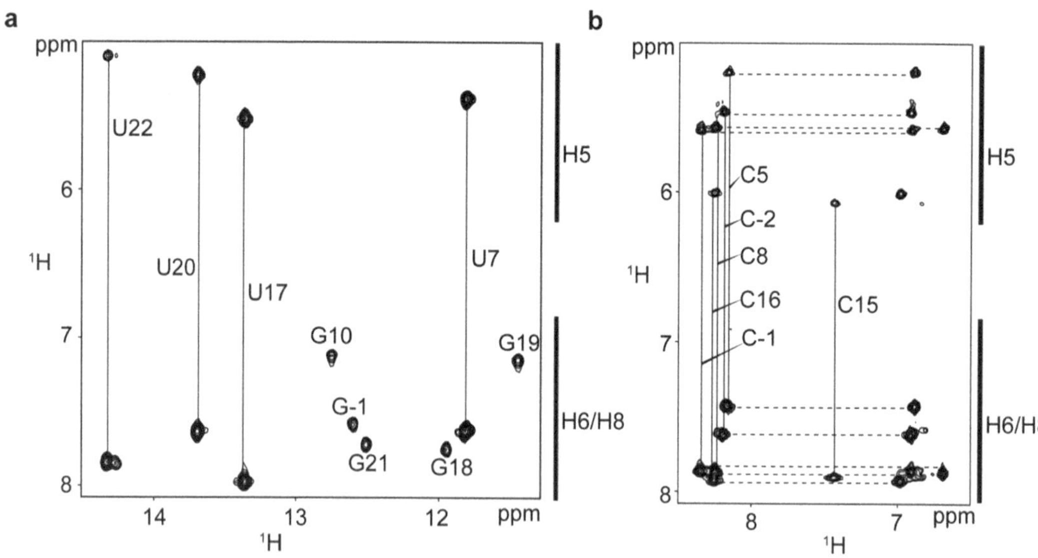

Fig. 5 Portions of the 2D H6/H5(C4N)H spectrum obtained for DsrA-SL1 showing the correlations from the imino/amino protons to the H6/H5 protons of pyrimidines and H8 proton of guanine

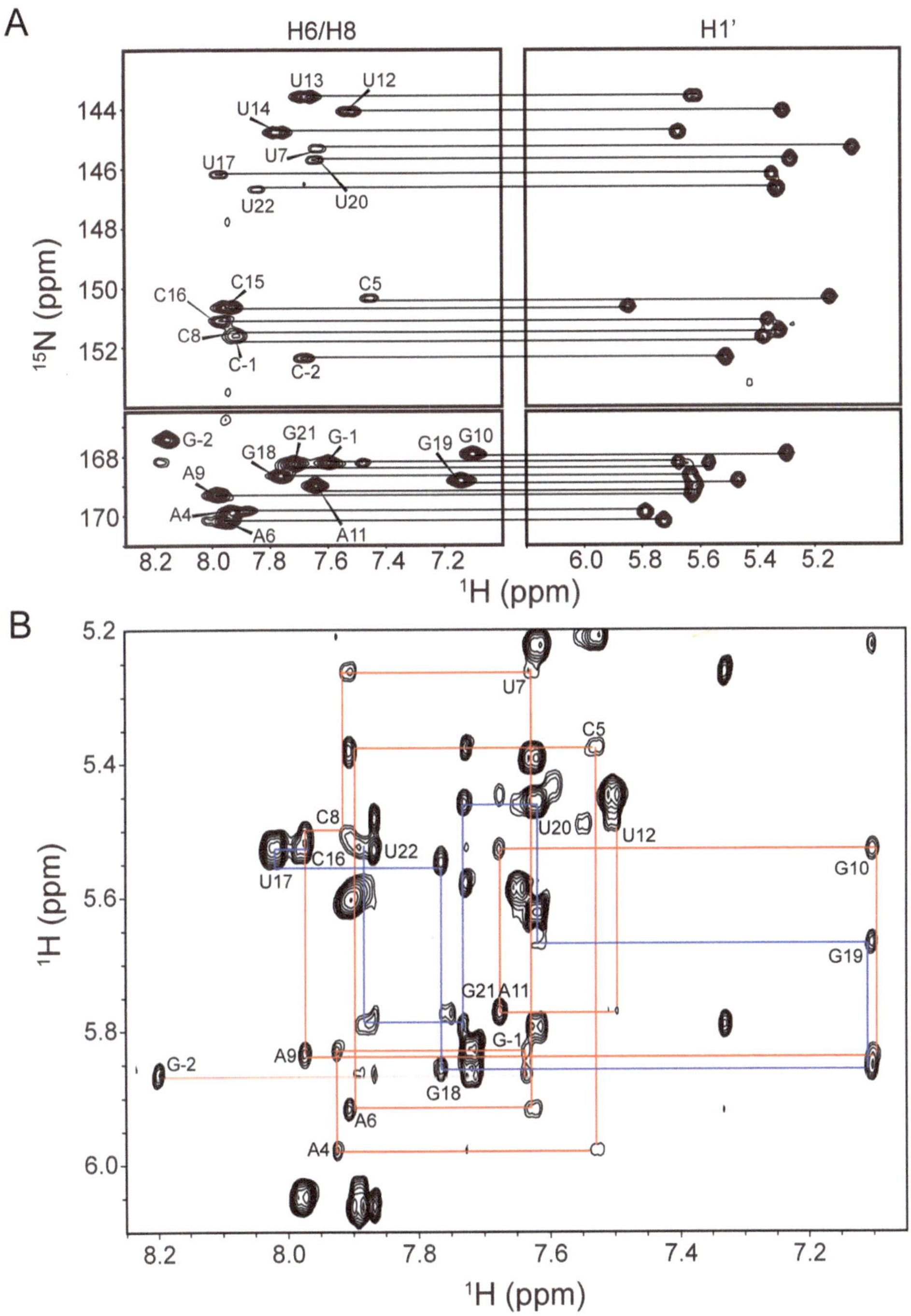

Fig. 6 Assign H1′ protons. (**a**) The HCN spectrum correlates the ribose H1' and H6/H8 protons via scalar interactions. (**b**) Sequential NOE connectivities between aromatic H6/H8 protons and ribose H1′ protons

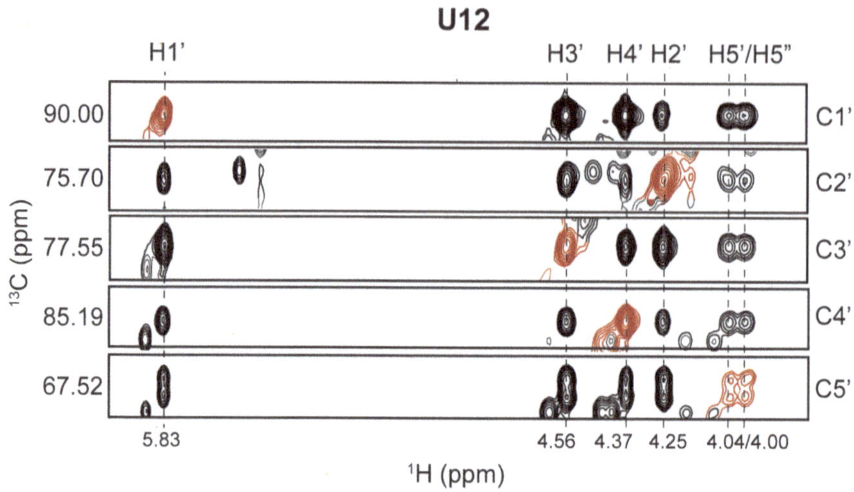

Fig. 7 Portions of HCCH-TOCSY spectrum showing the ribose connectivities of residue U12 of DsrA-SL1

2. Employ standard hydrogen-bonding restraints for all base pairs derived from 2D ^{1}H-^{1}H NOESY and ^{1}H-^{15}N HNN-COSY.
3. Employ cross-helix phosphate–phosphate distance restraints (with a 20% weighting coefficient) in the helical segments to limit the approach of the phosphate groups.
4. Employ torsion-angle restraints for the nucleotides of standard A-form geometry, allowing for ±25° deviations from ideality ($\alpha = -62°, \beta = 157°, \gamma = 48°, \delta = 83°, \varepsilon = -152°, \zeta = -73°$). Nucleotides with observable H1′-H2′ and H1′-H3′ correlations in 2D ^{1}H-^{1}H TOCSY are constrained as C2′-*endo* (south, $\delta = 142.5° \pm 37.5°$).
5. Once a set of restraints is assembled, calculate the structures with the program CYANA.
6. Refine the 20 final structures generated with CYANA in the ff99bsc0_chiOL3 force field using the sander module of AMBER14 (Fig. 8).

3.5 Specific Case of Modified RNA

Another particular case for NMR spectroscopy is to solve the structure of modified RNA. In contrast to the unmodified RNA, RNAs with chemically modified nucleotides cannot be transcribed readily in vitro readily by T7 polymerase. Therefore, commercially synthesis is used to obtain most of modified RNA samples. Modified RNAs usually are not marked by protonated-^{13}C and -^{15}N, and so the assignment for peaks mainly depends on a series of homonuclear 2D NMR spectra and the comparison with the analogous RNA without modification. RNA modifications always involve many kinds of chemical processes, including methylation, pseudouridylation, and so on. Here we take the RNA with $m^6{}_2A$-h45 (a stem-loop RNA with 24 nucleotides) as an example

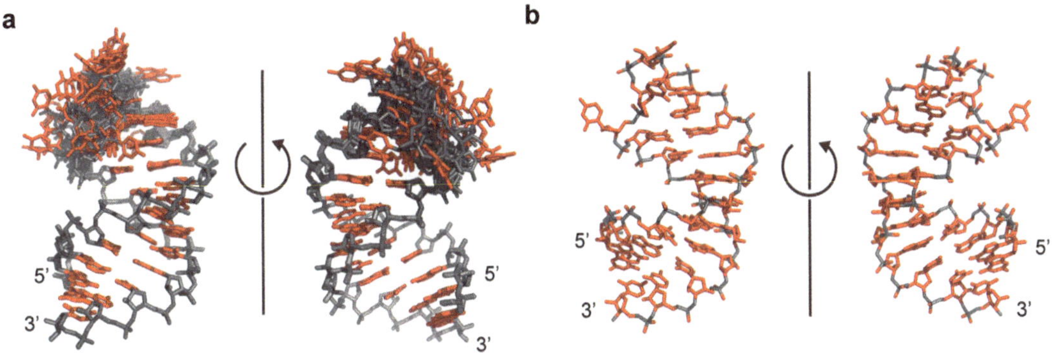

Fig. 8 NMR solution structure of DsrA-SL1. (**a**) Superposition of the 20 lowest energy structures. (**b**) The lowest-energy structure represented as a ribbon-and-stick model. The base and ribose are colored red and grey, respectively

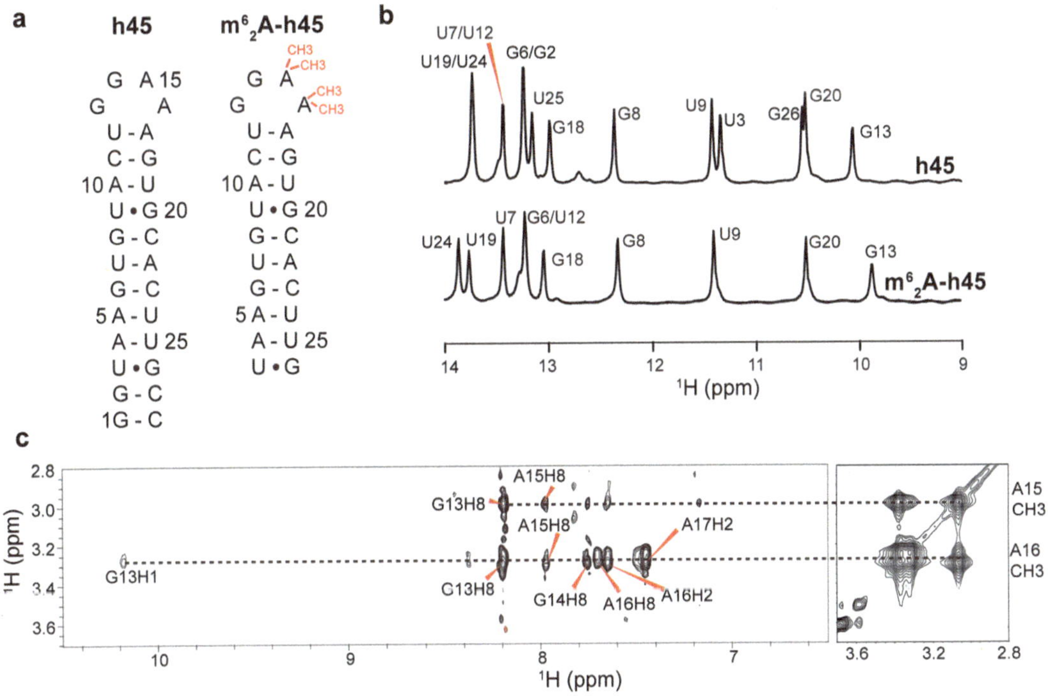

Fig. 9 (**a**) Secondary structures of $m^6{}_2A$-h45, renumbered as 3–24. The $m^6{}_2A$-h45 RNA contains four methyl groups in the "GGAA" tetraloop. (**b**) 1D spectra for h45 and $m^6{}_2A$-h45. Comparison of imino peaks showed the same peak positions for most of the base pairs in the stem region except the terminal two base pairs. The first nucleotide in the loop, G13, had a slightly different chemical shift and weaker intensity after dimethylation. (**c**) 2D NOESY spectra in H_2O for $m^6{}_2A$-h45. Methyl group region provides more information for modified atoms

(Fig. 9a). The methyl groups of this RNA are located at two adjacent adenosines in the loop region. Comparison of the 1D spectra of h45 and $m^6{}_2A$-h45 showed the same peak positions for most of the base pairs in the stem, but a slightly different chemical shift and weaker intensity for G13 after dimethylation, indicating that modified methyl groups in the loop region did not affect the stem region and did not cause the adenine to flip outside (Fig. 9b). The 2D NOESY spectra in D_2O of $m^6{}_2$A-h45 still showed a clear "walk" pattern in the base-pairing region, implying the A-form stem was also present in the $m^6{}_2$A-h45 structure. The chemical shift of methyl hydrogen atoms is located at approximately 3.0 ppm, which is a significant information for assignment of modified RNA. For $m^6{}_2$A-h45, only one diagonal peak could be observed on the 2D NOESY spectrum in H_2O for two methyl groups of $m^6{}_2$A15, yet two corresponding peaks were observed for $m^6{}_2$A16 (Fig. 9c). According to this, more NOEs related to methyl groups can be found. Certainly, the 2D DQF-COSY spectrum can provide more details for sugar pucker conformations of modified RNA, which is the same with unmodified RNA.

4 Notes

1. The addition of the five nucleotides 5′ of the T7 promoter sequence increases stability of the dsDNA and raises yields of product RNA.
2. RNA markers are unstable and expensive. We use DNA markers instead.
3. To determine the optimum transcription conditions, the concentrations of NTPs, DNA template, and T7 RNA polymerase also need to be optimized.
4. After the addition of EDTA, the white precipitate which is Mg-pyrophosphate should go back into solution.
5. For a new RNA, the buffer conditions, the annealing temperature, the time of incubation, and the method/speed of cooling can all be varied to identify the optimal annealing protocol.
6. The imino proton from a sheared G:A base pair appears at about 9–10 ppm. The two imino protons from U:U base pair appear at about 10.5–11.5 ppm.
7. Due to broadening from exchange with H_2O and rotation about the C–N bond, amino protons of RNA are more difficult to be observed than imino protons. Amino protons of the cytosine can be the most commonly observed, because they are often in the slow exchange regime when involved in a G:C base pair. They appear in the range of 6.5–8.5 ppm. The imino and amino protons fast exchange with H_2O, which depends on pH and temperature. To better observe these resonances, usually we use lower pH (6.5) and lower temperature (283 K).
8. An overlay of NOESY and DQF-COSY spectra helps to distinguish pyrimidine to purine in the region of the aromatic to H5/H1′ region.

Acknowledgments

This work was supported by Hefei National Laboratory for Physical Sciences at Microscale and School of Life Sciences, University of Science and Technology of China. We are grateful to Richard A. Lease (The Ohio State University, USA) for critical reading of the manuscript and many fruitful comments.

References

1. Morris KV, Mattick JS (2014) The rise of regulatory RNA. Nat Rev Genet 15:423–437
2. Marchanka A, Simon B, Althoff-Ospelt G, Carlomagno T (2015) RNA structure determination by solid-state NMR spectroscopy. Nat Commun 6:7024
3. Cruz JA, Westhof E (2009) The dynamic landscapes of RNA architecture. Cell 136:604–609
4. Dethoff EA, Chugh J, Mustoe AM, Al-Hashimi HM (2012) Functional complexity and regulation through RNA dynamics. Nature 482:322–330
5. Wu P, Liu X, Yang L, Sun Y, Gong Q, Wu J, Shi Y (2017) The important conformational plasticity of DsrA sRNA for adapting multiple target regulation. Nucleic Acids Res 45:9625–9639
6. Xue Y, Kellogg D, Kimsey IJ, Sathyamoorthy B, Stein ZW, McBrairty M, Al-Hashimi HM (2015) Characterizing RNA excited states using NMR relaxation dispersion. Methods Enzymol 558:39–73
7. Dieckmann T, Suzuki E, Nakamura GK, Feigon J (1996) Solution structure of an ATP-binding RNA aptamer reveals a novel fold. RNA 2:628–640
8. Furtig B, Richter C, Wohnert J, Schwalbe H (2003) NMR spectroscopy of RNA. Chembiochem 4:936–962
9. Ahmed YL, Ficner R (2014) RNA synthesis and purification for structural studies. RNA Biol 11:427–432
10. Scott LG, Hennig M (2008) RNA structure determination by NMR. Bioinformatics Methods Mol Biol 452:29–61
11. Milligan JF, Groebe DR, Witherell GW, Uhlenbeck OC (1987) Oligoribonucleotide synthesis using T7 RNA polymerase and synthetic DNA templates. Nucleic Acids Res 15:8783–8798
12. Keane SC, Heng X, Lu K, Kharytonchyk S, Ramakrishnan V, Carter G, Barton S, Hosic A, Florwick A, Santos J et al (2015) Structure of the HIV-1 RNA packaging signal_supplementary. Science 348:917–921
13. Duss O, Michel E, Yulikov M, Schubert M, Jeschke G, Allain FH (2014) Structural basis of the non-coding RNA RsmZ acting as a protein sponge. Nature 509:588–592
14. Furtig B, Richter C, Bermel W, Schwalbe H (2004) New NMR experiments for RNA nucleobase resonance assignment and chemical shift analysis of an RNA UUCG tetraloop. J Biomol NMR 28:69–79
15. Varani G, Tinoco I (1991) RNA structure and NMR spectroscopy. Q Rev Biophys 24:479–532
16. Pardi A (1995) Multidimensional heteronuclear NMR experiments for structure determination of isotopically labeled RNA. Methods Enzymol 261:350–380
17. Gallo S, Furler M, Sigel RKO (2005) In vitro transcription and purification of RNAs of different size. Chimia (Aarau) 59:812–816
18. Dingley AJ, Nisius L, Cordier F, Grzesiek S (2008) Direct detection of N-H[...]N hydrogen bonds in biomolecules by NMR spectroscopy. Nat Protoc 3:242–248
19. Wohnert J, Gorlach M, Schwalbe H (2003) Triple resonance experiments for the simultaneous correlation of H6/H5 and exchangeable protons of pyrimidine nucleotides in 13C,15N-labeled RNA applicable to larger RNA molecules. J Biomol NMR 26:79–83
20. Sklenar V, Dieckmann T, Butcher SE, Feigon J (1998) Optimization of triple-resonance HCN experiments for application to larger RNA oligonucleotides. J Magn Reson 130:119–124
21. Pardi A, Nikonowicz EP (1992) Simple procedure for resonance assignment of the sugar protons in C-13-labeled Rnas. J Am Chem Soc 114:9202–9203
22. Bax A, Clore GM, Driscoll PC, Gronenborn AM, Ikura M, Kay LE (1990) Practical aspects of proton carbon carbon proton 3-dimensional correlation spectroscopy of C-13-labeled proteins. J Magn Reson 87:620–627

Index

Veronique Arluison and Frank Wien (eds.), *RNA Spectroscopy: Methods and Protocols*, Methods in Molecular Biology, vol. 2113, https://doi.org/10.1007/978-1-0716-0278-2, © Springer Science+Business Media, LLC, part of Springer Nature 2020

FSC
www.fsc.org
MIX
Papier | Fördert
gute Waldnutzung
FSC® C083411

Kolibri 360 GmbH
Ferdinand-Jühlke-Straße 7
99095 Erfurt, Deutschland
produktsicherheit@kolibri360.de